Terje Tvedt is Professor of Geography at the University of Bergen and Professor of Political Science at the University of Oslo. He is an international authority on the history and politics of water and the author of a number of successful books and films on water issues, including *The River Nile in the Age of the British* (I.B.Tauris, 2004), which was shortlisted for the British Society for Middle East Studies Prize 2004. Among his other books are *The Nile: An Annotated Bibliography* (revised edition, I.B.Tauris, 2003), *Southern Sudan: An Annotated Bibliography* (revised edition, 2 volumes, I.B.Tauris, 2004) and, as series editor, *A History of Water* (I.B.Tauris, 2006 and forthcoming).

# The River Nile in the Post-Colonial Age

## Conflict and Cooperation among the Nile Basin Countries

Edited by Terje Tvedt

I.B.TAURIS
LONDON · NEW YORK

Published and reprinted in 2010 by I.B.Tauris & Co. Ltd
6 Salem Road, London W2 4BU
175 Fifth Avenue, New York NY 10010
www.ibtauris.com

Distributed in the United States and Canada Exclusively by Palgrave Macmillan,
175 Fifth Avenue, New York NY 10010



The boundaries, names and designations shown on the maps used in this volume do not imply acceptance or official recognition by the authors and the Nile Basin Research Programme



ISBN: 978 1 84511 970 6

A full CIP record for this book is available from the British Library
A full CIP record is available from the Library of Congress

Library of Congress Catalog Card Number: available

Designed and Typeset by 4word Ltd, Bristol, UK
Printed and bound in Great Britain by CPI Antony Rowe, Chippenham

FSC
www.fsc.org
MIX
Paper from responsible sources
FSC® C013604

# Contents


| | | |
|---|---|---|
| | Foreword by Tore Saetersdal | vii |
| 1 | About the Importance of Studying the Modern History of the Countries of the Nile Basin in a Nile Perspective<br>*Terje Tvedt* | 1 |
| 2 | Burundi and the Nile: Water Resource Management and National Development<br>*Pascal Nkurunziza* | 13 |
| 3 | Rwanda and the Nile: Water Plans and their Implementation<br>*Robert Baligira* | 31 |
| 4 | Unlocking Economic Growth and Development Potential: The Nile Basin Approach in Tanzania<br>*Honest Prosper Ngowi* | 57 |
| 5 | The Congo Nile: Water Use, Policies and Challenges<br>*Raphael M. Tshimanga* | 73 |
| 6 | The Nile Waters and the Socio-Economic Development of Western Kenya<br>*Mary C. Mwiandi* | 93 |
| 7 | Independent Uganda and the Nile: Hydroelectric Projects and Plans<br>*James Mulira* | 125 |
| 8 | Institutional Development and Water Management in the Ethiopian Nile Basin<br>*Yacob Arsano* | 161 |
| 9 | The History of the Nile Waters in the Sudan<br>*Fadwa Taha* | 179 |
| 10 | Egypt and Collective Action Mechanisms in the Nile Basin<br>*Hosam E. Rabie Elemam* | 217 |

11 Some Conceptual Issues Regarding the Study of Inter-state Relationships in River Basins 237
*Terje Tvedt*

Notes 247

Contributors 281

Index 285

# Foreword

This book is the product of the first research group of the Nile Basin Research Programme, convened at the University of Bergen in the spring semester of 2007 under the theme 'Nile Basin Developments in the Post-Colonial Period'. The contributors to the volume are from nine Nile basin countries, and their chapters reflect the academic traditions of Nile-related research in these countries. For any student of Nile development, this volume will be indispensable to a proper understanding of the complex modern history of the river basin, the politics surrounding it and the efforts that are now being made to jointly manage it.

The Nile Basin Research Programme began in March 2006. It is based at the University of Bergen and is funded by the Norwegian Ministry of Foreign Affairs. The programme is devised as a guest programme for researchers from the Nile basin countries.

Professor Tvedt is responsible for a number of books and television documentaries on the importance of fresh water and the politics of water in major river basins, especially the Nile. It was accordingly appropriate and important that he headed the first group of Nile basin researchers in Bergen. As leader of the group he is also editor of this volume and is to be thanked for his efforts and input on the programme.

Dr Martin Daly, author of a number of well-known books on the history of the Sudan, helped improve the style and grammar of the chapters. The Nile Basin Research programme decided to have new maps made for this volume. These have been drawn by senior technician Kjell Helge Sjøstrøm, Department of Geography, University of Bergen. The programme administrative coordinators, Mr. Bård Hekland and Mr. Torleif Markussen Lunde, are thanked for acting as assistants to the editor. Numerous specialists and colleagues have commented upon drafts and have contributed to the volume in a variety of ways, and are to be thanked for their efforts: Professor Anders Bjorkelo, Professor Gaspard Ntakimazi, Professor Charles Odiki, Dr Jockey Nyakanna, Dr Ahmed El Rashidy and Professor Ashok Swain.

Tore Saetersdal
Director
Nile Basin Research Programme

# 1

# About the Importance of Studying the Modern History of the Countries of the Nile Basin in a Nile Perspective

**Terje Tvedt**

Whether the River Nile will be an object of violent conflicts or of peaceful cooperation in the years ahead, the way its water is managed in coming decades will definitely have worldwide implications.[1] Struggle over the Nile's waters has had global political consequences in the past and could fan existing conflicts in the Horn of Africa and Somalia, threaten the peace agreements in the Sudan, and influence the power balance in the Middle East in the future. Yet, in terms of jurisdiction and development, the Nile has been cited as one of the few international river basins with legal arrangements for sharing the waters, and has at times been portrayed as a possible model for other international river basins.[2]

Literature on the River Nile is quite extensive.[3] The importance of the river and the efforts to control it during the British colonial period are relatively well documented and analysed,[4] but for the post-independence period there is no comprehensive and empirical description of the riparian countries' efforts to manage the watercourse.[5]

One premise of this study of the modern history of the Nile basin, and the imprint the river and attempts at modifying and sharing it have had on regional and national development, is that the only way to escape the power of history is to know it. The manner in which states, political leaders and ordinary people have adapted to and used the Nile are important in themselves, since these human–river relations form central elements of the history of every basin country. Conventional, 'water blind' history writing has tended to overlook the importance of the river and its tributaries in structuring development patterns and has failed to grapple with the regional political implications of powerful actors' Nile strategies. The deep connections between river and society have continuously created and recreated a particular arena for policy-making and diplomacy in the Nile valley, and impacted, as well, the history of each and every country. In order to grasp the modern history of the peoples of the Nile basin and the challenges of cooperation and optimal and rational utilization of this finite water resource, it is necessary to focus in detail on how developments in different countries followed

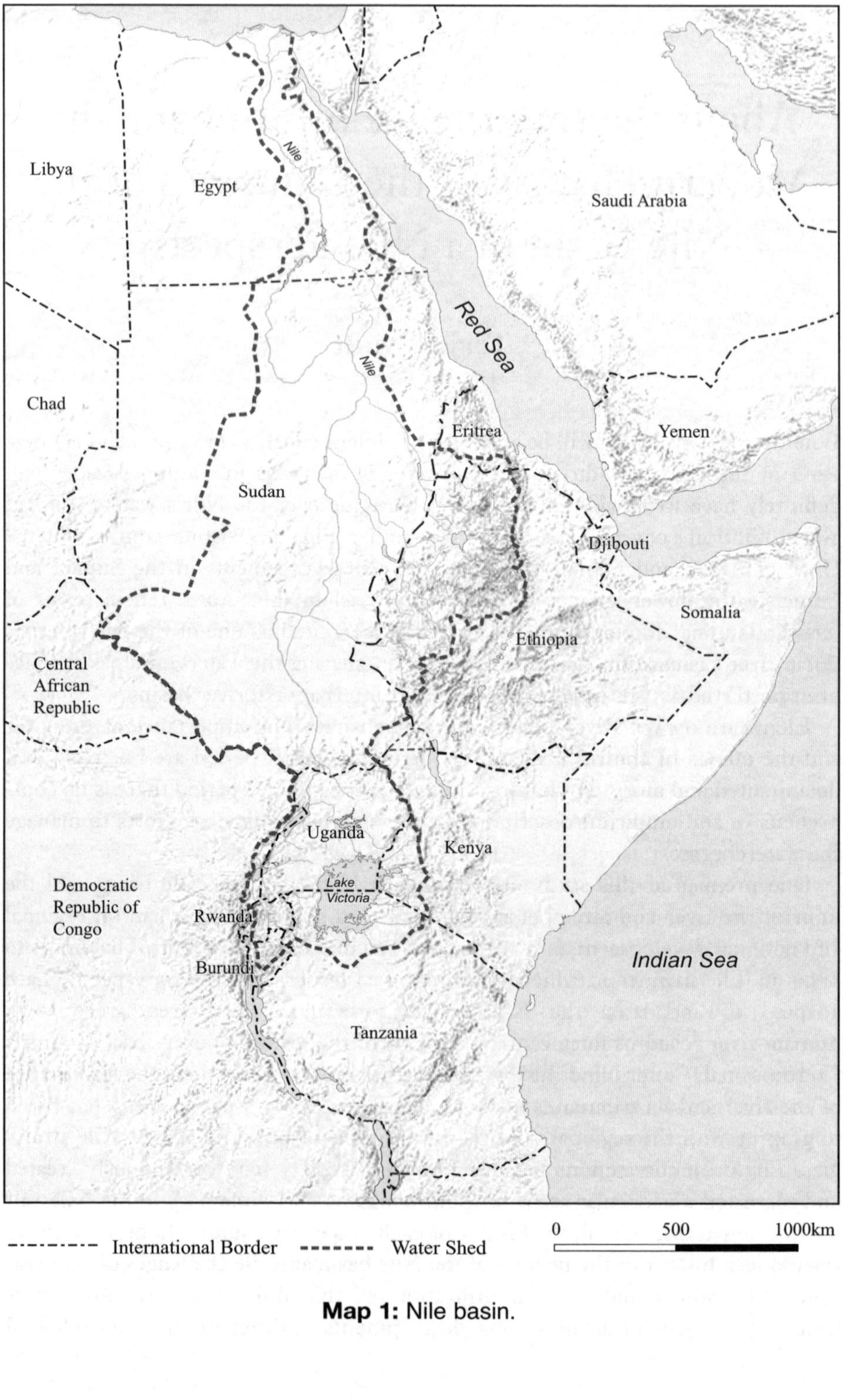


**Map 1:** Nile basin.

their own particular pattern, structured and made possible by historical, ecological and power configurations in the Nile basin.

## THE AGE OF THE BRITISH

The history of the Nile basin in the age of the British is one of water wars, of hydropolitics on a grand scale, and of a river empire – stretching from the Mediterranean to the heart of Africa. The British colonial period was also a time, at least in part of the Nile basin, of revolutionary pushes to modernize Nile control and thus the water economy as a whole, and hence (because of the importance of water and the Nile) to modernize other economic sectors of the society. From source to mouth the river was for the first time controlled from one centre; and from London, and by the British government. This was the moment in time when political leaders came to regard this widely varying resource as one hydrological and political unit, with far-reaching consequences for the peoples who for generations had been living along the banks of the river as if the river and its tributaries were local water courses.

British Nile policy, the development it stimulated and the conflicts it engendered had important effects locally and regionally, and, in certain cases, they had world historical importance. Global political events – such as the partition of Africa, the collapse of the League of Nations, the demise of the British Empire and the birth of such new states as Uganda, Kenya and a sovereign Sudan independent from Egypt – were all closely related to British Nile strategies, and partly a consequence of these strategies.[6] Britain's occupation of Egypt in 1882 was a *démarche* owing more to the Suez Canal than to the Nile, but its accession of Uganda in 1894 and conquest of the Sudan during the 'river war' of 1896–98 were first and foremost aspects of an imperial expansionist strategy for control of the Nile as a resource.[7] These military campaigns brought the kaleidoscopic procession of civilizations, peoples and the hundreds upon hundreds of language groups of the enormous Nile basin into the maelstrom of world politics – but under a kind of Pax Britannica. Conventional explanations of the partition of Africa and the race to Fashoda have one-sidedly focused on the role of European rivalry in forcing a supposedly reluctant Britain to occupy the Nile basin; that it was the fear of the French and the Italians that forced, so to speak, the British to move upstream. But what actually happened, and what is well documented in secret sources from the time, was that the British took control of Uganda and the Sudan because of a deliberate kind of water imperialism.

British expansion upriver was a rational imperial policy driven by a complex mixture of economic and political considerations that were influenced by how they understood the structuring capabilities of the Nile's geographical and hydrological characteristics.[8] Britain had two (partly conflicting) visions of Nile management which framed colonial strategy in this part of Africa and the Middle

East. On the one hand, it sought to secure more water for Egypt in order to increase agricultural production, and especially the profitable cotton production, and to enrich the country under its own benign oversight. Posing as the guardian of Egypt's lifeline, Britain could maintain its position at Suez. But on the other hand, upstream water projects under London's control might also be used as leverage *against* Egyptians, elite and fellahin alike, if they ever grew restive.[9] It was for these combined and partly conflicting motives that the British aimed at 'taking the Nile in hand'.[10] As part of this policy, London signed pacts and agreements with other European colonial powers active on the 'fringes' of the Nile basin, such as Germany, Italy, France and Belgium, curtailing their influence and instituting a regime of 'non-interference' with the hydraulic integrity of the river system.

The widespread understanding in the historical literature of the Sudan as a kind of 'buffer state' between European rivals,[11] more or less as Afghanistan was in Asia during the nineteenth century, is misleading. This way of conceiving the Sudan's importance disregards the geopolitical, structuring role of the River Nile. The Sudan should not be reduced to a 'buffer' between more important countries. On the contrary, it held nothing less than the very key to the planned development of the Egyptian irrigation economy. The Egyptian land-owning elite urged for a re-occupation of the Sudan (lost to the Mahdist uprising in 1885), conceiving the area more or less as an Egyptian province on the Nile. The central African lake area became crucial to imperial strategy because by controlling it London would at the same time control the headwaters of the White Nile, the most important tributary for the cotton crop in Egypt during the first decades of British rule, and thus the very lifeline of Egyptian summer cultivation. These regions of the Upper White Nile were primarily conceived of as aqueducts bringing water to the irrigated fields in the far north, and London allowed only very modest water development upstream. The territories south of Lake Victoria and east of Lake Albert were considered to be of marginal importance to the flow of the main Nile, and they were therefore also of marginal strategic importance to Britain. The British similarly thought it unnecessary to occupy Ethiopia for hydropolitical reasons – something that might have been difficult to achieve, anyway, due to both potential strong local military resistance and diplomatic opposition from other European powers. The technology available at the time did not make it possible to control and store for usage in the low season the violent and silt-laden flood waters of the Blue Nile, and its waters could thus contribute little to planned economic activities related to cotton growing in the summer, or *sefi*, season.

The British first invented the Nile as one planning unit in 1904.[12] Their Nile strategy led to an overall water policy, demanding limited development and modernization upstream in the basin.[13] Here the main aim was to maintain law and order without high expenditure, while in Northern Sudan and Egypt the British rulers invested in river control works, irrigation development and cotton production made possible by artificial watering. After the Egyptian Revolution in 1919, London also worked tirelessly to expand cotton cultivation in the Sudan, especially in the Gezira scheme, not least because the cotton that was produced here was produced in an area still firmly controlled by Britain and therefore it was regarded as being a more secure source of supply than post-revolutionary Egypt. The relative efficiency of British Nile policies in these early decades of imperial rule reflects the fact that they were implemented by a comparatively small group of political strategists in London, Cairo and Khartoum, in conjunction with an even smaller group of water experts, supported wholeheartedly by the government in London and enacted in a political–institutional environment where the most powerful regional stakeholders supported them.[14]

Gradually the coherence of British imperial policy became weaker, and different imperial interests started to conflict with each other. When the British had the Sennar dam built in the mid-1920s, by a stroke they initiated a development that turned the Sudan into a hydraulic state, also creating, along the way, an economic and political elite with a strong vested interest in ever-increasing Nile control. From a geopoliticial perspective, Foreign Office strategists in London, eager to weaken Egyptian influence in the Sudan, repeatedly stressed that whoever controlled the Sudan 'held Egypt at her mercy', because the Sudan lay astride the Nile upstream of Egypt. The dam and increased irrigation in northern Sudan were thus both in line with Lancashire's interests for more cotton and with geopolitical considerations, but at the same time British colonial officers, stationed in the country, started to view the Sudan and its development as important in itself. London never managed to win necessary Ethiopian support for plans to build a dam at Lake Tana in Ethiopia; project ideas they had developed at the very beginning of the twentieth century. Haile Selassie, the strong man in Ethiopia from the First World War to the 1970s, distrusted London's intentions, fearing that the dam was a plot in a larger plan aiming at the partition of Ethiopia. In return for permission to build a dam there, he wanted support from Britain for his claims to Eritrea against Italy, a promise he was not given. (I have elsewhere shown how the history of the Tana dam is part and parcel of important world history, related to the Italian occupation of Ethiopia in the mid-1930s, Britain's attitude to Mussolini's attack on Ethiopia, and the subsequent collapse of the League of Nations, the forerunner of the United Nations.)[15] In Uganda, British colonial administrators disliked what they considered a downstream bias in British Nile policies. From the 1940s on, especially, they started to put pressure on London to revise their Nile policy; they wanted a hydroelectric dam in Uganda to modernize the country. When they finally received Egyptian acceptance for the building of the Owen Falls dam, finished in 1954, the British introduced large water-control works in Uganda and turned an important new page in the history of that country. The crushing defeat of Britain in the Suez crisis of 1956 – after their political tactics related to their financing of Nasser's Aswan dam backfired and ended in the Suez war – was the death blow to Britain's Nile project.

The British also pioneered a system of river-sharing agreements. Some were negotiated from strength (e.g. in 1902 with Emperor Menelik II of Ethiopia), and some from relative weakness (notably the exchange of notes with Egypt in 1929, largely an effort to regain some trust and reputation in Egypt as the powerful provider of Nile waters, and largely lost through the so-called Allenby ultimatum in 1924, when the leading British politician in the region declared that the Sudan unilaterally would be able to withdraw more water from the Nile in the Sudan than what had been promised to Egypt). In the 1950s, when the approach of the 'wind of change' made it evidently clear that the days of the British Nile empire were numbered, London tried to reach a kind of basin-wide agreement with Egypt based on a new and more equitable Nile Valley Plan, giving more attention to the water needs of the upstream countries still under British control, but they failed.

A balanced interpretation and assessment of the British colonial period in this part of the African continent cannot be achieved without giving due emphasis to the River Nile and its economic and political importance, its physical characteristics and how these were understood and manipulated for the benefit of certain types of economic transformations by the British water planners and politicians. Moreover, knowledge of the British period on the Nile is also important, seen as an event in the very long history of relationships between man and river, because the empire's Nile discourse, Nile plans and Nile strategy developed and became very different from those of the past, and have had and will have long-lasting consequences.[16] The British authorities were the first to be in a position that enabled them to formulate and popularize a basin-wide techno-scientific planning concept of the river basin as a whole. It will also have long-lasting implications that their vision for Nile development and London's hydropolitical thinking were biased in favour of Egypt and from the 1920s also in favour of the northern riverain Sudan, both due to imperial strategic plans and visions and for economic reasons. Since the emergence of independent Africa, the Nile states have been at loggerheads over the sharing and control of the waters of the Nile. While the British Empire collapsed decades ago, the legacy of the British Nile empire has had important implications for the development of the entire Nile basin up until the present; thus justifying empirically, and from a new perspective, the term the 'post-colonial age'.

## THE AGE OF INDEPENDENCE

Gradual disintegration of colonial authority, and London's weakened ability to enforce one strategy for the basin as a whole, led to what from a long Nile historical perspective can be seen as the emergence of a more decentralized decision-making process when it comes to water projects, and less comprehensive and coherent Nile management plans. With independence a new era opened because suddenly, and for the first time in the long history of the river, nine sovereign states (ten, when Eritrea declared independence from Ethiopia) were responsible for using and sharing the Nile basin. Two historical processes coincided; new states were established just as it became technologically feasible to control the Nile waters on a much larger scale than before, and in places where it previously had been physically impossible to tame it. Post-colonial Nile developments were therefore from the outset influenced by aspirations of national pride and sovereignty and state power on the one hand, and considerations of technological feasibility and economic ability on the other hand. Respective national policies developed at very different paces, however, reflecting varying levels of development, degrees of political stability and specific geographical positions in the Nile basin, as well as varying legal positions as to the right to Nile waters.

Egypt has throughout the post-colonial period been by far the most important actor on the Nile, and has also been the state that has benefited most from the Nile arrangements institutionalized by the British Nile Empire. The two most formative events in this period in the Nile basin, influencing the whole region's economic development and hydropolitical power relations, took place under Gamal Abdel Nasser. The Egyptian leadership aimed at liberating the country from potential pressure and interference from upstream powers (i.e. at that time, the British) by the decision in the 1950s to build the new Aswan Dam. The dam was designed to hold two years of successive Nile flows. The idea was that it, as Nasser expressed it many times, should help turn Egypt into the 'Japan of Africa', and that it at the same time should tame the river within the borders of Egypt. The dam was finished in 1971 and became the foremost symbol of Nasser's vision of independent Egypt. It played a very important role in expanding power production and agricultural area. The dam has also served an important hydrodiplomatic aim since it has helped to bolster Egypt's claim that they have acquired rights to 55.5 billion cubic metres (bcm) of Nile water every year. Nasser's impact on the present Nile regime cannot be overestimated, also because he was instrumental in bringing about the 1959 agreement with the Sudan: the agreement 'For The Full Utilization of the Nile Waters' was signed at Cairo on 8 November 1959 and was in force by 12 December 1959.

The Egyptian government has since then determinedly pursued water control projects within Egypt's borders, while at the same time recognizing upstream countries' development of the Nile's waters as a potential national security threat. Egypt aimed for some time to enforce the principle of 'exclusive user rights', but that idea now belongs to the past. The old, deep-seated cultural notion that Egypt was entitled to the waters of the Nile could not be maintained in the modern post-colonial era, challenged partly by a growing body of international water law but also by a growing political strength in the upstream countries. In the 1970s and 1980s, Egypt stated its willingness to resort to military measures to secure its water supply. They threatened to use military force against anybody interfering with what has generally in Egypt been considered an Egyptian river, because Egypt had, it was argued, since the dawn of civilization been dependent upon the waters of the river while the up-stream countries have had no, or very weak, traditions of Nile control. Since at least the mid-1990s, however, Egypt has changed its policy and diplomatic strategy. The military option is now publicly discarded, regarded as neither viable nor justifiable. Egypt's official policy has been to foster cooperation and goodwill among the countries in the basin,[17] and it has followed this up by taking concrete measures; in a number of cases, Egypt has offered economic assistance. For example, after the Sudan Peace Agreement, Egypt offered support to build up an irrigation administration in South Sudan (under the British there was quite a strong division of what was then called the Egyptian Ministry of Public Works established in the swamp area of South Sudan, mainly concerned with planning the projected Jonglei Canal). Egypt's policies have been formulated within a context of conflicting aims: the need to control the Nile while sharing sovereignty over it; balancing the inherent vulnerability of a downstream country with military strength far exceeding that of the other Nile countries; dealing with a permanent food shortage through imports from Nile neighbours instead of from Argentina and Australia; and maintaining a status quo in water-sharing issues that is unsustainable for Egypt itself, for it needs more water than it has and looks to projects in southern Sudan for close to 20 bcm of extra water. Egypt's official policy has been to stress that regional cooperation is in Egypt's own national interest, and should not be dismissed by sceptics as a mere shift in tactics. In upstream countries rumours have often had it that Egypt instead has aimed at sustaining instability and weak governments there, so as to indirectly hinder them in developing the water resources. Egypt has pursued its Nile diplomacy by intentionally playing down deep-seated differences in conceptualizations and policies among the Nile basin states, especially between Ethiopia and Egypt. All the actors, including the leaders of Egypt, also know that extolling cooperation also sits well with the international donor community.

The position of upstream states has in general been described in this way: 'Diversion, overuse, contamination and flow delay are tactics that are available in accordance with one's position on the riparian totem-pole.'[18] But a riparian upstream position understood within such a perspective regards it simply as a static form of potential coercion, being bound to cause conflicts, and does not fit the actual historical developments of the Nile story. A much milder version of this attitude has been expressed in some upstream countries, nurtured by what has been conceived as a lack of progress on new Nile-sharing agreements. Among the questions that have been discussed are: To what extent should agreements reached during the British era be binding today? And what about the 1959 agreement between Egypt and the Sudan, which left these two parties to decide whether to hold negotiations with other riparian states about sharing of their waters? If this agreement is not renegotiated among all the stakeholders, is genuine cooperation even possible? Some actors have argued that it is possible to sustain and deepen cooperation, while at the same time the issue of water sharing should be circumvented by focusing on water harvesting potentials and economic ties more generally; however, other actors, and some upstream countries especially, have, more or less persistently, argued that the present Nile regime is outdated and reflects the power relations of the post-colonial period, and that it has to be changed – although, it has been in general underlined, in an atmosphere of cooperation and not confrontation.

The Sudan has for most of this period sided with Egypt on Nile issues. The Nile agreement from 1959 – legally expressing and confirming this specific hydrorelationship between the two hydraulic states in the northern parts of the basin – has enabled the Sudan to implement a number of Nile projects without facing vetoes or strong objections from Egypt. Influential political voices have lately argued that perhaps Sudan's interests lay more in cooperation with Ethiopia, since this could facilitate steps to improve flood control and protect the Sudan from the threat posed to its reservoirs by silt deposits from the Ethiopian tributaries.[19] However, Egypt has sought to deepen its ties with the Sudan through investing in land and irrigation in northern Sudan, and Khartoum and Cairo have had and most likely will continue to have common interests in the execution of the Jonglei project and other reclamation schemes planned in southern Sudan. The territorial integrity, political stability and future political alliances of the Sudan will be central to determining the overall development of relations within the Nile basin. The ambition of turning the country into the breadbasket of the Arab world, and the big water projects that will become necessary if this aim is to be reached, might, if they succeed, change the balance of power further in the direction of the Sudan and enable the Sudan to play a more decisive role on its own.

Since the 1980s the governments of Ethiopia and Egypt have repeatedly aired sharp differences over use of the Nile. The background is clear: the Ethiopian Nile tributaries contribute more than 80 per cent of the main Nile's flow as measured at Aswan, but Ethiopia is not free to exploit the tributaries within the borders of Ethiopia. Egypt and the Sudan claim that Emperor Menelik II in 1902 on Ethiopia's behalf promised not to use Nile water without the prior agreement of Britain and the Sudan. These veto rights endure, it is argued, with Cairo and Khartoum today. Ethiopian politicians have time and again declared that 'this inequitable state of affairs' cannot continue, and their legal experts have concluded that the agreement does not bar them from constructing works on the Nile.[20] In addition to the legal arguments, the Ethiopian government has also pointed out that existing and planned Nile projects in the deserts of Egypt and the Sudan are wasteful and irrational water management practices when viewed from a basin-wide perspective. They argue that water would be much more effectively stored on the Ethiopian plateau.[21]

In public debates in upstream countries it has often been suggested that the 'Nile treaties' have impeded their use of the river and thus barred national and local growth. The issue of Nile control in the different countries is much more complicated, however, and the lack of development cannot be explained simply as a function of Egyptian opposition to dam projects or to new water agreements. The hydraulic infrastructure of upstream countries is certainly undeveloped, but this also reflects political instability, poverty, poor state finances and other problems, not least a lack of real leadership and interest in and need for Nile waters.[22] Nevertheless, in a broader historical context, although not very much happened for decades during the post-colonial era in most of the countries, since the 1970s all basin states have slowly started to contemplate and implement Nile control works.

Disagreements among the Nile countries over sharing the waters have been real and deep-seated. The post-colonial era in the Nile basin has been a period marked by legal battles concerning the successors of the colonial Nile agreements and the degree to which these should be legally binding on what is described as non-contracting, sovereign basin states. While these water agreements have played a lesser role in barring upstream development than some have maintained,[23] the water-sharing issue has been and still is a major point of contention and will continue to be so. Water-sharing issues were, for example, omitted from the 2005 North–South peace accord in the Sudan, because the Nile issue would have threatened the agreement as a whole. Kenyan and Ugandan legislators have insisted on declaring the 1929 Nile waters treaty void. Ugandan President Yoweri Museveni dismissed these 'childish demands', and while blaming a misinformed subordinate for mistakenly announcing Kenya's renunciation of the water treaties, that country's water minister praised the excellent relations with Egypt. At the same time as leading politicians in upstream states have assured their intention and willingness to cooperate, the upstream states have declared that the Nile agreements concluded by Britain and Belgium are no longer applicable and should be revised. The positions of the different countries have in relation to this issue remained basically the same: Egypt wishes to secure the amount that they historically have regarded as acquired rights and hopefully even increase the water they can use. The Sudan shares the interest in some form of status quo but with slight modifications, and in the future some degree of convergence of views with Ethiopia might develop. The upstream states are, although to differing extents, in favour of new water-sharing arrangements and have taken the position that the existing Nile treaties cannot be maintained in the long run. Agreements signed between the Sudan and Ethiopia (23 December 1991) and Egypt and Ethiopia (1 July 1993) helped to weaken, at least in principle, extreme positions that based themselves upon doctrines safeguarding the interests only of one of the parties. All states now seem to agree that regulations for the use of Nile waters should be worked out in detail on the basis of the rules and principles of international law, although they do not agree on which particular doctrines should be given priority. Gradually the upstream countries have attempted to develop a Nile diplomacy arguing the principle of equitable uses, while refuting the claims of 'historical, natural or acquired' rights. Egypt's policy regarding the uninterrupted natural flow of the Nile, a legacy of the past and the British colonial era, has also partly been modified, and Cairo has been looking for some sort of compromise as an alternative to unilateral action from upstream states.

To understand and reconstruct the modern history and the recent development of the countries of the Nile basin it is necessary to incorporate the region's defining element, the River Nile, in the analyses. Many historical, diplomatic and political studies are produced about the region and its countries that fail to recognize the importance of the river and its waters, and some of the studies that do bring water politics into the picture suffer from a superficial understanding of Nile hydrology and Nile projects, and how they are interlinked. It must therefore be important to bring forth more information both about the Nile as physical space and structuring force in the different Nile states, creating certain ranges of economic and technological possibilities, as well as about the many concrete initiatives politicians and water experts have undertaken in order to exploit and control the river. This is what this book aims at doing.

# 2

# Burundi and the Nile: Water Resource Management and National Development

**Pascal Nkurunziza**

## INTRODUCTION

This chapter analyses water resources in Burundi in the context of the Nile basin, its management and utilization, and current threats of climate change. Because Burundi has an interest in the Nile Basin Initiative (NBI), some consideration will also be given to the former Kagera Basin Organization (KBO) and the projects of the Nile Equatorial Lakes Subsidiary Action Programme (NELSAP). Only a few of the many projects that have been formulated have been carried out, because of political conflicts between or among the member states of these organizations.[1] We will endeavour to show that Burundi is not running short of water; what Burundi lacks is the technology and expertise to meet the needs of its population for drinkable water, irrigation and hydroelectric production. Finally, we will analyse the impact of climate variability observed in the past (1964–99), as well as that projected by various meteorological studies for the period 2000–50.

## THE GEOGRAPHICAL AND HYDROLOGICAL CONTEXT OF BURUNDI

The republic of Burundi is a landlocked country bordered to the west by the Democratic Republic of Congo, to the south by Rwanda, and to the east by Tanzania. Its total surface area is 27,834 km$^2$ and it has a population of about 8 million.

It is from the Congo–Nile divide that various rivers spring and flow towards the Nile. These include the Ruvubu and its affluent, the Ruvyironza. The Kanyaru river, on the border between Rwanda and Burundi, is an affluent of the Nyabarongo, a river mainly in Rwanda. The source of the Ruvyironza is at Rutovu, the remotest source of the Nile. (The spot is marked by a pyramid built in 1938 by Burkhart Waldecker in reference to the Nile's significance to Egypt.[2])

In Burundi each of the two hydrographical systems, the Congo and Nile, is divided into sub-basins. The hydrographical system of the Congo comprises the Rusizi, Tanganyika and Malagarazi sub-basins. Within Burundi there are important lakes, including Lake Tanganyika in the Congo basin. It collects all the water from rivers to the west of the Congo–Nile divide and that of Lake Kivu via the Rusizi river. With a maximal depth of 1,433 m and an average depth of 700 m, Lake Tanganyika is the second deepest lake in the world after Lake Baïkal. Its length is 677 km and its average width 50 km. It is the largest reservoir of fresh water in Africa and the second largest worldwide. Other important lakes in Burundi include Lake Cohoha and Lake Rweru. The basin of the Nile is divided into the Kanyaru–Nyabarongo, Kagera and Ruvubu sub-basins. The Ruvubu river is 480 km long and has a basin of 12,300 km$^2$. It springs from the eastern part of the Congo–Nile divide at 2,500 m and joins

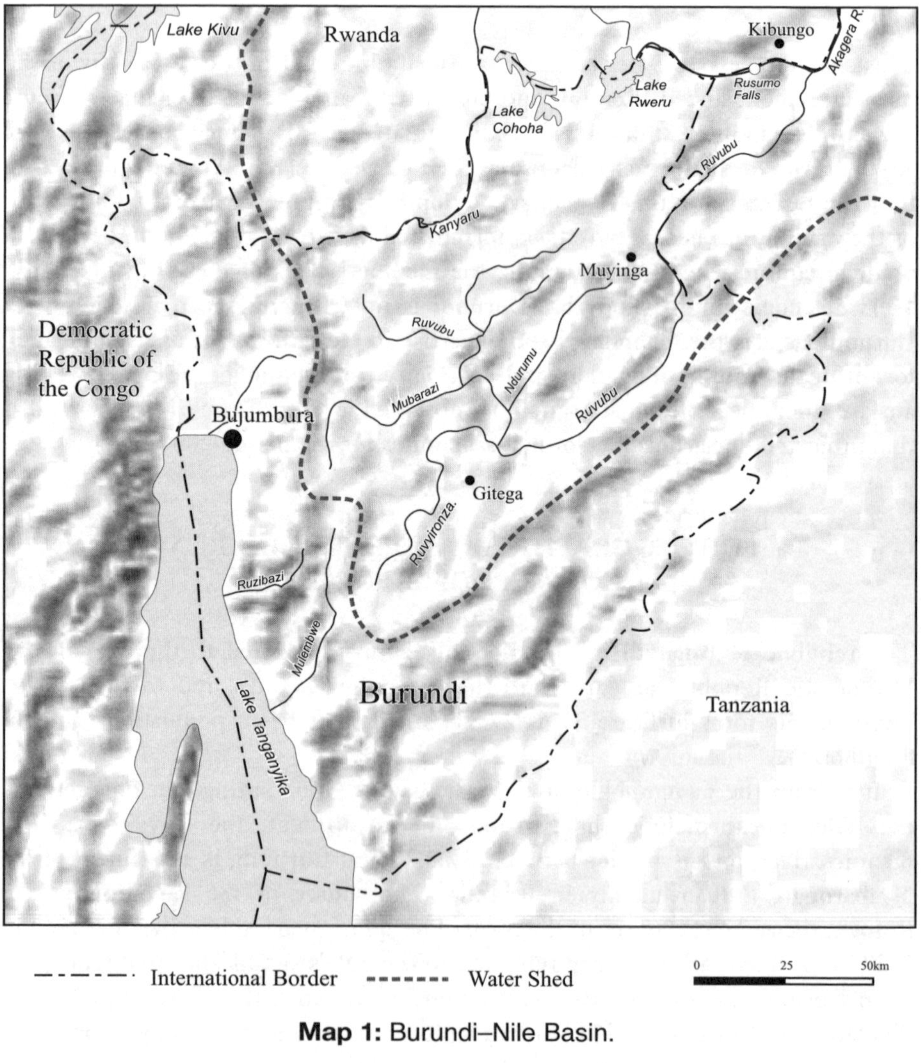


**Map 1:** Burundi–Nile Basin.

the Kagera river just above the Rusumo Falls, with an average slope of 150 m/km. The total flow from Burundi to the Nile is estimated at 2.6 billion cubic metres (bcm).

Burundi is compared to the downstream states a very rainy country, with an average rainfall of 1,274 mm a year. Three rainy zones are distinguished: the Congo–Nile divide's high mountain region (2,600 m) of 1,400–1,600 mm/year; the Central plateaus (altitude between 1,400 and 1,750 m), with average rains of 1,200 mm; and the lowland regions of the Imbo plain, the eastern (Moso region) and north-eastern (Bugesera region) depressions located at an altitude between 1,000 and 1,500 m, with average rains of 1,000 mm/year. Temperatures vary between 12 and 16°C in the Congo–Nile divide region, between 16 and 20°C in the Central plateaus, and between 20 and 24°C in the Moso and Imbo plains, as well as in the north-eastern depressions.

In their quest to discover the sources of the Nile, Europeans conducted many expeditions. In 1858, John Hanning Speke and Richard Burton arrived in Burundi. They skirted the coast of Lake Tanganyika and visited Resha, Kabezi and the delta of the Rusizi, and moved on to Uvira in the Congo. Henry Morton Stanley and David Livingstone came in 1871. The aim of their expedition was to gather information on the political and economic situation of Burundi and of eastern Africa in general. Oscar Baumann, a German, crossed Burundi in 1892 with a particular interest in finding the best way to the great lakes region.

As latecomers to the 'Scramble for Africa', the Germans obtained the Tanganyika territory, or what is now Tanzania, Burundi and Rwanda. They first settled in Burundi in 1886, at Kajaga near Lake Tanganyika, but because that site was malarial they moved on and established themselves at Usumbura, today's Bujumbura. During the First World War the Belgians chased the Germans from the territory, and after the war the League of Nations put Rwanda and Burundi under Belgian tutelage, while the Mandate for the Tanganyika Territory was awarded to Great Britain.

## BURUNDI AND THE NILE TREATIES AFTER INDEPENDENCE

There was no formal treaty on the management of the Nile between Burundi and the British. All the international agreements on the Nile concerned the needs and rights of downstream countries such as Egypt and the Sudan, rather than of those upstream; only the Anglo-Belgian treaty has retained the attention of independent Burundi. This treaty aimed to regulate the sharing of the Kagera river between the (British) Tanganyika Territory and (Belgian) Rwanda–Urundi. This treaty[3] stipulated that:

> Water diverted from a part of the watercourse situated wholly within either territory shall be returned without substantial reduction to its natural bed at

> some point before such water course flows into the other territory or at some point before such watercourse forms the common boundary.

After independence, Burundi sent a note to the United Nations in June 1964, giving its position on the treaty of 1934. This note reads as follows:

> The Government of the Kingdom of Burundi is prepared to succeed to bilateral agreements subject to the following reservations: (1) the agreements in question must remain in force for a period of four years, from 1 July 1962, the date of Independence of Burundi, that is to say until 1 July 1966; (2) the agreements in question must be applied on a basis of reciprocity; (3) the agreements in question must be renewable by agreement between the parties; (4) the agreements in question must have been effectively applied; (5) the agreements in question must be subject to the general conditions of the law of nations governing the modification and determination of international instruments; (6) the agreements in question must not be contrary to the letter or the spirit of the constitution … of Burundi.

With this announcement it was obvious that Burundi, like other upstream countries of the Nile basin, did not consider itself bound by colonial-era treaties. In any case, Burundi has never felt any pressure in the management of the waters of the Nile, either from the British or from any other foreign power. As a country where agriculture depends almost entirely on rainfall, Burundi posed no threat to the flow of the Nile, and the need to subscribe to regulation of the Nile outflow through treaties and conventions hardly arose.

## WATER RESOURCES AND THEIR UTILIZATION

Water is plentiful in Burundi but rather poorly distributed in time and space. According to published statistics,[4] the average rainfall is 1,274 mm/year, 68 per cent of which undergoes evaporation–transpiration; the remainder is divided into surface water (9 per cent) and underground water (23 per cent). The reserve of underground water as well as its exact distribution are still to be determined; underground waters not available at the source but as a 'deep resource' are generally present in the alluviums in the bottoms of valleys, such as in the sediments of the Rusizi and Kumoso plains.[5]

Lake Tanganyika constitutes the biggest reserve of fresh water worldwide (18,900 km$^3$), and of its surface area of 32,600 km$^2$ some 2,600 km$^2$ is in Burundi. Of the country's other lakes the most important are Lake Cohoha (59 km$^2$), with a volume of 5 bcm, and Lake Rweru (91 km$^2$), with a volume of 3.7 bcm.

## *Supply of drinking water*

Before independence and long after, the population consumed water directly from rivers and lakes. This practice was quite often dangerous. Hydrous diseases such as cholera, bacillary dysentery and bilharzias were endemic. It was only in 1939 that the first institution for production and distribution of drinking water was created by the colonial administration for both Burundi and Rwanda. In 1963, just after independence, this plant became a public company with an autonomous management, and was split into two branches: one for Burundi and the other for Rwanda. The company's mission was to supply water only to towns, especially to the capital.

According to the 2005 report by the Food and Agriculture Organization (FAO), total demand for water in Burundi is estimated at 0.288 $km^3$/year, whereas the country has internal renewable sources of water of some 10.06 $km^3$ – a demand of less than 3 per cent of the renewable water available. Moreover, 2.5 $km^3$ of water is imported from neighbouring countries. Including that amount, the demand is even lower, at about 2.4 per cent of the available volume. Put another way, the total real renewable water resources per inhabitant are 1,774 cm/year, whereas appropriation per inhabitant is 46 cm. According to a consensus developed by various experts in relation to the evaluation of the degree of water shortage, water is considered scarce when a person has less than 1,000 cm of water per year.[6]

Statistics for the year 2000 estimate domestic use at 49 mcm of water, or 17 per cent of the total amount used. According to surveys carried out by the Direction Générale de l'Hydraulique et des Energies Rurales,[7] the gross rate is 70 per cent and the net rate 42.9 per cent (the gross rate takes into account all infrastructure, functional or not, while the net rate considers only that which is operating). This difference is mainly attributable to the war, which destroyed much infrastructure, while other works have not been repaired owing to the lack of spare parts or to the lack of funding for their replacement.

With a population growth of 3 per cent, the amount of drinking water needed in urban areas doubles every ten years. From 22 mcm in 1990, consumption reached 40 mcm in 2000 and is expected to reach 70 mcm in 2010. In rural areas, the growth rate of demand for drinking water is 58 per cent every ten years. Demand increased from 170 mcm in 1990 to 293 mcm in the year 2000, and may reach 434 mcm in 2010.

## *Management of drinking water*

The management of water is the responsibility of various government ministries. The Ministry of Land Management, Environment and Tourism deals with the protection and conservation of water. A government agency, the Institut Géographique du Burundi (IGEBU), is in charge of gathering, interpreting and maintaining meteorological and hydrological data. The agency has 52 hydrometric stations on the main streams, 110 rain stations and 16 main climate stations distributed throughout the country.[8] The Ministry of Energy and Mining is in charge of production and distribution of drinking water in urban areas, as well as for water exploitation as a source of hydroelectric energy. Under the same ministry, the Direction Générale de l'Hydraulique et des Energies Rurales (DGHER) is in charge of supplying drinking water and electricity to rural communities.

The Ministry of the Interior and Public Security is responsible for Water Agencies in Local Administrations (Régies Communales de l'Eau), and for management and administration of hydraulic infrastructure and its maintenance in rural areas. Similarly, through the Services Municipaux d'Assainissement de la Ville de Bujumbura (SETEMU), the ministry is in charge of drainage in Bujumbura, the capital city. The Ministry of Agriculture and Livestock oversees all water use for agriculture and livestock, including irrigation techniques and marsh draining. Sanitation and hygiene at the national level are the responsibility of the Ministry of Public Health, while water management at the regional and international levels is the responsibility of the Ministry of External Relations. Many organizations of civil society as well as international organizations and NGOs are active in the domain of water, especially during recent crises.

With so many intervening, uncoordinated actors, conflicts of competence and interference in activities are common. A shortage of competent personnel limits good management of the sector.

### *Development of hydroelectric energy in Burundi*

Water and hydro-power will be crucial for the development of Burundi's standard of living. Until recently, that development was seriously retarded. Production of electricity at the national level has lately actually decreased following poor rains in 2004, a lack of spare parts, and concomitant decay of equipment in power stations. At independence the country was still supplied with electricity largely from a high-tension wire of 70 KV from the hydroelectric dam on the Ruzizi river in the (Belgian) Congo's South Kivu province. It was only in 1982, 23 years after independence, that Burundi got its first important hydroelectric plant, an 8 MW dam built by the Chinese on the Mugere, a tributary of Lake Tanganyika. Between 1982 and 1988, eight other plants were built to supply electrical power to urban centres. In 1989 a second hydroelectrical plant was built on the Ruzizi river (Rusizi II).

The Régie de Distribution d'Eau et d'Electricité (REGIDESO), which was created in 1939 by the Belgians, produced and distributed water and electricity in Rwanda–Urundi. Its head office was in Bujumbura but its managing office was at Bukavu in Congo. In 1962, REGIDESO became a state company and in 1963 it was split in two, REGIDESO for Burundi and another company for Rwanda. The Direction Générale de l'Hydraulique et des Energies Rurales (DGHER) is in charge of water and electricity supply in rural areas.

In 1994 the hydroelectric potential of Burundi was estimated at 1,371 MW. But, in terms of technical feasibility, only 300 MW could be achievable. In 1997 the installed capacity of electrical power was 43 MW, 32 of which came from hydroelectric dams, the rest from thermal power. Some 17 small hydroelectric dam installations have a total capacity slightly higher than 1 MW.

In order to meet its needs in electricity, Burundi derives power from Ruzizi I and Rusizi II, respectively of about 4 and 8.3 MW, accounting for about 30–40 per cent of consumption. In order to achieve the country's hydroelectric capacity, projects for 34 MW have been planned since 1997. These include dams on the Mpanda (10 MW), which would allow irrigation of 5,550 ha in the plain of Ruzizi, and a dam on the Kaburantwa (20 MW). Completion of these projects was scheduled for 2000 (Mpanda) and 2002–03 (Kaburantwa), but at the time of writing the projects have not even started yet.

Since 2005 another series of hydroelectric projects, on the rivers Jiji (15.5 MW) and Mulembwe (16.5 MW), have been projected. Massive new international projects – including notably Ruzizi III and at Rusumo on the Akagera river, with respective capacity of 82 MW and 62 MW – have been drawn up within the framework of NBI/NELSAP.

To summarize, the production of power in Burundi today is primarily hydroelectric and is provided by works at 27 dams, two of which are regional REGIDESO projects, with nine power stations with a total installed capacity of 30.9 MW representing more than 96 per cent of total consumption. Two of these dams (Rwegura with 18 MW and Mugere with 8 MW) produce 82 per cent of the total capacity. REGIDESO supplies mainly urban centres and the industrial sector. DGHER exploits five power stations for a total of only 0.3 MW, and private producers operate ten stations with an installed capacity of 0.7 MW. These serve rural communities.

In spite of heavy public investment agreed upon for the sector, the rate of electrification of the country (the percentage of households connected) remains very low – less than 2 per cent. The war worsened the situation: in the capital more than 2,000 connections were damaged. The embargo and freezing of external assistance also contributed to the collapse. In August 2000 a law liberalizing control of water and electricity was promulgated, allowing private investment in this sector, in the hope of speeding the recovery.

In any case, hydroelectric production represents only 0.6 per cent of the country's power. The energy sector remains dominated by traditional sources: firewood and charcoal account for 95 per cent. Consumption outstrips the production of wood. Non-electrified rural areas consume 76 per cent of this wood, or about 2.9 kg per inhabitant per day. Exploitation of peat reserves, estimated at 100 million tonnes (with 57 million exploitable), might slow deforestation. At present only one company exploits peat, producing no more than 12,000 tonnes a year or a miniscule 0.04 per cent of the energy consumed in Burundi.[9] By comparison, imported oil products represent 2.5 per cent of the total energy consumed.

**Table 1:** Managed and unmanaged marshes in the Congo and Nile basins of Burundi in 1990 and 1995.

| *Catchment area* | *Area (ha)* | | | | | |
|---|---|---|---|---|---|---|
| | *Total 1990* | *Managed marshes 1990* | *Unmanaged area 1990* | *Total 1995* | *Managed area 1995* | *Unmanaged area 1995* |
| **Congo Basin** | | | | | | |
| Ruzizi | 2,056.92 | 1,098.29 | 9,58.63 | 2,557.55 | 1,365.51 | 1,192.04 |
| Tanganyika | 2,443.93 | 1,185.48 | 1,258.45 | 3,638.00 | 2,064.68 | 1,573.32 |
| Malagarazi | 2,3517.80 | 9,891.54 | 13,626.26 | 23,517.80 | 11,294.69 | 12,223.11 |
| Total Congo Basin | 28,018.49 | 12,175.31 | 15,843.34 | 29,713.35 | 14,724.88 | 14,988.47 |
| **Nile Basin** | | | | | | |
| Ruvubu | 59,284.07 | 33,284.52 | 25,999.55 | 61,635.17 | 43,037.44 | 18,597.73 |
| Kanyaru | 17,277.36 | 5,555.44 | 11,721.92 | 18,157.30 | 5,838.38 | 12,318.92 |
| Kagera | 74,25.02 | 1,250.81 | 6,174.21 | 7,425.07 | 1,489.01 | 5,936.06 |
| Total Nile Basin | 83,986.45 | 40,090.77 | 43,895.68 | 87,217.54 | 50,364.83 | 36,852.71 |
| Total Congo + Nile Basin | 112,004.94 | 52,266.08 | 59,739.02 | 116,930.89 | 65,089.71 | 51,841.18 |

*Source:* Ministère de l'Energie et des Mines, Direction Général de l'Eau et de l'Energie, PNDE, April 1998 (modified by the author).

### *Irrigation in Burundi*

Up to now, irrigation has occupied a marginal position in Burundi. According to FAO statistics, the irrigated area in 2005 amounted to 21,430 ha, or 1.6 per cent of the cultivated area. The average annual increase during the period 1985–2000 was 2.7 per cent. Analysis of Burundi's capacity for irrigation reveals a dichotomy in terms of the country's topography.

The irrigable part, between 700 and 1,000 m of altitude, comprises the northern part of the Imbo plain drained by the Ruzizi; the central part of the Imbo, drained by the Mpanda; and the southern Imbo plain, a region of palm trees and savanna around Lake Nyanza.[10] The total irrigable area is 95,000 ha. The other part – made up of the central, north and north-east plateaus – is in the Nile basin. It is a hilly region intersected with marshlands, at 1,100–1,700 m of altitude.

Marshlands cover 120,000 ha, or 4.3 per cent of the total surface of Burundi (lakes excluded).[11] Marshes play an important role during the long dry season (July–September), as sponges that store water and retain fine mineral material eroded from surrounding hills. Underground water feeds small streams that flow into these marshes. According to the last inventory on marshlands conducted in 1998,[12] a classification can be made on the basis of two specific criteria: the nature of soil and hydrology. From a soil point of view, marshes are classified on the basis of the composition of organic material. Those with less than 20 per cent of organic material are called mineral soil; those with 20–40 per cent are organic soil; and the peat soil has more than 40 per cent of organic material. From a hydrological point of view, marshes are classified into marshes with temporary, long and permanent flooding.[13] In combining soil nature and hydrological criteria, we obtain nine classes of marshes to which is added a tenth particular category, that of floating peat marshes. Marshes may also be classified according to altitude: lowland marshes (at less than 1,000 m of altitude); medium (1,000–1,400 m); intermediate (1,400–1,900 m); and highland marshes (1,900 m and above).[14]

Most marshes are peat bogs, especially between 1,650 and 2,300 m altitude. In the Nile basin, there are many marshes in the Akagera Nyabarongo sub-basin, where the natural region of Bugesera is characterized by papyrus up to 5 m high. These marshes spread over 11,200 ha, half of which are cultivated.

Traditionally, marshes serve as livestock grazing lands and as a reserve of raw materials for house building and handicrafts such as pottery and the weaving of mats and baskets, and the manufacture of rope, fishing nets, umbrellas, and so forth. In the past, Burundians knew how to exploit marshes as a source of kitchen salt.

What is the status of the marshes today? Rapid population growth (3 per cent) has increased pressure on cultivable land, especially on marshland. Since 1990, marsh cultivation has increased from about 50 per cent to almost 100 per cent. Marshes are very sensitive ecosystems. Water is almost level with the surface and uncontrolled cultivation can generate incommensurable ecological imbalances.

The outcome may be the total drying out of the marsh and the loss of cultivable land. In the past, marshes were cultivated during the dry season only, using the reserve of water stored during the rainy season. But today cultivation takes place even during the rainy season, and there is a risk to the water balance. When the surplus of water during the rainy season is drained away to keep only the water necessary for the growth of plants, the consequence is that no water is left for the dry season. The ecological balance having been destroyed, there will be negative repercussions on the fauna and the flora, as well as ultimately an impact on food production for the human population.

Marshland management is very important, and the lack of harmonization may easily lead to tension among peasant farmers. Marshes fall under the responsibility of the Ministry of Land Management, Environment and Tourism. Inventory and planning for management of new areas is through the 'Direction du Génie Rural et de la Protection du Patrimoine Foncier'. It is this office that develops policy for the protection of land. An inventory of marshes was conducted in 1999, and a general scheme for the management of marshes was drawn up.[15]

### *Demand for water in the sub-basin*

According to data gathered by the Ministry of Energy and Mining,[16] it appears that marshes in the Nile basin make up 75 per cent of all marshland in Burundi. With the high growth of the population, exploitation of marshes is likely to increase with time. A scheme for the management of marshes was first produced in 1990.[17] This allowed for a better programme in the exploitation of marshes and has been a guide for the government in its policy and decision-making about the production of subsistence crops in the marshes. The same scheme mentions the existence of 25 sub-catchments, 13 of which are tertiary in the main basin of Lake Tanganyika and 12 tertiary in the main basin of the Nile, mainly in the drainage area of the Ruvubu river.

### *Need for water in the industrial sector*

Burundi has a very low level of industrialization. Most industry is located in the capital city, Bujumbura, in order to exploit the energy infrastructure, water distribution network, transport network, macadam roads, international airport, and other modern facilities absent elsewhere. An inventory of 80 industries and their needs for water has been made. The 'Plan Directeur National de l'Eau' made another inventory in 1998 and revealed that there are 180 industries in the country, mostly of small production units. The volume of water needed by these units is very small. Even in projections for the year 2010, only 7,000 cm per day of water will be the overall need for water in all these industries; 95 per cent of this demand will be in Bujumbura, and accounted for mainly by a brewery, a textile complex and the University Hospital.

### *Water pollution*

The most important factors in water pollution are manmade, biological and industrial. Industrial pollution, even despite the low level of industrialization, may be more important than pollution caused by households. Heavy metals, colouring, oils and detergents have been identified in rivers and channels. The sanitation services in Bujumbura cover only 38 per cent of the capital. The remaining wastewater is discharged into rainwater channels flowing ultimately into Lake Tanganyika.[18] The other urban centres have no purification stations or sewage systems. In rural areas, the most important source of pollution is human sewage. With increasing population and a lack of sanitation infrastructure, this pollution leads to outbreaks of infectious diseases such as cholera and dysentery.

## FISHERIES AND FISH CULTURE

In Burundi, modern fish culture can be traced back to the year 1980. In 1991 the government, with the assistance of the UNDP (through the 'Direction des Eaux, Pêches et Pisciculture') and the American Peace Corps, tried to re-launch this sector. An assessment revealed that there were only 40 ha of fish ponds in place. Fish farming was difficult to master because it involved a lot of new techniques, including water-quality control and fish feeding. Another limitation is that water temperatures in high attitude regions (above 1,750 m) are too low for the reproduction and growth of tilapia, the most widely farmed species in Africa.

Fisheries are almost exclusively limited to Lake Tanganyika, which constitutes a potential source of some 23,400 tonnes of fish per year. Most of the fish caught are the sardine-like *Stolothrissa tanganicae* (locally called *ndagala*) and *Limnothissa miodon* (*lumpu*) (90 per cent of the catch); *Lates stappersi* (*mukeke*) (9 per cent of the catch) and catches of the larger *Lates* make up the rest. The northern lakes, which are in the Nile basin, allow a much lower yield because of their small size. The potential catch from Lakes Cohoha and Rweru is 700–1,500 tonnes per year. Of the smaller lakes, Kanzigiri has a potential of 50–70 tonnes per year, Gacamirindi 19–25 tonnes and Rwihinda 34–42 tonnes.

### *Burundi and the Kagera basin*

The Kagera basin encompasses almost 60,000 km$^2$ with a population estimated at about 14 million people (of the 75 million total population of the countries in the basin).[19] The Kagera Basin Organization (KBO) came into existence after the failure of the East African Community. Burundi and Rwanda were immediately interested in the organization because of their landlocked situation. The two countries saw the solution to their problems of transport in access to the port of Dar-es-Salaam. The KBO was officially inaugurated in Bujumbura in May 1977 by Burundi, Rwanda and Tanzania. Uganda joined later. The organization was formally dissolved in 2004.[20]

Such an organization had in fact been mooted since 1969, when the UNDP had started lobbying with Tanzania, Burundi and Rwanda. It then proposed a technical committee in order to produce a regional plan. In 1976 the technical committee presented its work to the governments. It was clear that these countries faced similar socio-economic problems and had to work together in order to solve them. These problems included deforestation and drastic erosion, flooding, desertification, disturbance of hydrologic regimes and infectious diseases. Recognizing the scope of these problems, the idea arose of creating a community capable of planning for, and joint exploitation of, resources for the benefit of all.

The objectives of the KBO encompassed the areas of transport and telecommunication, energy, health, agriculture, livestock and the environment. Technical studies were initiated of the possibility of navigation on the Kagera. These concluded that a distance of 580 km between Lake Rweru (bordering Burundi and Rwanda) and Lake Victoria is navigable. Today, this project is still valid because it would allow avoiding or reducing costly road transport. But the only project implemented that members could be proud of was in telecommunication.

The Rusumo hydroelectric project is still in the pipeline, scheduled to be implemented within the framework of the Nile Basin Initiative. The KBO could not afford it: member states were unable to raise the 30 per cent of the total amount required by donors as their own contribution. This in turn was the result of mutual mistrust. Projects for agriculture and livestock, and for creation of a Polytechnic Institute, failed for similar reasons.

### *The Nile Basin Initiative*

Many experts[21] have argued that the Nile Basin Initiative (NBI) offers a framework for dialogue, consultation, analysis and the peaceful resolution of conflicts. It can also, if it is well understood by political decision-makers in the respective countries, become a long-term crucible for the integrated development of the basin countries. It could become a framework for the prevention of water conflicts. It groups the following ten countries: Burundi, the Democratic Republic of Congo, Egypt, Eritrea, Ethiopia, Kenya, Rwanda, the Sudan, Tanzania and Uganda.

This organization underwent a long period of gestation, dating back at least to discussions at Geneva in 1993. Formal establishment took place in 1998. Donors represented by the World Bank encouraged the project by making available $3 billion for the first phase of the project. The NBI programme was very ambitious; here we limit ourselves to a discussion of projects developed within the framework of the Nile Equatorial Lakes Subsidiary Action Programme (NELSAP).

## BURUNDI AND THE NILE BASIN

What does Burundi really represent in the Nile basin framework? Since the ten countries are gathered for the common goal of managing the Nile watercourse, what is the contribution of Burundi in this management?

Burundi covers 6 per cent of the total surface area of the Nile basin countries and represents 22 per cent of the Kagera basin area; 75 per cent of the Burundian marshlands are located in the Kagera basin. These represent an important reserve of water and help feed local populations through agriculture. The Ruvubu river contributes 2.6 mcm of water to the flow of the While Nile. Since we know that the White Nile represents 14 per cent of the total water of the Nile (84 bcm),[22] we have an idea of the importance of the Ruvubu. In fact, 22 per cent of the average annual flow of the White Nile comes from Burundi.

**Table 2:** Distribution of the Nile Basin and percentage area within the Kagera basin.

| *Country* | *Total catchment area* | *% of catchment area within the Nile basin* | *% of area within the Kagera basin catchment's area* |
|---|---|---|---|
| Burundi | 22,152 | 5.90 | 22.2 |
| Kenya | 56,834 | 15.14 | — |
| Rwanda | 25,208 | 6.71 | 33.2 |
| Tanzania | 164,296 | 43.76 | 34.8 |
| Uganda | 85,294 | 22.72 | 9.8 |
| RD-Congo | 21,680 | 5.77 | — |
| Total | 375,464 | | 100 |

*Source:* Nkokoni, *Regional development planning.*

**Table 3:** Land distribution by national divisions and Kagera basin.

| *Country* | *Area in km²* | *% in the Basin* |
|---|---|---|
| Burundi | Total: 25,680 | |
| | Kagera: 13,260 | 51.6 |
| Rwanda | Total: 26,338 | |
| | Kagera: 19,876 | 75.5 |
| Tanzania | Total: 945,000 | |
| | Kagera: 20,765 | 2.2 |
| Uganda | Total: 236,000 | |
| | Kagera: 5,820 | 2.5 |
| KBO member State | 1,233,018 | |
| Kagera Basin | 59,721 | |

*Source:* Nkokoni, *Regional development planning.*

So far, the water of the Ruvubu river is not used for irrigation. With a watershed of 12,300 km$^2$, however, the Ruvubu has obvious irrigation potential. Such exploitation could be made possible through expertise from members of the NBI. From the energy point of view, the part which belongs to the Nile basin contains 15 hydroelectrical dams whose capacity is 26 per cent compared to the current hydroelectrical capacity in Burundi.

### *Involvement of Burundi in NELSAP projects*

The Nile Equatorial Lakes Subsidiary Programme (NELSAP) is a component of the NBI. It includes countries of the Kagera basin as well as those bordering it. The objectives of NELSAP include investment in energy projects, interconnection of electrical networks in the region, management of the waters of the Nile, fisheries, agriculture development, and the control of the water hyacinth.

Among the ambitious objectives of NELSAP are the eradication of poverty, promotion of economic growth and establishment of environmental indicators. Within the NELSAP framework many projects are under way, notably including the promotion of hydroelectric energy. A number of studies have evaluated the potential of the Kagera sub-region in this regard. These studies were conducted in two phases, in February and November 2005, and were financed by the World Bank, the Canadian Agency for International Development (ACDI), and SNC-Laval International in collaboration with Hydro-Quebec International. The first phase covered Burundi, Rwanda and western Tanzania; the second incorporated the Congo, Kenya, Tanzania and Uganda.

The study estimated that demand for electricity in the six countries would increase from 2,800 MW to 16,000 MW between 2005 and 2020. By 2020, all options for electric development without serious social and environmental impact would have been exploited; projects after 2020 will carry some social and environmental risks. After consultation with the governments, the following recommendations were made:

- the Bujagali and Rusumo hydroelectric projects and Kivu methane gas project should be implemented immediately;
- the Kabu 16, Kakono and Ruzizi III hydroelectric projects, as well as a geothermal project and gas project in Tanzania, should be implemented in the near future;
- the electrical grids of the member states should be connected immediately; and
- measures must be taken to rehabilitate the hydroelectric dams in the Congo and restore its eastern electrical transmission network.

Burundi is very interested in these projects because it has a serious energy deficit. Implementation of the Kabu 16 project (20 MW) on its territory, and connection to the electrical network from the Rusumo dam (82 MW), should help, although how much will depend on Burundi's share from Rusumo, which has yet to be determined.

## CONJUNCTURE PROBLEMS: CLIMATE CHANGE AND ITS EFFECTS

A relatively rich literature describes the consequences of continuing climate change. Catastrophic predictions of famine, drought, ecosystem disturbances and a consequent recrudescence of diseases such as malaria, schistosomiasis and onchocercosis have been made. The sub-Saharan region could experience malnutrition on a scale requiring its permanent dependence on food assistance.[23] In this scheme of things, the waters of the African great lakes region might already be in a pre-drying phase.

To be sure, Lakes Victoria, Kivu, Cohoha, Rweru, Gacamirinda and Rwihinda, among others, lose important volumes of water every year. This has an impact on the ecosystems of the lakes beyond the decrease in their flow and level of water. Climate warming affects physical and chemical properties of the lakes and consequently reduces their pelagic productivity.[24] Such impacts may be more important than those caused by the direct action of man. In terms of health, the increase in temperature provokes the implantation of malaria in higher altitude regions, as has been observed in Burundi and neighbouring Rwanda;[25] scientists predict a new distribution of this disease at a global level if the tendency of temperatures to rise persists.

Whether because of climate change or temporary changes, the water levels of the lakes have been declining. Lake Kivu, which supplies the water for the hydroelectric dams of Ruzizi I and Ruzizi II, has experienced a 1 cm per year decrease in average level.[26] This has already resulted in a huge decrease – some 50 per cent – in the amount of power produced at Ruzizi. As a consequence, the towns of Bukavu, Uvira and Goma are more frequently plunged into darkness, except at strategic installations such as hospitals, officials' houses, utilities and some big companies. Specialists think that this deficit in the production of electrical power could worsen in the future as a consequence of continuing deforestation in the region and the progressive warming of the Earth.

In spite of projects in gestation for the exploitation of resources in the framework of the Nile Basin Initiative in general and NELSAP in particular, studies need to be carried out in order to establish energy forecasts and eventually to take appropriate measures in order to counter the fluctuation in the flow of rivers and lake water as a consequence of climate change. These might significantly change energy forecasts and the estimated volume of water needed for irrigation. Various scenarios are already available and often predict catastrophes for countries in the region, including Burundi. Perhaps most relevant to the present discussion, changes in the flux of rivers in Burundi are likely to have a real impact on the flow of the White Nile.

*Fluctuation of rainfall and temperature in Burundi*

In the framework of the United Nations Conventions on Climate Change (CCNUCC), the report of the National Action Programme of Adaptation of Water Resources to Climate Changes[27] provides information on climate fluctuation observed during the period 1961–99. IGEBU has published climate change scenarios for the next half-century (2000–50).[28]

For the period 1961–99, analysis of the rainfall at the Gisozi station reveals a cyclic character with a periodicity of more or less ten years between positive and negative disparity from the normal. Positive and negative disparities have coincided respectively with the increase and decrease of the level of Lake Tanganyika, which reached its highest level in 1964 (777.1 m) and its lowest level in 1950 (772.8 m).[29] Air temperature analyses indicate a persistent increase during the same period.

The outcomes of such meteorological disparities are better expressed with reference to calamities such as drought and floods, which in turn have caused famine and displacement, and increased mortality among human and animal populations. The modern history of Burundi is marked by periods of famine consistent with climatic cycles: 1917, 1923, 1931, 1943, 1961, 1984–85, 1999–2000 and 2005–06.[30]

Taking as a reference the changes of the period 1961–99 and extending the tendencies up to 2050, Burundi is likely to face a progressive reduction of the flow of rivers and the levels of lakes as predicted by IGEBU (Figure 1). Indeed, temperature projections for this period indicate a persistent increase of about 0.4°C every ten years; the total increase would be of 1.9°C by 2050. The average flow of the most important rivers, the Rusizi and Ruvubu, would decrease during this

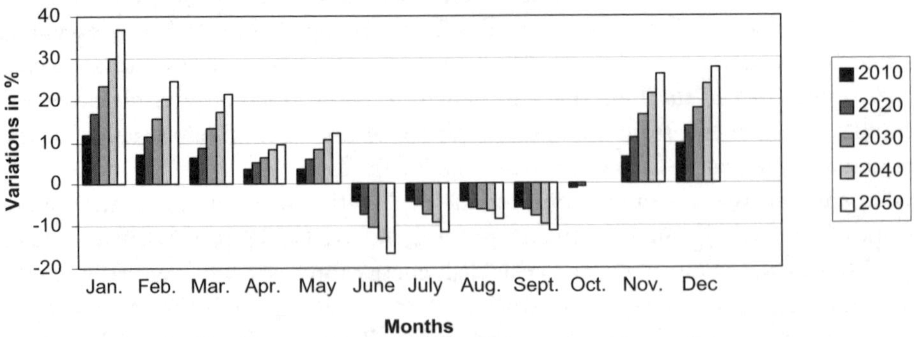


*Source:* IGEBU, 2001.

**Figure 1:** Projected rainfall variations for the period 2010–50.

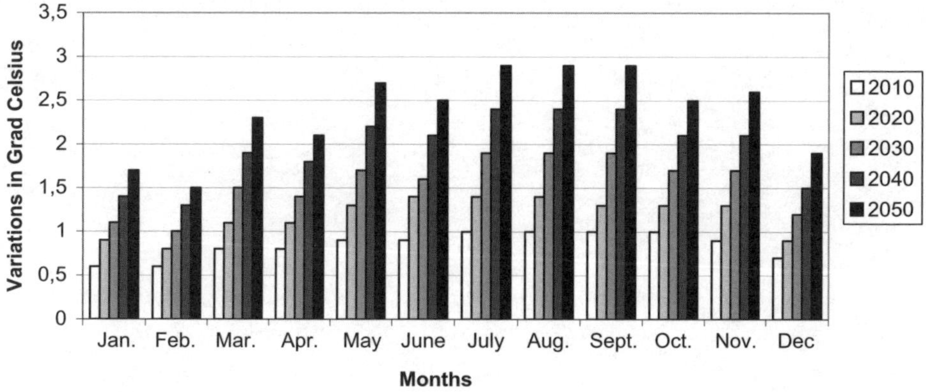


*Source:* IGEBU, 2001.

**Figure 2:** Temperature variations 2010–50.

period by 24 per cent and 30 per cent respectively. It must be pointed out that alternate scenarios call for an increase of flow and lake levels during the same period.[31]

The effects of these reductions and increases of rainfall and temperature could be huge. If current tendencies are maintained, the flow of the Ruvubu river would, in dry periods, decrease by 30 per cent (from 2.6 $km^3$ to 1.8 $km^3$ per year), which in turn would mean a loss of 0.8 $km^3$ per year for the Nile. In higher rainfall periods, the flow of the Ruvubu would increase by about 44 per cent (to 3.74 $km^3$), meaning an increase of 1.14 $km^3$ for the Nile.

Climate change might affect the life of populations not only in Burundi, but also all over the Nile basin. The repercussions (already visible today) will be felt in energy production, agriculture and food production, and availability of drinkable water.

# 3

# Rwanda and the Nile: Water Plans and their Implementation

**Robert Baligira**

This chapter provides a historical analysis of Nile water use in Rwanda since independence. It examines the water resources, catchment delineation and hydrological characteristics of the Rwandan Nile basin, emphasizing the water resources of the sub-basins that contribute to the Kagera river. After analyzing Britain's policy to control the headquarters of the White Nile, and the legal instruments applied to Rwanda during the colonial period, we shall trace the history of water's institutional development through successive regimes since 1962, a period that can conveniently be divided into three phases: 1962–90, 1990–94 and 1994–2006. Finally, we shall explore Rwanda's role in Nile basin cooperation and how this has affected water planning and economic development within the country itself.

## RWANDA DURING THE COLONIAL PERIOD

The colonial era in Rwanda began in 1897 with the arrival of the German officer Von Ramsay, who appeared at the royal court and proposed an alliance that the then Queen Mother Kanjogera immediately accepted on behalf of the king, a minor.[1] The Germans retained the monarchy and adopted a policy of indirect rule. In 1916, during the First World War, Belgium occupied both Rwanda and Burundi, and in 1919 her position there was confirmed by the Treaty of Versailles.[2]

Important changes took place under Belgian rule. Rwanda was organized into *chefferies* and *sous-chefferies* with the regrouping of royal administrative entities created since the seventeenth century, and the abolition of the old structure of chiefs of the land, chiefs of pastures and chiefs of the army. The effect of these changes was revolutionary.[3] From 1926 the Belgian administration adopted agricultural reforms to prevent a recurrence of the famines that had occurred in 1904, 1906, 1910, 1917–18 and 1924–25. Both political and administrative reforms aimed at economic change: the obligatory growing of food crops and the introduction of new crops for export.[4] A *Fond du Bien-être Indigène* was established to combat the food shortages and malnutrition brought about by periodic drought. An ironic result of the reforms was a reduction in the amount of arable land, which in turn engendered changes in agricultural methods; because traditional farmers had not managed their land by shifting cultivation and fallowing, soil exhaustion had resulted. The introduction of Christianity brought similarly revolutionary changes in thought and practice: massive conversions in the 1930s led to cultural disintegration and the rejection of ancestral cultural identity.[5]

After the Second World War, Rwanda was awarded to Belgium by the United Nations Trusteeship Council. But events during the war had aroused Africans' political consciousness, and demands for independence and the abolition of social and political inequalities ensued. In Rwanda, revolt broke out in November 1959 and continued sporadically thereafter, until Rwanda became independent in July 1962.

## PHYSICAL CHARACTERISTICS

Rwanda is among the ten countries of the Nile basin. Located in the south-west Victoria basin, it shares borders with the Democratic Republic of Congo to the west, Uganda in the north, Tanzania in the east, and Burundi to the south. The country has a total area of 26,338 km$^2$ and is divided into two main basins by the Congo–Nile ridge of some 20–50 km;[6] some 17 per cent of Rwanda lies in the Congo basin, while the rest is drained by the Nile system. Mean precipitation varies from 700 mm to 1,600 mm per year, with two rainy and two dry seasons; rainfall is least in the eastern part of the country and greatest in the west. The mean temperature is 18°C. Sometimes called the 'country with the thousand hills' because of its location in the Albertine Rift (the western part of the Eastern Africa Rift), Rwanda is one of Africa's most important sites of biodiversity, encompassing mountains, active volcanoes, hot springs, alpine vegetation, temperate forest and savanna.[7] In the Virunga volcanic chain the slopes of the highest volcano, Karisimbi (4,507 m), are still covered by glaciers.

The main tributaries of the Nile in Rwanda are the Nyabarongo, Akanyaru and Akagera, all of which empty into Lake Victoria, together with which they provide 90 per cent of the water resources available in Rwandan territory. Some studies have claimed the Ruvironzo river in Burundi as the most remote source of the Nile, giving it a total length of 6,611 km. But the 2005 expedition of McLea, MacIntyre and McGregor seems to have proven that the ultimate source in fact emerges at 2,428 m in the south-eastern part of the Congo–Nile Ridge in Rwanda, whence a filament of water springs up to form the Nyabarongo and Akagera, thus extending the Nile to 6,718 km.[8]

Information obtained through generic functions of the geographical information system and remote sensing classifies the Nile-basin part of Rwanda as three sub-catchments:

(i) The Akanyaru Haut sub-catchment is in the south-western part of the basin. That region of Rwanda and the watersheds of northern Burundi are drained by the Akanyaru river. This sub-catchment has an area of 2,800 km$^2$, an elevation between 1,325 and 2,500 m, rainfall varying from 1,050 mm to 1,390 mm, and average evaporation estimated at 1,555 mm.

(ii) The Nyabarongo sub-catchment is entirely in Rwanda and drains, among many streams, the Rukarara in the south-west and the Mukugwa flowing from the north; the Nyabarongo river crosses the country in an arc from south to north to south. The range of the sub-catchment's elevation is 1,400–4,500 m, its mean average rainfall is 1,100–1,750 mm/year, and evaporation averages 1,032.38 mm per year on the hillsides and 824.09 mm in the marshland.[9] Part of the Akanyaru river joins the Nyabarongo at Kanzenze, and another tributary originates in the mountain lakes of Burera and Ruhondo in the north. The Nyabarongo flows through the Lake Rugwero depression before connecting with the Akagera river.

(iii) The Akagera sub-catchment is located in the eastern part of Rwanda and collects water from Lake Rugwero, the Ruvubu river from Burundi, and the Muvumba river to Lake Victoria. The Akagera river itself flows through many marshlands of papyrus and small lakes. The Kagera basin and Upper Nile catchment area are particularly important in any consideration of Lake Victoria's water utilization. Other minor streams also contribute to the Akagera.

We should note that generally the hydrography of the region is dense and that the total flow of the country's basins is about 5 billion cubic meters (bcm) per year.[10]

## POPULATION

Rwanda is one of the most densely populated countries in Africa, with 326 inhabitants per km$^2$ in 2002 (as against 283 in 1991 and 191 in 1978). The population increased rapidly from 2 million in 1952 to about 5.7 million in 1987, and some 8.13 million in 2002. The Nile basin is more densely populated (with 328 people per km$^2$) than the Congo basin (with 314 per km$^2$).

One factor accounting for this rapid growth appears to have been an increase in the quantity and quality of cultivated food crops. With the adoption of new farming practices from about 1850, Rwandan society was transformed. Irrigation as well as the introduction of phaseolus beans, sweet potatoes (*ipomea batatas*) and water yams improved the food supply. Notable irrigation structures were recorded along the upper and middle Nyabarongo in Budaha and Ndiza, as well as north of Lake Muhazi.[11]

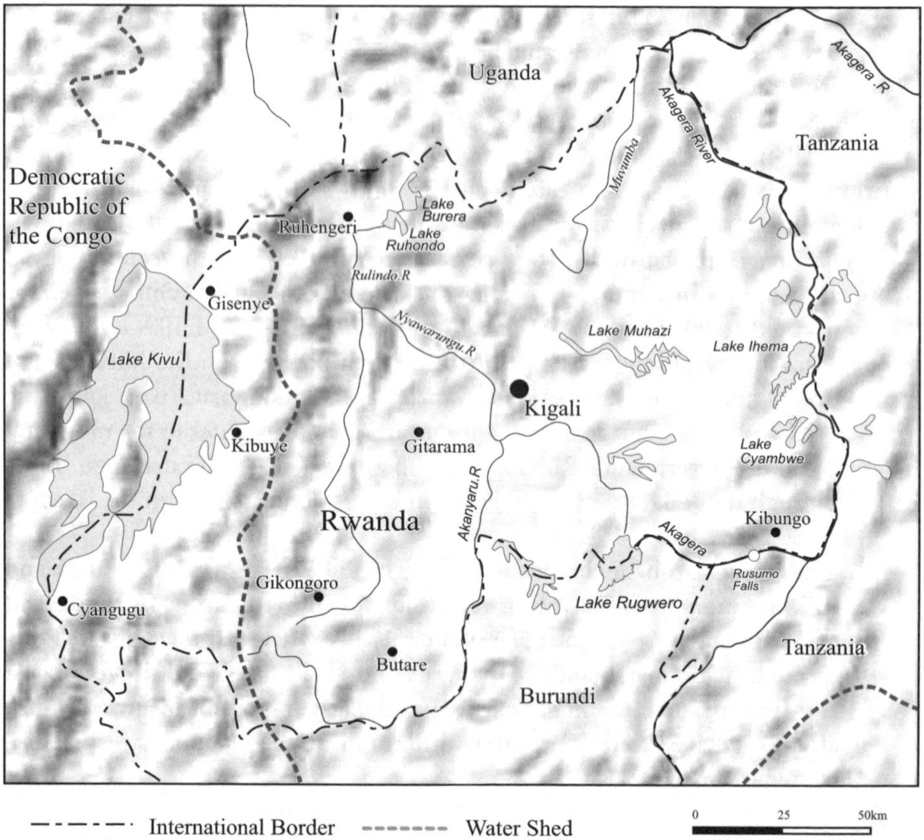


**Map 1:** Rwanda.

Although infant mortality has remained quite high, the rate of population increase was estimated at 2.6 per cent in 2002 and fertility at 5.9 per cent. During the period 1991–2002, population growth decreased (1.2 per cent) compared to the period 1978–91 (3.1 per cent). This rapid population growth has been attributed to the lack of family planning during the war of 1990–94.[12]

The Rwandan population is mostly young, with 67 per cent less than 20 years old; 52.3 per cent are female. The illiteracy rate remains rather high: 36 per cent of 15-year-olds are illiterate; only 4 per cent of women can read and write. Of the whole population, 60 per cent have had primary education, only 5.8 per cent secondary, and a mere 1 per cent have reached the tertiary level.[13]

## THE ECONOMY

Rwanda ranks among the poorest nations of the world. About half the population is under 16, and 90 per cent live in rural areas.[14] The economy is dominated by agriculture, which in 2003 represented about 45 per cent of GDP in real terms (based on 1995 prices); services accounted for 36 per cent and the industrial sector 19 per cent.[15] Since 1995 the country has made remarkable progress in rebuilding. Microeconomic performance has continued to improve despite the effects of the war and genocide of 1994. A large number of internally displaced people and returning refugees have been resettled, and progress has been made in rebuilding capacity for economic management, restoring macroeconomic stability and initiating key structural reforms.

Agriculture remains the most important sector of the economy, with 81 per cent of men and 93 per cent of women actively involved in it.[16] Coffee and tea are the major export crops. From 2001 to 2004, tea production rose from 18,268 tonnes to 29,400 tonnes; in 1998 coffee exports earned about $49 million. But agricultural production in general is based on small family farms that produce mainly for their own consumption (more than 80 per cent). Integration with the market is weak, and mainly local.[17]

The cultivated land area increased from 704,418 ha in 1990 to 826,344 ha in 2004, when crop production reached 6,099,310 tonnes, as against 4,795,071 tonnes in 1990; the peak year was 2002, with 7,098,608 tonnes. Food-crop production fell in 2003 by 4 per cent,[18] mainly because of lower-than-average rainfall. Only a few crops have been important export earners. In 2004, Rwanda's 28,000 tea farmers produced some 11,300 tonnes for export.[19] However, the production of coffee, traditionally the main export crop, has continued to decline, mostly on account of international price fluctuations and the drought; coffee for export was estimated at 19,000 tonnes in 2004, and efforts began (e.g. by the Maraba Coffee Cooperative in Butare) to produce 'full-washed' coffee, of better market potential. In 2004 commercial production of pyrethrum (a kind of chrysanthemum) was estimated at 1,250 tonnes. In recent years the government has developed a strategy with the objective of increasing rural income, enhancing food security, and moving away from subsistence agriculture to a market system.[20]

Industrialization in Rwanda really began only in 1978. Statistics show that by 2004 industry constituted about 20 per cent of GDP, mostly from manufacturing and construction, but employed less than 2 per cent of the population. Value added through manufacturing remains relatively low (about 7 per cent of GDP in 2003). Most manufacturing is small, with capital investment of less than $1 million; there is some medium-sized enterprise in beer and soft drinks, cement and textiles. Apart from coffee and tea, which constitute the base of a promising agro-industry, other agriculture activities are far from being industrialized. All other manufacturing units produce for domestic consumption and mostly for import substitution. Most factory production processes imported raw material.[21]

Access to electricity in Rwanda is very low.[22] There are four hydropower stations in the country – Ntaruka (11.3 MW), Mukungwa (12.5 MW), Gihira (1.8 MW) and Gisenyi (1.1 MW) – and one thermal station at Gatsata (2 MW).[23] Demand outpaces supply, and electricity is imported from Rusizi in the

Democratic Republic of Congo and from Uganda. The independent electrical production sector (including thermal energy and photovoltaic installations) is not well developed. The most recent figures show that the industrial sector accounts for 20 per cent of electricity consumption, and domestic consumers and the commercial sector account for 40 per cent each.

Biomass remains the major source of energy, contributing up to 95 per cent of the national needs. Wood products and plant residue provide most energy consumed at a household level, in industry and in handcrafts. Rwanda has therefore experienced intensive deforestation and environmental degradation. Some 80 per cent of imported petrol products are used for road transport.[24] From early 2003 to late 2006 fuel prices increased from 370 Rwandan francs (RwF) per liter to 572 RwF, with consequent effect on almost all other prices and on the currency, which lost over 20 per cent of its value in 2002–03 alone.

## THE DEVELOPMENT OF WATER INSTITUTIONS: HISTORICAL AND POLITICAL CONTEXT

Since the nineteenth century many laws and regulations relating to water have been promulgated, based on principles and methods still applicable today. Customary water rights are somewhat insignificant and what little remains of them today is unlikely to impede new legislation. Some scholars associate the weakness of customary water rights in Rwanda to the abundance of water resources, which meant that a few simple management principles were sufficient.[25] In any case, water in Rwanda is commonly regarded as the right of everyone.[26] Traditionally, water has a sacral character; the host offers the water of God in the absence of milk or beer, and an earthenware jar must be maintained full for this purpose. Water commanded respect; the rivers must not be polluted with refuse.

Information about water management in pre-colonial times is still scanty. All land and water in theory belonged to the king (or *mwami*), but rights to use were allocated to families or clans and passed from one generation to the next through inheritance. Chiefs of lands (*umunyabutaka*) and cattle (*umunyabukenke*) consulted, laid down rules and arbitrated disputes, for example over watering animals and rights in marshlands. Customary users of the resource always had priority over newcomers or other applicants.

From 1899 the colonial administration introduced land laws providing for the registration of title, demarcation of boundaries, and individual ownership. The water sector too was governed by a series of regulations that were modified and supplemented over time. The codes of the Congo Free State were applied to Rwanda after Belgium occupied it in 1916.[27]

The first ordinance relevant to our discussion dates from 1 July 1914, was amended in November 1936, July 1940 and March 1942, and related to the pollution and contamination of surface and groundwater. This ordinance obliged territorial authorities to determine zones of protection for springs, lakes, rivers, or parts of rivers used or potentially useful for drinking water. Under terms of legislation adopted in May 1952, each province would have a water commission,[28] and provincial governors could grant concessions for water use. Services for hydro-electric energy development, hydro-agricultural management and hydro-meteorological data management were established in 1956.

### *The first republic, 1962–73*

Rwanda achieved independence from Belgium in 1962, amid rising tensions between republicans and monarchists. Even before independence, violence had led many people to seek refuge across the international borders; refugees who had fled in 1959 repeatedly invaded the country, and the government of Gregoire Kayibanda resorted to punitive actions against Tutsis in 1959–61, 1963–64 and 1973. Violence rose to the level of physical elimination of the opposition. In 1973 northern politicians and soldiers united to bring down Kayibanda and his party. These developments occurred in a context of catastrophic violence in neighboring Burundi, where Bahutu were victims of genocide.

During this chaotic first republican period, little was achieved in terms of water development. Only two important pieces of legislation were enacted: one relating to cooperative companies adopted in November 1966, and another in April 1971 that modified the mining code of 1937. While the first meteorological service was created in 1966, the water sector continued to be managed by different institutions. Under terms of an agreement signed in 1966, the International Association of Rural Development (AIDR)[29] was the only non-governmental organization (NGO) that dealt with water supply in rural areas. Another organization[30] created during the colonial era was in charge of urban water supplies. Between 1965 and 1970, several projects for land reclamation and resettlement in Icyanya, Mayaga and Bugesera were implemented.[31] Resettlement of some 30,000 families was followed by construction of new hydraulic systems, but the agricultural sector remained dependent on rainwater. The only works executed for water management in agriculture involved marshland drainage; reclamation of the first 450 ha of the Nyabogogo began in 1970.[32]

### *The second republic, 1973–94*

Whereas the first republican regime was preoccupied with domestic affairs, its successor sought to open up the country. Rwanda experienced a period of distinct modernization, manifesting itself in relations with the outside world, urban growth, and increased investment and business activity. Diplomatic relations were established with many countries. President Habyarimana traveled frequently and reciprocated as host; Kigali held the sixth Franco-African

Conference in 1979. Rwanda was a co-founder of the Communauté économique des Pays des Grands Lacs (CEPGL) in 1976 and of the Organisation pour l'aménagement et le développement de la rivière Akagera (OBK) in 1977. Important investments in infrastructure, including a road network and telecommunications, were made.

Beginning in 1970, a series of international environmental and water conferences concentrated on basic resource needs and rights to water.[33] The 1977 Mar del Plata statement set the goal of meeting basic needs. In 1986 the United Nations General Assembly adopted the Declaration on the Right to Development, Article 8 of which declared: 'States should undertake, at the national level, all necessary measures for the realization of the right to development and shall ensure, inter alia, equality of opportunity for all in their access to basic resources.'[34] In Rwanda the launching in 1980 of the worldwide International Decade of Water provided the context for important administrative reforms regarding water management. In 1984 the General Directorate of Water (DGW) was created in the Ministry of Public Works, Energy and Water, with responsibility for managing water resources, planning and monitoring projects, and assisting communes and user associations. Overall management of water resources, including meteorology, hydrology and water for agricultural irrigation remained outside the realm of DGW responsibilities.

During this period, legislation created Electrogaz[35] as a publicly-owned company for producing, transporting and distributing electricity, water and gas. And by a presidential decree of 15 May 1987, communes took ownership of water supplies in rural areas. The perceived need for reform, however, related only to drinking water. A study carried out in 1985 by a consultancy proposed an administrative and financial arrangement for water supply management in rural areas.[36] The study recommended that users form self-financing water associations under communal control; fees collected from users were supposed to support the operation and maintenance of any hydraulic water supply networks that were installed.

Despite the difficulties encountered during the second republican period, including political instability in neighboring Uganda (from 1970 to 1980) and Burundi (1966 to 1988), important results were achieved. In the 1980s the World Bank and other organizations considered Rwanda prosperous; in 1987, Rwanda's debt was only 28 per cent of GDP, one of the lowest percentages in Africa. The country's economic performance was rather good in spite of the inherent handicaps of its landlocked status, demographic pressure, and lack of raw materials. One of the five poorest countries in 1976, Rwanda had become by 1990 one of the least poor.[37] Its record in creating infrastructure – especially in terms of road construction, postal communication, a telecommunication network and the extension of the electrical grid – was impressive.[38]

## TOWARDS CRISIS, 1990–94

The collapse of world coffee prices in the mid-to-late 1980s marked the beginning of a crisis in Rwanda. Between 1986 and 1992 coffee prices decreased by 75 per cent, contributing to a four-fold increase in the debt:service ratio. Economic downturn and government corruption led to dissatisfaction with the Habyarimana regime and its policies. The crisis peaked in 1990, when the first measures of a structural adjustment program were enacted to enable the government to reduce its deficit. This hit hardest the peasants who comprised 90 per cent of Rwanda's population.

Severe devaluations, removal of price controls and retrenchment in the civil service resulted in further economic deterioration. The consequences for salaries were rapid and dramatic. Purchasing power declined as the cost of imported goods increased. The crisis affected all sectors of the economy. The Habyarimana regime and its one-party state were seen more and more as an obstacle on – rather than the road to – further development, a view propagated mainly by urban politicians from the opposition and by the Rwandese Patriotic Front.[39] The international donor response was to employ 'positive conditionality' to promote democratization through support for the justice system, the free press, and local human rights organizations. The fall of the Berlin wall and the end of the Cold War also influenced political development in Rwanda, as international aid was increasingly linked to democratization, good governance and respect for human rights.

Apart from other inherent economic limitations, the increasing population of an already densely populated country had led to a situation in which the average peasant family possessed no more than 0.7 ha of land. Under prevailing crop patterns, families faced increasing difficulty in meeting their own needs. Severe droughts in 1989–90, 1991 and 1993, and diseases affecting two staple crops (cassava and sweet potatoes), resulted in food shortages and malnutrition affecting at least half a million people; in 1989 there was famine in the south of the country.[40] Widespread government corruption and a diversion of budgetary resources for military expenditure after the invasion of RPA forces from Uganda in October 1990 further complicated the situation. Military spending quadrupled from 1989 to 1992, when it amounted to 7.8 per cent of GDP, and subsidies to the coffee sector amounted to 46 per cent of export receipts in 1992. Among the consequences of these pressures was severe damage to the 'social safety net'.

Despite a major increase in the provision of aid to Rwanda by external donors, the water sector was affected negatively during the economic crisis. All important water supply projects were stopped in October 1990 owing to insecurity. Two major projects were canceled outright. A rural water supply project supported by the Japanese government and designed to cover a total area of 2,667 $km^2$ in Kibungo Prefecture was suspended. And the foreign experts working in the 'Alimentation en eau potable des laves' project in Gisenyi and Ruhengeri were forced to leave the region.

## AFTER 1994

Since the genocide of 1994 the international community has mobilized substantial financial support for Rwanda's recovery. Projects were carried out on an emergency basis to repair and rehabilitate damaged infrastructure. From 1994 to 1998, over 120 NGOs were registered in Kigali, many of them involved in water supply and sanitation. The first projects financed by the World Bank included an assessment of hydraulic infrastructure for further rehabilitation.[41]

A policy of decentralization adopted in 1996 brought changes in the legislative and institutional context of water. By 2004, sector policies had been developed for the environment, agriculture, health, wetlands, energy, land tenure and drinking water.[42] A series of decrees created a legal framework, and new institutions such as RURA[43] and REMA[44] were created for public utilities and the environment respectively. Under terms of an administrative reorganization, Rwanda today is composed of four provinces, and the number of districts has been reduced from 105 to 30, changes designed to decentralize the decision-making process. The new water law submitted to Parliament in 2006 emphasizes an integrated water-management approach within basins. Assistance from NGOs and the private sector has been solicited on all levels, mainly in strategic planning, financing, project execution and program evaluation. Decentralization gives local communities the right to delegate management to associations or to private operators.

## WATER RESOURCES IN RWANDA

### *Rainfall*

Rainwater in Rwanda occupies an important place in agriculture and hence in the national economy. While historical rainfall data are reliable, they misrepresent rainfall patterns. The first data describing rainfall date to 1907 and relate to the station at Save, near Butare.[45] By 1990 there were 195 weather stations in operation. The Meteorology Unit is the main agency in charge of rainfall data. Specialist agencies of the UN have contributed to development of the unit, focusing on capacity building and the supply of equipment for data gathering. The unit has 143 stations, of which 49 were established during the colonial era, 23 in 1962–73, 69 in 1973–94, and only two since then. Currently, of six automatic and ordinary rain-gauge stations that are operational, five are in the Nile basin catchment zone. Mean monthly statistics for the period 1970–2005 indicate that rainfall can vary between 23.704 and 36.87 bcm per year. The maximum monthly precipitation ever recorded between 1930 and 2006 was 281 mm one April, and the minimum for the same period of 1930–2006 was zero. Most rainfall in Rwanda – 82 per cent – is lost through evaporation. The remaining water represents 4.3 to 6.6 bcm per annum.

**Table 1:** Physical characteristics of lakes located in the Nile basin catchment.

| *Lake* | *Elevation, m* | *Surface area, km²* | *Maximum depth, m* | *Mean depth, m* |
|---|---|---|---|---|
| Bulera | 1,862 | 55.0 | 173.0 | 100.0 |
| Ruhondo | 1,764 | 28.0 | 68.0 | 40.0 |
| Muhazi | 1,443 | 34.0 | 13.0 | 10.0 |
| Mugesera | 1,374 | 54.0 | 3.8 | 3.0 |
| Bilira | 1,350 | 6.0 | 6.5 | 5.5 |
| Sake | 1,350 | 20.0 | 4.3 | 4.0 |
| Nasho | 1,300 | 13.5 | 6.3 | 4.2 |
| Rwampanga | 1,300 | 9.5 | 7.0 | 5.2 |
| Cyambwe | 1,290 | 19.2 | 6.7 | 4.1 |
| Ihema | 1,290 | 90.0 | 7.0 | 4.5 |

*Source:* Kanangire K. Canisius, 'Etude hydrobiologique des eaux rwandaises: une nécessité pour une gestion durable', *Actes du VIème Congrès du Nil 2002*, Kigali (with some modifications by the author).

### *Surface water*

Surface water in Rwanda is affected by geographical and temporal rainfall distribution. The most important flows are recorded during the great rainy season. For the Nile basin catchment, the highest river discharge occurs during April and May; the lowest water levels are in August and September. Irregularity is more marked in the Congo basin, where the strongest flows are in November or December. Rwanda's permanent rivers cover 7,260 ha and the discharge measured at different hydrological stations indicates the following values: for the Nyabarongo at Kigali station, 78 $m^3/s$; for the Nyabarongo at Kanzenze, 100 $m^3/s$; for the Akagera at Rusumo, 232 $m^3/s$; and for the Akagera at Kagitumba, 256 $m^3/s$.[46]

Lake water covers nearly 6 per cent of the total surface of the country, or approximately 1,500 $km^2$, of which 65 per cent comprises the Rwandan part of Lake Kivu in the west of the country. Some 30 other lakes belong to the Nile basin catchment. These are categorized as lakes of the north, the central region, Lake Muhazi, lakes of the Bugesera Depression, lakes of the south-east, and lakes of the Akagera national park. The Bulera and Ruhondo lakes in the north have relatively high depths (up to 100 m), while the other lakes range in depth from 5 to 7 m and hence are greatly vulnerable to pollution.[47]

Since 1990 the hydrological monitoring network has changed and only two hydrological stations, at Rusumo-Akagera and Butare Bridge on the Nyabarongo, remain in operation under the support of the FAO. The situation for water-level measurement seems better, with monitoring on Lakes Bulera, Ruhondo and Cyohoha, where water is used for electrical power generation or for drinking.

### *Groundwater*

Groundwater in Rwanda is categorized as springs emerging on hillsides; shallow aquifers with a depth up to 60 m; and deep wells up to 200 m. Springs supply more than 86 per cent of the country's drinking water. This is normally of good quality, often drinkable untreated, and thus cheap. The ground-water potential and principal hydro-geological characteristics of Rwanda were detailed in the early 1990s, when a geophysical survey was conducted to locate the water table with a view towards meeting demand in the eastern part of the country.[48] Studies showed that the water table of Rwanda's volcanic ground roughly represented a regular flow of 35,000 $m^3$ per hour at least over 750 $km^2$ of surface.[49] All survey data have been collected in a database (called Aquarium) developed in 2005.

Information on the number of springs in the country as a whole, and in the Nile basin catchment, is incomplete. Official data record only the administrative location (sector, district and province) of springs. During the colonial period some 10,200 springs were documented in this way. A more recent inventory by UNICEF listed 22,300.[50] In 2001–06 the project on Water and Sanitation in Rural Areas financed by the World Bank recorded some of these with a global positioning system. A 2002 survey in the north-east revealed widely varying rates of discharge, from 0.05 to 43.23 $m^3/h$.[51] Taking into account the mean flow of 0.4 l/s, we may estimate water potential of springs of as much as 290 million $m^3$/year.

Research has established that groundwater in Rwanda has two main origins.[52] The first category includes rainwater infiltration into soil. This water is generally of low mineral content. The second type, formed by condensation, has a high mineral content and concentration of chlorine, and is found at greater depths. Groundwater exploitation in Rwanda dates from 1955–59, when 300–400 wells were sunk in Byumba, Kigali Ngali and the town of Kigali. These wells were drilled to a depth of 8–12 m, had casings of 80 cm, and produced a flow of 1–6 $m^3/h$.[53]

More recently, a number of well-sinking projects have been undertaken. In 1985, 'Alimentation en eau potable de la région Orientale du Rwanda', a project assisted by the Japanese International Cooperation Agency, made the first drillings to depths of 30–100 m. This study conducted a drillings survey, and used the Schlumberger method and electromagnetic EM 34 to obtain longitudinal profiles of bedrock resistance. By 1991 the project had sunk 72 wells in the Eastern Province, 80 per cent of which are still operating. In 1998 the Lutheran World Foundation (LWF) launched a campaign to sink 75 wells in the former province of Umutara, resulting in 58 positive drillings. In 2002 the 'Projet Développement des Ressources Communautaires et d'Infrastructures d'Umutara' (PDRCU) carried out 58 drillings, of which 35 were positive. In 2003–04, Médicins Sans Frontières (MSF Belgique) constructed 23 wells on the Island of Nkombo in Cyangugu Province. In July 2004, Electrogaz contracted with the Foraky company to carry out test drillings along the main rivers around Kigali and in Ruhengeri

Province, in order to investigate the nature and thickness of the alluvial deposits and analyze the possibility of groundwater exploitation in the bedrock. The reports of these studies contain water-quality analyses as well as pumping test parameters.[54]

Apart from these boreholes, drilling of large diameter and low depth has been carried out with private financing. These are primarily limited to the alluvial deposits; water is withdrawn for domestic use, mainly for cleaning. In some cases this water is used for animals. The wells operate without official cognizance, so we have no data on their number, location, design, flows, and so forth. Other boreholes of relatively low depth were drilled by NGOs during the emergency period of 1994–98.

## WATER RESOURCES: PLANS AND DEVELOPMENT

During the six decades of German and Belgian colonial occupation, economic development was of little importance. Colonial policy aimed to safeguard traders. The first plans for water use were developed only after independence, with increased economic activity in several fields. Most planning focused on supplying water for domestic use and on the reclamation of marshes. Figure 1 shows how the general water demand for agricultural crop production has increased with population growth: from 1962 to 2005 demand almost doubled, from 1.71 bcm to 3.53 bcm.

### *Agriculture*

Hilly Rwanda is unsuited to mechanized agriculture. Farming along the hillsides as well as in bottomlands and marshes depends on rainwater, and the two rainy seasons allow two harvests a year. The first crop is harvested from June to September, immediately after the great rainy season; production is endangered if the cropping period is delayed by prolonged drainage from the river banks. The second crop, from November to March, is threatened if the rainy season involves floods or if the great rainy season begins too early. In general, during the great rainy season flooding precludes agricultural activity in the valleys.

The Belgians introduced commercial agriculture and compulsory labor during the 1920s and 1930s. The colonial government started the *paysannat* system under which slightly larger rural settlements were established, along main roads, with each family receiving 2 ha of land for housing and agriculture. This policy, which aimed at intensifying agricultural production and ensuring wider distribution of land, continued after independence.

Through the 1940s the western highlands were the only region densely populated and with intensive farming. Abundant rainfall and fertile soils were especially conducive to agriculture. Farmers settled first along the upper ridges of

**Table 2:** Number and location of boreholes.

| *NN* | *Project* | *Location* | | *Number of boreholes* | | |
|---|---|---|---|---|---|---|
| | | *Basin* | *Provinces* | *Positive* | *Negative* | *Total* |
| 1 | Electrogaz | Nile | Kigali and Ruhengeri | | | 15 |
| 2 | JICA | Nile | Kibungo | | | 93 |
| 3 | AAR | Nile | Kibungo | 4 | 7 | 11 |
| 4 | LWF | Nile | Umutara | 33 | 58 | 91 |
| 5 | PDRCU | Nile | Umutara | 23 | 35 | 58 |
| 6 | SOCINCO 1962 | Nile | Kigali | | | 17 |
| 7 | AIDR:RW/1870/81 | Nile | Kigali | | | 2 |
| 8 | BURGEAP 1974 and WAKUTI 1975 | Nile | Kigali | | | 2 |
| 9 | GKW Consult | Nile | Kigali Ngari | | | 21 |
| 10 | ZOA | Nile | Kigali | | | 71 |
| 11 | MsF Belgique | Congo | Cyangugu | | | 23 |
| Total boreholes | | | | | | 404 |

Data produced by Robert Baligira, 2007.

the hillsides, where soils were more fertile and cultivation was easier than it was farther down or on the steeper slopes and in the marshy valleys. Pressure on cultivable land was intense, and as early as the mid-1950s a mass exodus from the western part of the country was under way, which culminated in a government-sponsored resettlement program (*paysannat*) of grand proportions during the 1960s and 1970s.[55] Spontaneous migration towards the eastern prefecture of Kibungo continued into the early 1980s.

The first five-year plan (1966–71) for economic and social development aimed at transforming the agricultural sector, in particular by draining marshlands to increase production.[56] During this period the government adopted a strategic plan to study marshland drainage and the development of 28,000 ha in the Nile basin catchment. To ease resettlement, water resource plans focused on supply for domestic use rather than on agriculture, which would continue to depend on rainfall. The plan nonetheless took note of the need for agricultural diversification. In 1967 sugar cane occupied only 245 ha. Rice planting started in 1963, and by 1970 some 700 ha were cultivated in the Nyabugogo valley. Cotton cultivation was impractical away from the plain of Bugarama (in the Congo basin), and its production continued to decline from 332 tonnes produced on 719 ha in 1965 to 257 tonnes on 280 ha in 1970. The first modern irrigation program started in 1970 with the inauguration of a sprinkler system using water from Lake Rumira to irrigate the soya plantation. Although this pilot project allowed an increase in crop production of as much as 700 per cent under dry conditions, it was not widely copied elsewhere.

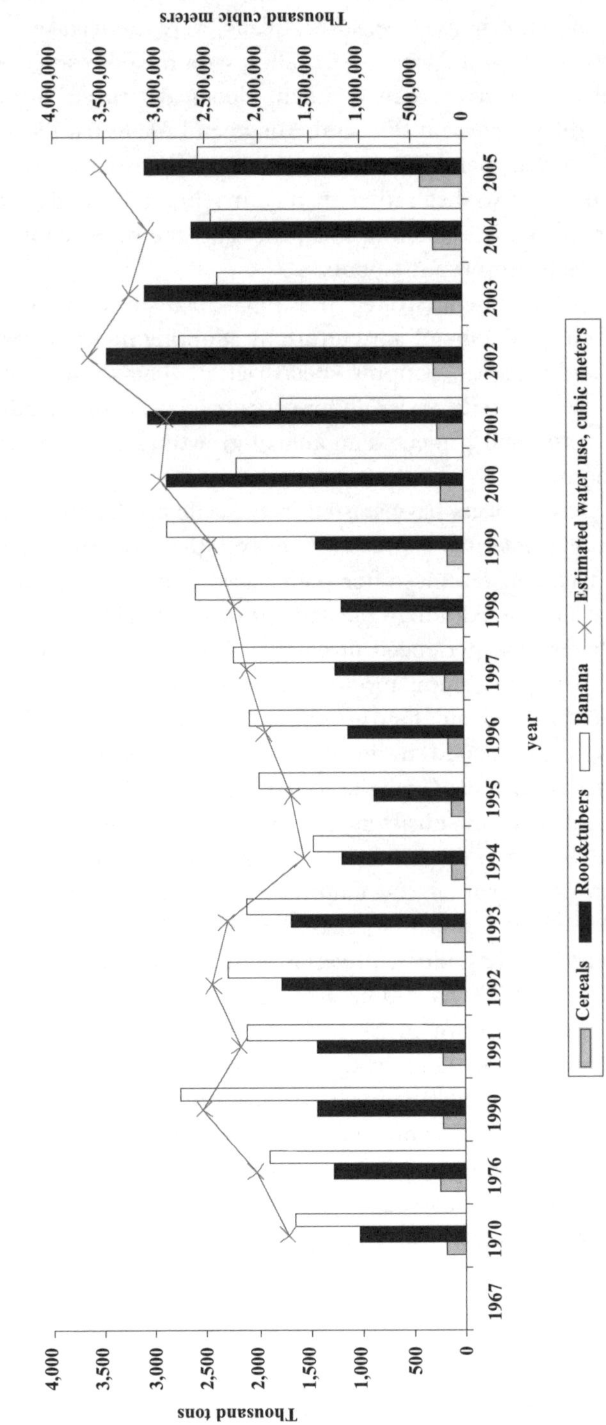


*Source:* Robert Baligira.

**Figure 1:** Chronological crops production and estimated water use.

Beginning in 1983, with the participation of the population through the community work program known as *Umuganda*, plans were undertaken for land drainage in the Nyabarongo and Akanyaru marshes, with mixed results. Attempts to modify the Nyabarongo river regime failed; floods destroyed the drainage works. Although the 1988 report on the Nyabarongo and Akanyaru Development Master Plan concluded that marsh development had had positive results,[57] these consisted of drying up the marshes rather than controlling the floods. The water level in drained marshlands has fallen dramatically and the water content of the soil there no longer meets crop requirements.

The Agricultural Development Strategy for 1998–2003 shifted policy towards maximizing the economic benefits of agriculture by adopting more intensive farm production systems and regional cropping specialization. The strategy assumed a rebalancing between sectors, with fewer and more intensive farms shedding labor to a growing urban sector, and projected an annual growth rate of 5.6 per cent in the agricultural economy.[58]

Since 2006 many new projects have started in the Nile catchment zone, where agricultural development potential is greater. One example is the Bugesera region, which has recorded frequent famines after poor harvests in the wake of drought and inadequate water control. Although blessed with considerable water resources (lakes and rivers), Bugesera experienced prolonged drought from 1998, resulting in food insecurity and massive population movements. On several occasions the government appealed for food aid for the region. A response came from Lux-Development,[59] which cooperated in implementing a project to pump some 14 million $m^3$ of water from Lake Cyohoha to irrigate 240 ha for crop production and facilitate the transition from subsistence to market agriculture.[60]

AQUASTAT, the FAO's information system for water and agriculture in Rwanda, reported in 2005 that of the country's 165,000 ha of marshland the area under controlled irrigation had increased from only 3,500 ha in 1996 to 8,500 ha in 2000. Marshlands under cultivation without any previous irrigation works were estimated at 94,000 ha.[61] The area of paddy fields had grown from 450 ha to 5,000 ha, and there are plans to increase that area to 62,000 ha by 2020.[62] Hillside irrigation of 6,300 ha in the Nile basin catchment and 200 ha in the Congo basin is also planned,[63] subject to creation of rainwater harvesting facilities in those regions. The Vision 2020 plan foresees hillside irrigation of 10,000 ha, mainly in the area characterized by rainfall deficit; development of 40,000 ha of marshlands; and construction of 140 dams. This would enable the country to increase water supply for food production, with priority for food safety and poverty reduction.

### *Hydropower*

At independence Rwanda had five hydropower stations, two of them in the Nile basin. Between 1965 and 1970 energy production increased from 10.5 GW to

23.4 GW per hour. Currently, the country's electricity comes from six hydropower stations, of which four (Ntaruka, Mukungwa, Gihira and Gisenyi) belong to Rwanda. The Ntaruka hydroelectric power station alone, between Lakes Burera and Ruhondo, produces about three-quarters of the country's electricity.[64] These hydroelectric power stations are exploited by Electrogaz.[65]

Energy demand has tended to increase with urbanization. (The population of the capital, Kigali, increased from only 6,000 at independence to 763,325 people, according to the national census, in 2002.) Overall energy consumption grew from 10.5 GWh in 1965 to 236.25 GWh in 2003. In 2000, demand in 2010 was forecast at 315 GWh at peak times.[66] After a decrease of 2.15 m in the levels of Lakes Burera and Ruhondo since 1995 resulted in a shortage of electricity, two large generators with a capacity of 12.5 MW were imported in September 2004 to supplement the output of Mukungwa and Ntaruka.[67] Under the terms of a 1995 agreement, the Uganda Energy Bureau undertook to supply 5 MW of electrical power to Rwanda, delivered at the Rwanda–Uganda border.[68]

Electrical power production, import and export over the period 1965–2003 are shown in Figure 2. The general trend is of decreasing levels of upstream storage, and a concomitant decline in the amount of energy produced by the hydropower plants of the Nile basin, mainly because of insufficient rainfall and mismanagement.

Various studies have addressed energy demand and the need to reduce the expensive import of oil. A Hydroelectric Master Development Plan commissioned before the 1990–94 war could not be implemented. An assessment of the hydropower potential of 100 sites has been conducted in order to develop small and medium-scale projects on Rwandan Nile tributaries. Table 3 provides details of plans up to 2020.

Rwanda's energy policy has emphasized hydropower development based on a combination of hydraulic potential and methane. The minimal amount of extra capacity needed to close the energy gap is estimated at 125 MW. Meeting this need implies an electrical consumption growth rate of 9.6 per cent per annum and electrification of 30 per cent of the country, which would increase access to energy from 6 per cent of the population to 53 per cent. Of the big hydroelectric stations, Rusumo Falls presents the best option for expansion, though environmental

**Table 3:** Planned hydropower projects 2005–20.

| *Rivers* | *Name of plant* | *Installed power (MW)* | *Nominal discharge ($m^3/s$)* |
|---|---|---|---|
| Rukarara | Rukarara | 9 | 13.50 |
| Nyabarongo | Nyabarongo | 27 | 1.84 |
| Kagera | Rusumo | 60 | 197.00 |

*Source:* MININFRA /Unité Energie (2005).

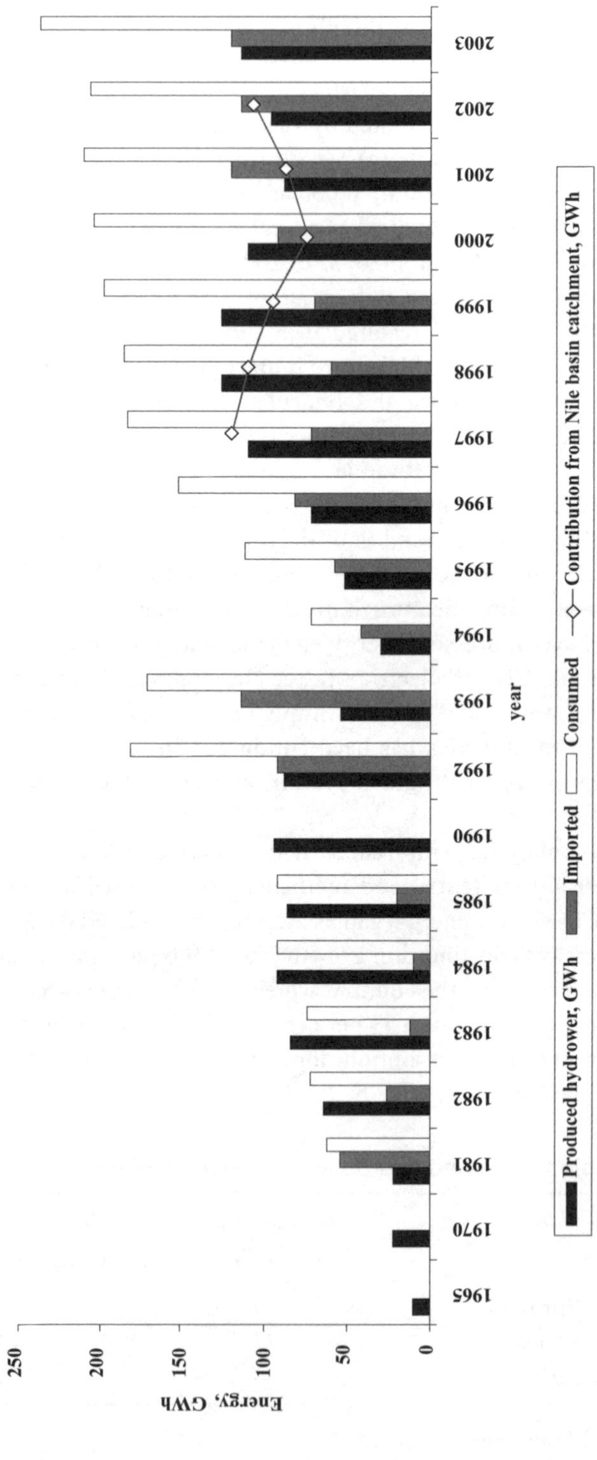


*Source:* Robert Baligira.

**Figure 2:** Relation between energy produced, energy consumed and the contribution from the Nile basin catchment.

aspects of hydropower potential and sediment transport issues have raised concerns. Weather uncertainty requires very careful attention to both national demand estimates and the dimensions of new works.

### *Drinking water*

Domestic water supply has developed more quickly than other sectors. In Rwanda, water for drinking purposes comes from springs, rivers, lakes and groundwater. Since independence, the supply and improvement of water in rural and urban areas has been entrusted to separate agencies. The third General Census estimated the country's safe drinking water requirements at 82 million $m^3$ per year, and access to drinking water at 54 per cent in urban areas and 44 per cent in rural areas. In the countryside, two delivery systems can be identified: the simple conveyance of water by gravity; and complex methods using pumps to lift water to users. In 1971 the country had five gravity supply systems of some 168 km and 17 pump systems of 48 km. All of these were financed by Belgium or through church or other bilateral donations.[69] By 1985 there were 292 simple water supply systems for rural areas, 243 of these of the gravity flow type, and six with purification facilities; some 3,939 km of piping and 2,459 supply connections supported the systems, which supplied 44,000 $m^3$ daily to an estimated 555,000 households.[70] Currently, of 809 water supply systems in Rwanda, 698 (nearly 86 per cent) are fed by one or more springs.[71]

Water distribution in urban areas is the responsibility of Electrogaz, which supplies most of the 12 major towns, including nine in the Nile basin. Of the water distributed in Kigali, 67 per cent comes from the Yanze river and groundwater from the Nyabarongo valley, and 10 per cent from Lake Mugesera via the treatment plant at Karenge; the rest comes from springs around the town.[72]

Access to drinking water in Rwanda is therefore very uneven. In many places people must travel great distances for water from lakes and rivers that are in any case often polluted. A survey carried out in 2004 revealed that some 61 per cent of the population used only 8 liters or less of water per day, which is far below the standard established by the World Health Organization.[73] The same report noted that fully 41 per cent of people suffer from diarrhea and that 83 per cent of children under the age of five suffer from water-borne diseases. Poor sanitation and the very low level of women's participation (1.36 per cent) in water and sanitation activities are seen as the main constraints on development.

### *Industry*

Rwanda has little modern industry, and what there is consumes very little water. Industrial activity is concentrated in the Nile basin catchment, where some 72 per cent of the country's industrial water use takes place.[74] Processing industries (e.g. soap, soft drinks and beer, mineral water) tend to locate in urban areas, where water supplied by Electrogaz to industries is equivalent to 6.67 per cent of its total output.[75] The agro-industry currently under development will certainly influence water demand in the future. Water use by the industrial sector (including tourism), now about 300,000 $m^3$ per year, is expected to triple by 2020.[76]

### *Navigation*

Water transport is relatively insignificant in Rwanda except on Lake Kivu. On most of the rivers and lakes in the Nile basin, however, dugouts and other boats of rudimentary manufacture are still used. Navigability studies will be undertaken within the framework of the Vision 2020 plan.

### *Fish farming and aquaculture*

Fish farming in Rwanda dates to 1950. This involved three species of tilapia, with good results in some places. Since 1980 piscicultural development has intensified. A study led by USAID and FAO showed that rural fish farming was both feasible and potentially profitable. FAO thereupon framed a four-year project, the main object of which was to create 2,400 ponds supporting the production of 180 tonnes of fish per year.

Research was undertaken in 1983 to identify the best techniques for adaptation to local conditions, and in 1985–86 a suitable technology was applied. Total potential production is estimated at 7,400 tonnes per year. In contrast, recent national production has been between 2,000 and 3,500 tonnes; the total catch, brought in by about 5,500 fishermen, accounts for only 0.3 per cent of GDP, but the sector is thought to generate about 35,500 jobs. Lake productivity is estimated at nearly 40 kg/ha per annum, compared to the African average of 150 kg.[77] But the constraints on piscicultural development are daunting: a lack of qualified technical supervisory staff, low local demand for fish, the expense involved in pond construction, and finally a lack of clarity in planning and implementation.

## RWANDA AND THE NILE

British imperial strategy called for control of the Nile from its mouth to its sources. The British regarded the Nile basin as a single unit and planned to establish control over its waters both technically, through a series of dams, and legally, through treaty arrangements with or on behalf of riparian polities. Bilateral treaties were concluded with various European powers from 1885 onwards to regulate water utilization. In July 1890, London reached an agreement with Germany that for the first time officially recognized the Nile Valley as a British 'sphere of influence'. For British strategists it was not important to control the whole of Lake Victoria, but only those parts crucial to supplying the Victoria Nile.[78] In

1894, London reached an agreement with King Leopold II of Belgium that gave the Congo international recognition as an independent state, and thus helped to thwart French ambitions on the Nile.

The legal position of successor states with regard to the international treaty commitments undertaken by colonial regimes is complicated. Did independence involve a change in circumstances so fundamental as to render colonial-era treaties untenable? In the event, most of the independent Nile Valley countries renounced such agreements, in some cases invoking the Harmon doctrine, which holds that a state has the right to do whatever it chooses with the water that flows through its boundaries, regardless of the effect on any other riparian state. On the other hand, competing principles of international law have allowed lower riparian states, especially Egypt, to argue that the Nile water treaties are binding in perpetuity.[79]

### *Nile treaties and agreements applied to Rwanda*

International agreements during the colonial period and since independence affect Rwanda's position regarding the Nile waters. The earliest was an Anglo-Belgian convention of May 1906, which came into effect only in August 1924 when by decree all agreements and treaties entered into by Belgium before its occupation of Ruanda–Urundi in 1916 were made applicable. The Anglo-Belgian agreement of November 1934, which came into effect in May 1938, was the most important of the colonial era. It modified the boundary between the Tanganyika Territory and Ruanda–Urundi, and defined the parties' rights in respect of the water of those rivers and streams that formed parts of that boundary. Regarding the Nile waters, Article 1 of the agreement stated that water 'diverted from a part of a river or stream wholly within the Tanganyika Territory or Ruanda–Urundi shall be returned without substantial reduction to its natural bed at some point before such river or stream flows into the other territory, or at some point before such river or stream forms the common boundary between the two territories'.[80]

When Rwanda became independent, the following declaration was made:

> The Rwandese Republic undertakes to comply with the international treaties and agreement concluded by Belgium and applicable to Rwanda which the Rwandese Republic does not denounce or which have not given rise to any comments on its part. The Government of the Republic will decide which of these international treaties and agreements should in its opinion apply to independent Rwanda, and in so doing will base itself on international practice.[81]

Since the 1934 agreement was never specifically renounced, it remains valid as far as the government of Rwanda is concerned, and therefore applies in Rwanda to water pollution monitoring and the mechanisms for water-flow measurement in the main tributaries of the Nile.

### *The Kagera river and the Nile basin*

The Kagera river is the main feeder of Lake Victoria.[82] The river takes water from streams in Rwanda, Burundi, Tanzania and Uganda. Its catchment area lies between latitudes 0°45' and 3°55' south, and longitudes 19°15' and 31°51' east; the basin is approximately 345 km long and has a maximum width of about 275 km, with a total area of 59,675 km$^2$.

In Rwanda, the main tributary of the Kagera is the Nyabarongo, which has a length of 305 km from its sources to its mouth at Lake Rugwero. The basin of the Nyabarongo has slopes ranging from 237 to 147 cm/km in its upper reaches for a distance of 30 km, and then reduces to 71 cm/km for a distance of 220 km. The river passes through large areas of swamp, and connects lakes in the lower reaches: the most important are Lakes Rugwero, Sake and Mugesera, which provide good natural regulation. The main tributary of the Nyabarongo is the Akanyaru, which originates in Rwanda and Burundi and has a catchment area of about 5,285 km$^2$. The total area of the Nyabarongo catchment is 18,600 km$^2$, including the Akanyaru basin.

In Burundi, the Ruvubu river joins the Kagera 496 km south-east of the Rwandan border. The Kagera basin's high slope in its upper reach is some 150 cm/km for a distance of 30 km, then 42 cm/km for a distance of 60 km, then 110 cm/km for 160 km, and finally 26 cm/km for 57 km. From Burundi, the total catchment area is estimated at 12,300 km$^2$. The Kagera's middle reach begins from 2 km downstream of its confluence with the Ruvubu before Rusumo Falls. The total length of this reach, to Kyaka Ferry, is 128.7 km. Along this reach the Kagera passes through steep cliffs at Rusumo Falls, with a drop of 12–15 m and a width of 25 m, before the valley widens and the river enters a large area (some 1,070 km$^2$) of swamp. The Kagera connects a number of lakes, of which the most important are Lakes Rwamanga, Ngoma, Rwehinkama, Nasho, Ihema, Kivumba, Hago, Rwankanjunju, Rwanya, Rushwa and Nyaruwale, with a total area of about

**Table 4:** Hydrological condition of the Kagera basin, by country.

| *No.* | *Country* | *Kagera catchment area, km²* | *Mean precipitation, mm* | *Area within the basin, %* | *Estimated rainfall volume, bcm* |
|---|---|---|---|---|---|
| 1 | Burundi | 13,060 | 1,180 | 22.50 | 15.41 |
| 2 | Rwanda | 20,550 | 1,182 | 35.00 | 24.29 |
| 3 | Tanzania | 20,240 | 1,100 | 22.50 | 22.26 |
| 4 | Uganda | 4,520 | 1,000 | 8.00 | 4.52 |
| Totals | | 58,370 | | | 66.48 |

*Source:* C.P. Gasarasi, *The Process of Regional Integration within the Organization for the Management of the Kagera River Basin*, Dar es Salaam 1979 (with modifications).

**Table 5**: Kagera basin characteristic parameters.

| *No.* | *Designation* | *Sub-catchments* | | | | |
|---|---|---|---|---|---|---|
| | | *Nyabarongo* | *Akanyaru* | *Kagera* | *Ruvubu* | *Mwisa–Nkono* |
| 1 | Surface area, $km^2$ | 13,099 | 5,329 | 10,401 | 12,263 | 17,257 |
| 2 | Basin length, km | 161.3 | 105.3 | 179.1 | 188.1 | 216.1 |
| 3 | Basin width, km | 90.0 | 45.7 | 63.3 | 85.8 | 90.9 |
| 4 | Perimeter, km | 1,228 | 630 | 985 | 967 | 1,323 |
| 5 | Rainfall, mm | 1,100–1,750 | 1,050–1,390 | 1032.38–824.09 | 1,150 | 1,100 |
| 6 | Compacticity index C==0.28P/A^1/2 | 3.004 | 2.416 | 2.704 | 2.445 | 2.819 |
| 7 | Aspect ratio, d=L^2/A | 1.986 | 2.081 | 3.084 | 2.885 | 2.706 |
| 8 | River length, km | 309 | 125 | (134–220) | 258 | (178–64) |

200 $km^2$. The size of the Middle Kagera basin from Lake Rugwero to Murongo Ferry has been estimated at 13,175 $km^2$.

The Kagera's lower reach runs from Kyaka Ferry to the river's mouth at Lake Victoria. Along this reach, with a mean land slope of 18 cm/km, there are two main tributaries: at 5 km upstream of Kyaka Ferry the Mwisa river, with a mostly swampy catchment area of 2,050 $km^2$ with heavy papyrus growth, joins the Kagera; and at 58 km the Ngono river from Tanzania, with a catchment area of 2,600 $km^2$, joins.[83]

We divide the Kagera basin into five sub-catchments coinciding with the main tributaries of the Kagera river: the Nyabarongo sub-catchment, including the Nyabarongo river, which the Digital Elevation Model (using data from the Shuttle Radar Topography Mission) confirms as the most remote source of the Nile; the Akanyaru; the Ruvubu; the Kagera; and the Mwisa–Ngono sub-catchment. Information on the sub-catchments and their hydrological characteristics is given in Table 5.

The gross area of the Kagera river catchment is 58,370 $km^2$ and the total length of the Kagera, assuming the Nyabarongo river as its farthest source, is 801 km. The average annual flow recorded at Kyaka Ferry during the period 1940–71 was 184 $m^3/s$; the maximum flow was estimated at 540 $m^3/s$ and the minimum at 103 $m^3/s$.

### *Kagera basin cooperation*

Kagera basin cooperation began in 1968, after the East African countries, seeking to relieve the areas surrounding Lake Victoria, increased the outflow at the Owen Falls Dam by 125 per cent, which caused flooding that killed tens of thousands of people.[84] Consequently, the World Meteorological Organization proposed a hydro-meteorological survey of the lake plateau financed by the UNDP. The project was later joined by the Sudan and Egypt, which advocated planning for the entire Nile basin. In August 1977, Burundi, Rwanda and Tanzania established the Organization for the Management and Development of the Kagera River Basin (KBO). Their agreement was amended in May 1981 with the accession of Uganda.

The objective of the KBO was the integrated development of the water and land resources of the Kagera basin. This would involve provision of additional hydropower to Burundi, Rwanda, Tanzania and Uganda, and subsequently stimulate both agricultural and industrial development in the region. Kagera development plans included construction of an 80.5 MW hydroelectric dam at Rusumo Falls on the Rwanda–Tanzania border; a 2,000 km railway system and telecommunications network linking member countries; road construction; agricultural and irrigation projects; river transport; and a polytechnic institute. Three important pilot irrigation projects in the Rusumo catchment (4,450 ha), the Kyaka Kakono (16,800 ha) and Bugesera (1,000 ha) would boost the agricultural economy of the basin.[85]

Most of the projects initiated by the organization were not executed before its activities officially ceased in 2004. This failure was owed to political instability in the region; to the role of external actors in KBO planning, particularly foreign financial institutions and donor agencies; to the lack of harmonization in KBO's development plans and those of its member states; and to their disagreement over the distribution of costs and benefits.[86]

### *Rwanda and Nile basin cooperation*

After independence Rwanda joined several water-resource organizations. In 1967, with the support of the UNDP, Rwanda became a member of an early Nile basin project called Hydromet, which was intended to foster joint collection of hydro-meteorological data; Hydromet operated until 1992. From February 1971, Rwanda participated in the negotiations to establish the Kagera Basin Organization,[87] and in 1977 hosted its headquarters. In 1992–93, Rwanda, with the riparian countries of Egypt, the Sudan, Uganda, Tanzania and Zaire, founded the Technical Co-operation Committee for the Promotion of the Development and Environmental Protection of the Nile Basin (TECCONILE), to focus on a development agenda. Also in 1993, the first in a series of ten Nile 2002 Conferences, supported by the Canadian International Development Agency (CIDA), was held to provide an informal mechanism for the exchange of views among riparian countries and with the international community. Within the framework of TECCONILE, a Nile basin action plan was prepared in 1995 with support from CIDA.

In 1997 the World Bank agreed to a request by the Council of Ministers of Water Affairs of the Nile Basin States (Nile-COM) to lead and coordinate donor support for their activities. The Bank, UNDP and CIDA thus began operating as 'cooperating partners' to facilitate dialog and cooperation among the riparian states. In 2006, Rwanda became a member of the East African Community (EAC), an organization intended to boost trade, security and political stability of a region defined as including also Uganda, Kenya, Tanzania and Burundi.[88] The EAC has established a number of projects for the sustainable development of the region under the Lake Victoria Environmental Management Program (LVEMP II). The member states are concerned about environmental threats, including infestation by water hyacinth, which has brought social, economic and environmental problems to the lake basin since its first appearance in the late 1980s.

Rwanda is engaged in the general negotiations on the basin of the Nile (the Nile Basin Initiative, or NBI), and in the parallel regional action plan of NELSAP (the Nile Equatorial Lakes Sector Action Program). Currently Rwanda hosts the NELSAP secretariat, with its main office in Kigali. NELSAP has the global objectives of poverty reduction, reversal of environmental degradation and socio-economic development within the Lake Victoria basin. Twelve NELSAP projects have been adopted, targeting investments in agricultural development, fisheries development, water resources management, water hyacinth control, and hydropower development and transmission interconnection. NELSAP is already implementing the Kagera, Mara and Sio-Malaba-Malakisi management programs within the Nile Basin Initiative.

The ten Nile basin countries have agreed to cooperate in developing the river, while sharing the substantial socio-economic benefits it provides and promoting regional peace and security. The riparian states have agreed on a shared vision to 'achieve sustainable socioeconomic development through the equitable utilization of, and benefit from, the common Nile Basin water resources', and to development of a Strategic Action Program aimed at translating this vision into concrete activities and projects.[89] The strategic action program of the NBI thus has two arms: a 'Shared Vision Program' and a 'Subsidiary Action Program'. The main task of the Shared Vision Program is creation of an environment conducive for investment through a set of basin-wide activities and projects to include proposals for international financing of projects to develop a cooperative framework; confidence building and stakeholder involvement; socio-economic, environmental and sector analysis; development; and investment planning and applied training.

The Subsidiary Action Program consists of measures to be undertaken by groups of countries and comprises actual joint development projects at the sub-basin level involving two or more countries. To help identify subsidiary action programs, two sub-basin working groups of countries were established, one involving the countries of the main and Blue Niles, the other of the main and White Nile countries, including Rwanda. The sole responsibility for joint projects rests with the concerned riparian working group, but all Nile basin states may participate within a basin-wide framework.

In Rwanda the NBI cooperation framework will definitely impose itself on future national legislation. Despite significant progress in administrative reform and in establishing international cooperative management and development bodies, it is too early to evaluate their impact on water-resource management. Obvious areas of concern are the instability of these institutions, management problems that include overlapping responsibility and duplication of duties, and conflicts between and among the various authorities involved. Population growth, consumption practices and patterns, diversion of water resources, and climatic and environmental conditions have moreover contributed to the beginning of water scarcity in the basin, and Egypt, Kenya, Rwanda and Burundi have already been classified as water-scarce states. Access to the waters of the Nile is therefore now a security issue, and the matter of rights and obligations is at the center of things.

# 4

# Unlocking Economic Growth and Development Potential: The Nile Basin Approach in Tanzania[1]

**Honest Prosper Ngowi**

*The Nile has intrigued people, historians and poets since the days of Cheops up to the present day and will continue to be at the heart of regional economy, politics and culture in the decades and centuries ahead.*[2]

## INTRODUCTION

The Nile basin countries are sleeping giants. They are endowed with substantial human and physical assets which, combined with proper financial and technological resources, have the potential to spur high levels of economic growth and development. In none of the riparian countries is the potential greater than in Tanzania.

Academic literature on Tanzania[3] as a Nile riparian country and on its participation in various cooperative arrangements in the Nile basin is not extensive. Partly for this reason, the potential and actual contribution of such cooperation to the country's economic growth and development has been neglected. None of the 63 plans and reports on Tanzania in the context of the Nile[4] has focused on the country's participation in Nile basin arrangements, nor indeed have any of the thousands of titles registered in a recent comprehensive bibliography.[5] In this chapter we therefore set out to identify, document and discuss Tanzania's participation in various Nile basin initiatives in its bid to attain higher levels of economic growth and development.

A significant conclusion of this work is that, contrary to widely held views, the colonial inheritance of the Nile Waters Agreement of 1929 has not been a real barrier to Tanzania's development. On the other hand, the potential of the country's various levels of participation in five major Nile basin and sub-basin cooperative development initiatives – Hydromet, Undugu, the Technical Cooperation Committee for the Promotion of the Development and Environment Protection of the Nile Basin (TECCONILE), the Kagera Basin Organization (KBO) and the Nile Basin Initiative (NBI) – has hardly begun to be tapped.

## TANZANIA AS A NILE BASIN COUNTRY

Only a small, relatively remote part of Tanzania lies within the Nile basin, and the country has many other dependable sources of water. But that small part is vitally important to downstream states: Lake Victoria and the rivers that flow into it. Of the surface area of the lake, 49 per cent lies within Tanzanian territory, while Uganda and Kenya occupy 45 and 6 per cent respectively. Tanzanian rivers that flow into the lake make up about 61 per cent of the long-term average river discharge into Lake Victoria's catchment area; Kenyan and Ugandan rivers make up about 37.5 and 1.5 per cent respectively. The main rivers from Tanzanian catchments are the Kagera (33.5 per cent of the total discharge), Simiyu (5 per cent),

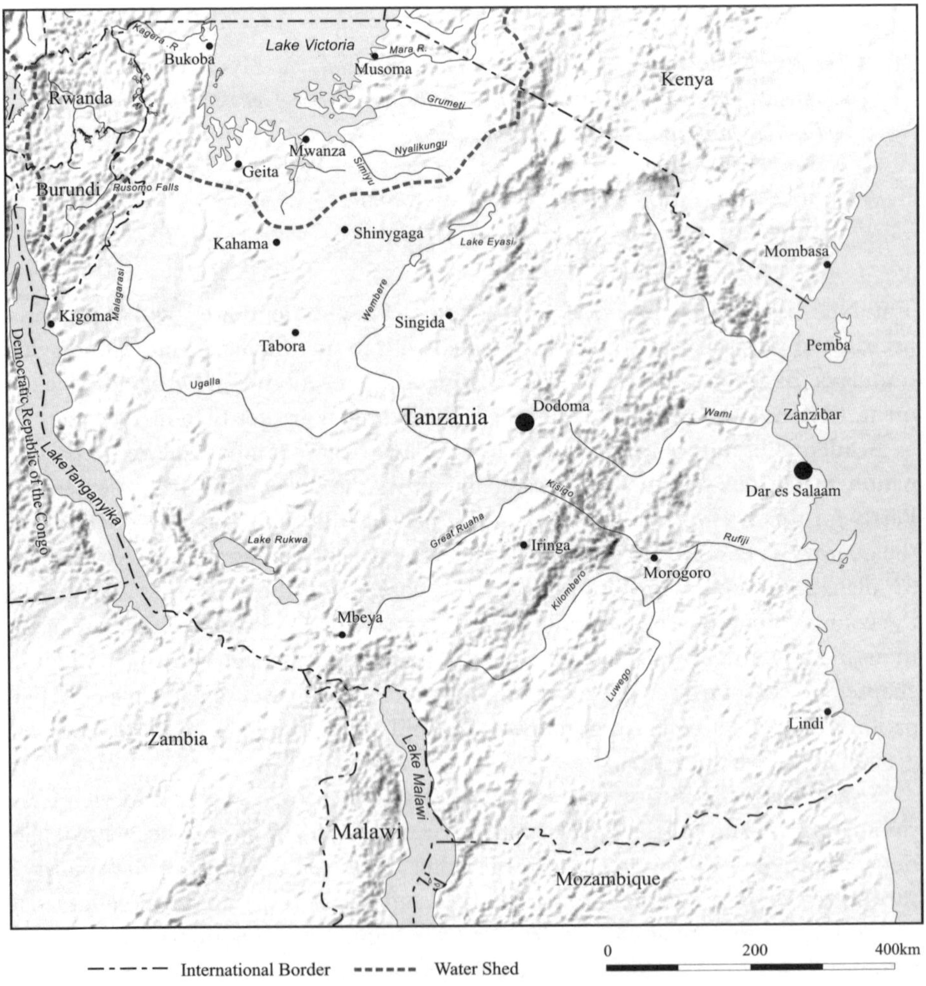


**Map 1:** Tanzania.

Mara (4.8 per cent), Isanga (3.9 per cent), Biharamulo (2.3 per cent), Grumeti (1.5 per cent) and Magogo Moame (1.1 per cent).

The Kagera is the single largest contributor of river water to the lake. By comparison, Kenya's biggest contributor is the Nzioa (14.8 per cent), and Uganda's the Katonga (0.7 per cent). To the extent therefore that Lake Victoria and the rivers feeding it are important for the Nile, Tanzania is an important riparian state. While the White Nile contributes much less than the Blue Nile to the total flow of the main Nile, the White Nile crucially provides a steady water flow. This is very important for downstream countries during Ethiopia's dry season, when the contribution of the Blue Nile drops.

The country's encounter with the rest of the Nile basin is in the area drained by the Kagera (represented by Bukoba and Biharamulo districts in Map 1), the Mwanza and the Mara (Musoma district). It is possible that water from other regions bordering Kagera, Mwanza and Mara flows to these regions, thereby forming part of the Nile. This is among areas needing further research.

In any case, Tanzania is hardly dependent on its Nile basin water resources in the way the Sudan and Egypt are. But more than half of the country receives average annual rainfall of less than 800 mm, and central and northern Tanzania, including areas immediately south of Lake Victoria, are dry for about seven consecutive months a year. Most river flows in these areas are not continuous but intermittent. The rest of the country receives more than 1,000 mm of rainfall per year, and rivers in these parts are perennial, some of them experiencing frequent floods.

## SOURCES OF WATER IN TANZANIA

Tanzania has many sources of water. The country has seven large lakes, parts of three of them – Victoria, Tanganyika and Nyasa (Malawi) – covering some 7 per cent of the country's land area. About 50 per cent of surface runoff water is derived from rivers flowing directly to the Indian Ocean, including the Pangani, Wami, Mkondoa, Ruvu, Rufiji, Ruaha, Kilombero, Mbarangandu, Matandu, Mbwemkulu, Lukuledi and Ruvuma. The remaining 50 per cent is divided into surface water drainage into inland basins with no outlet to the sea (Lake Rukwa, the Bubu Depression complex, Lake Eyasi and Lake Manyara).

The rivers Meri, Simiyu, Mara and Kagera flow to Lake Victoria. The Malagarasi drains into Lake Tanganyika, and the Songwe and Ruhuhu drain southwards into Lake Nyasa and the Zambezi river basin system. The country has about 5,439,000 ha of lakes and swamps, which represent 5.8 per cent of the total land surface.

Groundwater is a major source of supply for many areas in Tanzania. It is the most viable alternative supplement in the central and northern parts of the country's drier regions of Dodoma, Singida, Shinyanga, Tabora, Mwanza, Mara, Arusha, Coast and Southern Kilimanjaro.

A major turning point in relation to utilization of the Nile waters came soon after the country's independence in 1961 with the declaration of the Nyerere Doctrine of the Succession of States. This doctrine, which takes its name from its author, Tanzania's first president, refuted the notion that independent states are bound by treaties made by their former colonial masters. Among such treaties was the 1929 Nile Waters Agreement.

Tanzania nonetheless gave all colonial-era treaties a two-year grace period, during which time they could be renegotiated. Agreements not so renegotiated would be considered terminated unless otherwise binding under international law. Tanzania's specific objection to the Nile Waters Agreement was that the requirement to secure the prior consent of Egypt before undertaking economic activities such as irrigation, power works or other similar projects on Lake Victoria or in its catchment area was clearly incompatible with its status as an independent sovereign state.[6]

Tanzania's unique position in the Nile basin has thus raised a number of issues. Should the country insist on making use of the Nile waters to unlock their economic potential while it has such abundant alternative resources to achieve more or less the same goals? To what extent are other water resources actually developed or capable of development? In what parts of the country would Tanzania's share of the Nile waters be most feasibly exploited for attaining higher economic growth and development levels? Are other stakeholders, especially the downstream countries, willing and able to finance the development and exploitation of Tanzania's other water resources so that the Nile waters are entirely reserved for them?

Such questions may be provocative even to mention, and difficult to answer. It suffices for our purposes to note that the vital importance of the Nile waters to downstream countries necessarily increases interest, for them as well as for Tanzania, in her water resources outside the Nile basin. This alone can form a basis for cooperation between Tanzania and the other riparian countries. Such cooperation, moreover, would involve more than the geographical proximity of water sources and their points of use. Issues of technological, financial, environmental, political, and social feasibility and desirability may be equally or even more important.

## WATER ADMINISTRATION

Today, after various reorganizations and changes in nomenclature, the administration of water resources in Tanzania in general is the responsibility of the Ministry of Water and Livestock Development. The ministry is in charge of coordinating water-resource development policy, rural and urban water supplies, sewerage and drainage, drilling and dam construction, water resource institutes, central stores, a central water laboratory, river basin development, water quality and pollution control, and water boards.

Tanzania has four major water-resource management institutions. These are the National Water Board, which is a financially and administratively autonomous organization, financed from water-user and water-pollution charges and accountable to the minister responsible for water; the Basin Water Boards, also financially and administratively autonomous, financed from water-user charges but accountable to the National Water Board; the Catchments and Sub-catchments Committees, which are similar to basin boards but carry out functions delegated to them by basin water boards; and Water Users Associations, which are legally constituted bodies drawing their membership from water users in a particular locality.

## GENERAL ECONOMIC POLICY AND NILE UTILIZATION

Since independence Tanzania has undertaken various initiatives for economic growth and development, famously including the Arusha Declaration of 1967, which ushered in the policy of *Ujamaa*, or African socialism. These initiatives had little or nothing to do with the Nile waters. Since the mid-1980s, moreover, Tanzania has been in the midst of many far-reaching reforms in the management of its economy and in other spheres of life, the main objective of which has been to create an environment conducive to development of the country's natural resources. This policy is rapidly turning Tanzania into a market economy, with mixed results. But it is fair to say that this new emphasis on the private sector has, like the earlier ideology of *Ujamaa*, had little impact thus far on Tanzania's status as a Nile basin country.

## TANZANIA'S NILE BASIN APPROACH

Despite the low priority given to Tanzania's potential benefits as a riparian state, the country has participated in various Nile basin initiatives, some in relation to general development policies.

### *The 1929 Nile Waters Agreement and Tanzania's position*

The Nile Waters Agreement of 1929 between Egypt and the Anglo-Egyptian Sudan has received a lot of attention. Controversy surrounding the agreement is the obvious result of its favoring Egypt at the expense of all upstream states. But these at the time were all under one degree or another of European influence or control, and were powerless to object.

We have already noted that Tanzania's reaction to the 1929 Nile Waters Agreement came in what would be known as the Nyerere Doctrine. In keeping with this Tanzania, after independence, allowed each colonial-era treaty a two-year grace period during which it could be renegotiated. When no such invitation was received, therefore, Tanzania declared that the Nile Waters Agreement no longer bound her; it was incompatible with Tanzania's sovereign status to recognize an Egyptian right to veto projects on Lake Victoria or in its catchment area. This has remained Tanzania's official position ever since, and is one shared by all the riparian states except Egypt and the Sudan.[7]

Therefore, although Tanzania has not used its Nile water resources efficiently, it has opposed barriers to its doing so. Whether this is an indication of its aims or a position based solely on principal and without practical implications remains the great unanswered question.

### *Recent Nile negotiations*

The Nile River Basin Commission (NRBC) negotiation committee issued a draft agreement in 2004, covering wide issues with regard to the use, development, protection, conservation and management of the Nile river basin and its resources. The draft affirmed the importance of the Nile for the economic and social welfare of the peoples of the basin states, and stated general principles for use of the river's resources.

These principles include the protection, use, conservation and development of the Nile basin and its waters through cooperation between the states; adoption of sustainable development principles; commitment to 'subsidiarity' (planning and implementation at the lowest appropriate level); equitable and reasonable use of resources; prevention of significant harm to other riparian countries; the right for each country to use, within its territory, the waters of the Nile in a manner consistent with the other basic principles in the agreement; protection and conservation of the basin and its ecosystems; prior notification to other riparian countries of the possible effects of planned measures; regular reciprocal exchange of information on existing measures and on the condition of water resources; environmental impact assessments and audits; peaceful resolution of disputes; and the need for existing agreements to conform with the NRBC's framework.[8]

After subsequent negotiations the target date for signing a new Nile treaty was December 2006,[9] but this was postponed and, at the time of writing, a treaty has still not been signed. And while recognition of principles may be seen as a victory for Tanzania and other upstream states excluded from the ambit of the 1929 Nile Waters Agreement, the real issue of the moment has been and is likely to remain the actual amounts of water to which each riparian state is entitled.[10]

## PROJECTS USING NILE WATERS IN TANZANIA

Meanwhile, in conformity with its position on the 1929 Nile Waters Agreement, Tanzania is using its Nile waters. The most notable projects are described below.

### *Kahama water project*

The Kahama Project – also called the Lake Victoria Project and the Shinyanga Project, depending on whether the water's source, the administrative district or the district benefitting is referred to – is designed to pump water from Lake Victoria to the Kahama district in the Shinyanga region for domestic use. According to a report of the ministry of water and livestock development,[11] this project began in January 2003, with financing from the government.

The idea for such a project dates from 1910, when the German colonial government recognized the potential of water from Lake Victoria for irrigation purposes, transport and hydroelectric power production in the dry regions of Mwanza, Tabora and Singida. In 1956 the British colonial government conducted a feasibility study for drawing water from Lake Victoria for irrigation purposes, but the scheme was not implemented because the soil in the target areas was very salty and hence unfit for irrigated agriculture.

Subsequent plans ended in similar disappointment. In 1971–73 the government of Tanzania, in collaboration with NEDECO of the Netherlands, conducted a feasibility study on drawing water from Lake Victoria for the town of Shinyanga and some villages in Mwanza for domestic use. No report on the outcome of this study has been located.

In 1999–2000 the government, in collaboration with NEDECO, DHH, Holland Water Group, Aquanet, and the local company M-Consult, conducted another feasibility study, for a project to draw water from the lake to the Ilalambogo Mountains in Misungwi district in Mwanza and then distribute it by gravity to Kahama and Shinyanga. Tanzania's application for funds from the donor community, including Spain and the World Bank, was unsuccessful.

In January 2001, President Mkapa directed the ministry of water and livestock development to investigate the possibility of implementing government-financed water projects. The ministry hired GIBB (E.A.) Ltd of Kenya, and the local Netwas Ltd and Don Consult Ltd, to conduct feasibility studies and to prepare tender documents. The project is under way with four contractors, as summarized below. The consulting and supervising contractor for all five was Norconsult International, in collaboration with Norconsult (T) Ltd and MMK Project Services Ltd. The total value of the contracts is Tsh. 201.07 billion and the total project cost is estimated at Tsh. 225 billion. The ministry is planning to establish an agency, the Kahama–Shinyanga Clean Water Authority, to run the project.

### *China Civil Engineering Construction Corporation (CCECC)*

The CCECC was contracted for the period 15 March 2004–30 November 2006. Its project included installation of water-pumping machines and the provision of electricity for them; construction of a water tank and the laying of a 3.3 km pipeline from Lake Victoria to the water filtration point; construction of a station at Ihelele village capable of cleaning up to 80,000 l per day; construction of a 35,000 $m^3$ (35,000,000 l) water storage tank in the Mabale Mountains, and installation of water pumping machines there; building a 25 km electricity line (of 33 kV) from Bulyanhulu electricity station to Ikangala; and construction of 3.7 km of roads. Most of these projects had been completed by July 2007, at a cost of Tsh. 26.38 billion.

### *Sinohydro Corporation of China*

Sinohydro was contracted to build two projects. The first involved laying a 27.7 km pipeline from the Mabale Mountains to Mwawile village and a 21.7 km pipeline from Mwawile village to Mwamashimba ward, and then connecting ten villages to the pipeline through local off-takes. Implementation began in February 2005 and was completed in November 2006 at a cost of Tsh. 44.54 billion. The second contract was for laying a 62.1 km pipeline from Solwa village to Kahama town; laying a total of 48 km of branch pipelines to Lyabukande, Kiungu and Nduku wards; construction of a tank with a 250 cm water tower and 300 cm break pressure tank; construction of a 198 km water-supply system in Kahama town; and connecting 20 villages to the pipeline. Implementation started in March 2006 and completion was scheduled for September 2007, at a cost of Tsh. 62.5 billion.

### *China Civil Engineering Construction Corporation (CCECC)*

The CCECC was contracted to lay a 63.7 km pipeline from Solwa to Shinyanga; build an 18,000 cm water tank at Old Shinyanga and four water towers; lay a water supply system in Shinyanga town; and connect nine villages to the pipeline. Implementation began in September 2005 and was scheduled for completion in July 2007, at a cost of Tsh. 54.16 billion.

### *National Contracting Company Ltd of Saudi Arabia*

The National Contracting Company committed to expand the Bulyanhulu electricity station and construct a 28 km electricity line (33 kV) from Ikangala village to Ihelele village. Implementation began in August 2006 and was scheduled for completion in September 2007, at a cost of Tsh. 6.36 billion.

The total project capacity is to be 80,000,000 l of water per day for the 'normal' needs of 450,000 people in Kahama and Shinyanga towns, 54 villages, and livestock up to the year 2017. There are plans to expand the water machine capacity to 120,000,000 l per day in 2017 so as to cater for the needs of over 1 million people by the year 2027. The ministry is investigating the possibilities of another phase of the project, to expand delivery to other villages and to Igunga, Nzega and Urambo towns in Tabora region. This would mean using more water from Lake Victoria.

Problems already facing the project include frequent power rationing; insecurity caused by theft of property; water hyacinth growth in the Isangya river at its source in Ihelele; and inadequate funds set aside for completion of the various elements, leading to contractors threatening to stop work. Potentially more serious is the impact of changing water levels in Lake Victoria, but the government has insisted that the project's source channel was designed for the lowest recorded level of the lake.

Although the project may be seen as an example of Tanzania's refusal to honor the 1929 Nile Waters Agreement, if the water is for domestic use only then the project's impact on the water balance in the wider Nile basin may be negligible or even none at all. And while media accounts characterized the project as in 'contravention of the two treaties colonial Britain signed with Egypt and Sudan controlling the use of water from the lake', Tanzanian authorities have remained avowedly sensitive to concerns in Egypt about wide-scale use of water from Lake Victoria without consulting Cairo.[12]

### *Use of Lake Victoria waters in the mining industry*

The use of Lake Victoria water in the mining industry is another indication that Tanzania is determined to realize its resource potential. It is worth noting that gold and diamond exports, mainly from Mwanza and Shinyanga, are among the country's major sources of foreign exchange. Official statistics indicate that diamonds and gold contributed 90–94 per cent of Tanzania's total mineral exports in 2001–02.

Water from Lake Victoria is used in the world-class gold and diamond mines in Mwanza and Shinyanga. The Kahama Mining Corporation Ltd has a water supply project through a 47 km pipeline from Lake Victoria to its site at Bulyahulu in Shinyanga region.[13] The $3.4 million water pipeline also provides clean water throughout the day to over 30,000 people in surrounding villages. The Geita Gold Mine has water pumping stations and a 22 km pipeline from Lake Victoria to its plant site, with 11 take-off points freely available to villagers along the way.[14] Other mines in Mwanza and Shinyanga that are likely to use water from the lake are the Ashanti Gold Mining Company at Geita Mwanza, which started operations in 2000, and the venerable Williamson Diamonds (Mwadui) at Kishapu Shinyanga, which started its operations in 1940.

## TANZANIAN INVOLVEMENT IN COOPERATIVE EFFORTS

Tanzania has participated in various initiatives to establish and implement forms of cooperation among the Nile basin countries. Among the most notable are the Hydro-meteorological Survey of Lakes Victoria, Kyoga and Albert (Hydromet); Undugu; the Nile Basin Integrated Development (NBID); the Technical

Cooperation Committee for Promotion of the Development and Environmental Protection of the Nile Basin (TECCONILE); the Kagera Basin Organization (KBO); and the Nile Basin Initiative (NBI).

The aim of all of these has been to facilitate cooperative and mutually beneficial approaches to unlocking the resources of the basin for economic growth and development. Some initiatives have involved the active participation of certain countries, while other countries have chosen to act as observers or not participate at all. Indeed, the NBI may be the only institution so far to have brought all the riparian states together. What follows is analysis of the extent to which Tanzania has participated in the initiatives and of the extent to which participation has provided opportunities to attain higher economic growth and development levels.

## THE HYDRO-METEOROLOGICAL SURVEY OF LAKES VICTORIA, KIYOGA AND ALBERT (HYDROMET)

Hydromet was established in 1967 by Egypt, Kenya, the Sudan, Tanzania and Uganda, with the participation of the United Nations Development Program (UNDP) and World Meteorological Organization (WMO). Ethiopia became an observing member in 1971.

Its main purpose was to study, analyze and disseminate to member countries meteorological data on the equatorial lakes and rivers. This involved evaluation of water balances in the Lake Victoria catchments in order to control and regulate the lake's level and flow of water. Hydromet remained in operation for 25 years, during which useful meteorological data were gathered.

Hydromet conducted comprehensive studies,[15] including hydrological surveys of three river basins, the Kagera (which drains parts of Uganda, Tanzania, Rwanda and Burundi), the Nyando (Kenya) and the Kafu (Uganda), in order to assess their potential for development. The study of the Kagera basin used discharge data extending as far back as 1940 at Kyaka Ferry in the Kagera region in Tanzania. It found that the Kagera basin would support a variety of uses, including hydropower production, swamp reclamation, irrigation schemes, tourism and fisheries.

The upper reaches of the Nyvarongo, Akyanaru and Ruvuvu, tributaries of the Kagera, have very steep slopes. They are therefore very suitable for construction of hydroelectric power dams. Natural regulation of the river by lakes and swamps, however, is cause for concern because of potential water loss to evaporation. A comprehensive plan involves diversion of the flood flows of the Nyavarongo through Lakes Tshohoha and Regwere to reduce evaporation losses while draining the land along the Nyvarongo; use of Lake Rushwa and the Kishanda valley further to reduce evaporation losses in the lakes and to reclaim the swamps; a dam and hydroelectric power station construction at Kyansoro Falls, 240 km upstream from the mouth of the Kagera; low dam construction at Kakono on the lower

Kagera, for irrigation; control of the Ngono river for reclamation and irrigation; construction of a power plant at Rusumo Falls, where headwater control might also help in reducing evaporation losses by draining some of the swampy area upstream; and construction of dams with power plants on the upper reaches of the Ruvuvu, Nyavarongo and Akyanaru rivers.

### *Undugu*

The earliest reference to the idea of *undugu*[16] may date to a speech by President Mobutu of Zaire during the General Assembly meetings of the United Nations in October 1973. It was not until November 1983, however, that the first meeting of the organization took place, in Khartoum. The founding members were Egypt, the Sudan, Uganda, the Democratic Republic of Congo (Zaire) and the Central African Republic. Ethiopia, Kenya and Tanzania had observer status. The objective of the forum was to cooperate in the fields of environment, telecommunications, electric power, trade and water-resource development. The forum concluded after its tenth ministerial meeting, held in Addis Ababa in 1993.

Undugu had little to show for its decade of existence. Tanzania's observer status appears not to have affected its status in the organization or any benefits stemming from it.

### *TECCONILE*

The Technical Cooperation Committee for Promotion of the Development and Environmental Protection of the Nile Basin (TECCONILE) was formed in 1992. Initiated by Egypt, with funding from the Canadian International Development Agency (CIDA), it was meant to fill the void left by the defunct Hydromet. The founding members were Egypt, the Sudan, Rwanda, Tanzania, Uganda and the Democratic Republic of Congo. Kenya and Ethiopia chose to be observers.

In the short term, TECCONILE aimed to assist member states in developing national master plans and their integration into an action plan for the entire Nile basin, and to develop the infrastructure and build the capacity and techniques required for optimal use of the basin's water resources. Long-term objectives included conservation and negotiating equitable sharing of water resources. Some scholars have argued that, during the first three years of its existence, TECCONILE did little; by 1998 the organization's record was limited to a modest contribution towards the Nile Basin Action Plan.[17]

TECCONILE aimed at building the capacity and techniques required to manage the basin's water resources. Such management will be crucial for the economic growth and development of Tanzania. In order to have adequate water resource management, there is a need for appropriate capacity and techniques. Thus far, however, Tanzania has little to show in this regard from its membership of TECCONILE. Periodic drought has continued to result in food crises: a notable recent example occurred in 2001. The country needs to produce more of its own food, and for this it must turn to irrigated agriculture. This, in turn, will depend to some degree on training in proper water-resource management. Another continuing problem is the relatively poor availability of clean, safe drinking water: as late as 2003 only about half the population was so served, and until this situation is rectified the country will remain seriously at risk from water-borne diseases.[18]

The question of equitable entitlement to trans-boundary Nile basin water resources has always been the bone of contention among the riparian states. Tanzania's membership of TECCONILE is of the greatest importance, therefore, to downstream states. At the very least, the benefits of joining TECCONILE enjoyed by the other member countries, especially those that have relatively closer geographical proximity to Tanzania, should in turn have positive spillover effects in Tanzania.

### *The Kagera Basin Organization (KBO)*

The Kagera Basin Organization's roots may be traced to 1969, when the UNDP, in consultation with the governments of Burundi, Rwanda, Tanzania and Uganda, recommended that a technical committee be established to coordinate orderly regional planning in the basin. The KBO was established by those countries in 1977 (with Uganda in an observer role until it decided to become a full member in 1981).

The aim of the KBO was to 'operationalize' the concept of regional planning. To this end it would conduct studies for implementing 14 priority projects in transport and communication, energy, agriculture, and information and training. Specific projects to be implemented were the Rusumo Falls Hydroelectric Power Project; Phase II of the Tsetse Fly and Trypanosomiasis Control Project; and rehabilitation of the Miramira Hill–Ntungamo–Ishaka and Mutukula–Bukoba–Biharamulo–Lusahanga roads. In the event, none of these important projects was realized for lack of funding. KBO faced a number of other problems too, not least political and ethnic conflicts within and between member states, and the organization was dissolved in 2004.

Nonetheless, some of the development plans sponsored by the KBO remain of interest for future reference. Studies funded by the UNDP, World Bank and others were carried out by foreign firms including Norconsult, Tractebel Energy Engineering, and Tractionel Electrobel and Engineering. More than 30 such plans and reports have been documented.[19] They include studies on technical and economic justification of the interconnection of networks linked to the Rusumo Falls hydropower plant;[20] institutional and tariff studies for Rusumo Falls hydroelectric power cooperation;[21] a development program for the Kagera basin;[22] environmental impact studies of the Rusumo Falls project itself,[23] and its hydroelectric scheme;[24] and pre-feasibility studies of the Kagera river basin development.[25]

Among the critical issues arising from discussion of these studies is the failure of implementation. That the KBO was unable to raise adequate funds to finance projects raises questions about donors' financing models. Chief among those questions might be why study phases of projects were commissioned when the financing for implementation was not yet secured. This practice has led to a number of 'white elephant' studies. Ideally, financing for the whole KBO project cycle (from feasibility to implementation and monitoring and evaluation) should have been assured from the start.

### *Nile Basin Initiative*

The Nile Basin Initiative (NBI) is the most important and, currently (2007), the only Nile-basin-wide initiative of its kind. It includes the Shared Vision Program (SVP) and Subsidiary Action Programs (SAPs).

The SVP is comprised of grant-based activities to foster trust and cooperation and build an enabling environment for investments. The SVP has eight projects under way. These are the Applied Training Project, which aims at strengthening the individual and institutional capacity needed for managing water resources in an integrated manner; the Confidence-Building and Stakeholder Involvement Project, which is working to increase the involvement of a broad range of stakeholders in public awareness of the NBI's programs and future investments; the Regional Power Trade Project, which aims at trans-boundary power projects in the basin; a Socio-economic and Benefits Sharing Project to explore options to enable the Nile basin countries to work together to develop and manage the resources of the Nile for their mutual and sustainable social, economic and environmental benefit; the Trans-boundary Environmental Action Project, which provides a strategic framework for environmentally sustainable development; the Efficient Water Use for Agriculture Project, to address the technical, economic and institutional issues related to water management, productivity and efficient use; the Water Resources Management Project, which aims to ensure that water resources in the basin are developed and managed in a way that is equitable for NBI countries, optimizes mutual benefits and is sustainable; and lastly the Shared Vision Coordination Project, which oversees the effective implementation of the other projects.

The SAPs are vehicles for member countries to engage in concrete activities for long-term sustainable development, economic growth and regional integration. The objective of the SAPs is thus to translate the Shared Vision into action, realizing trans-boundary development opportunities within the agreed basin-wide framework. There are two SAPs, one for the Eastern Nile region (Eastern Nile SAP–ENSAP) and the other in the Nile Equatorial Lakes region (Nile Equatorial Lakes SAP–NELSAP). ENSAP includes Egypt, Ethiopia and the Sudan, and has as its goal the development of the water resources of the Eastern Nile basin in a sustainable and equitable way to ensure prosperity, security and peace. NELSAP's mission is to contribute to the eradication of poverty, to promote economic growth, and to reverse environmental degradation in the Nile Equatorial Lakes region, including Tanzania.

NELSAP's portfolio consists of seven projects grouped under two sub-programs. These are the Natural Resources Management and Development sub-program and the Power Trade and Development sub-program.

The Natural Resources Management and Development sub-program includes the lakes Edward and Albert fisheries project; river basin management projects; the Regional Agriculture Program; and water-hyacinth abatement in the Kagera River Basin Project. The projects of most direct relevance to Tanzania are described in more detail below.

The river basin management projects are three pre-investment programs involving the Mara, Kagera and Sio–Malaba–Malakisi rivers. They seek to establish sustainable frameworks for joint management and development in each of the three shared river basins within overall Nile basin cooperation, prepare investment plans based on the resource endowments of the basins, undertake feasibility studies of promising investment projects, and provide funding for small-scale investment projects such as water supply, irrigation, transport and watershed management and feasibility studies of large-scale investment projects.

The Swedish International Development Agency (SIDA), the European Union (EU) and the Norwegian Development Agency (NORAD) have jointly funded three schemes. The four-year, $5.183 million Mara Transboundary IWRM project, with its office at Musoma in Tanzania, began in January 2006. The four-year Kagera Transboundary IWRM project is funded with $7.942 million, is headquartered in Kigali, and also began operation in January 2006. The four-year Sio–Malaba–Malakisi Transboundary IWRM project, with a budget of $5.184 million, is located at Kakamega in Kenya.

The Regional Agriculture Program aims to promote opportunities for cooperation in the Nile basin through private investment, public–private partnerships and enhanced trade. The main focus is in the field of high-value crops and produce. It will identify steps to increase food security through increased investments and income generation. The project for water-hyacinth abatement in the Kagera River Basin Project area aims to eliminate hyacinth infestation's adverse effects on environment, health and socio-economic activities.

The Power Trade and development sub-program consists of three main projects. The first one is the Rusumo Falls Hydroelectric Power and Multipurpose Development Project, which is embedded within the Kagera basin development framework. Its objective is to ensure productive multi-purpose use of water and energy resources with investment in sustainable livelihoods in the project area. The expected funding for the project is $5 million. The second project is the Regional Transmission Feasibility Interconnection, a feasibility study for six power transmission lines connecting Burundi, the Democratic Republic of Congo, Kenya, Rwanda and Uganda. Lastly, the Strategic Sectoral Social and Environmental Assessment of Power Development Options, funded by a grant of

$1.8 million from SIDA, has prepared a NELSAP power master plan. The plan will assist the NELSAP riparian countries in selecting the best power supply options and regional transmission interconnections. It will also facilitate the participation by multilateral and private financing institutions in development.

# 5

# The Congo Nile: Water Use, Policies and Challenges

**Raphael M. Tshimanga**

## INTRODUCTION

'From the Mountains of the Moon the Egyptian Nile takes its rise. It cuts horizontally the Equator in its course north. Many rivers come from these mountains and unite in a great lake. From this lake comes the Nile, the most beautiful and greatest of the rivers of all the earth.' This description, by a seventeenth-century Arab geographer cited by Henry Morton Stanley,[1] refers to the watershed of the Rwenzori Mountains, known today as the Congo Nile.

The Congo Nile is the only watershed within the Democratic Republic of Congo (DRC) that flows northwards into the Nile; the Congo and Shiloango basins, oriented westwards, discharge ultimately into the Atlantic. Although the Congo Nile's importance in terms of the internally renewable water is less significant than that of other basins of the Nile region, its importance in the history, current regional cooperation and future socio-economic development of the DRC should not be overlooked.

The Congo Nile has played a prominent role in the geopolitics of Africa. It served as a highway for the trading of goods between the Arabs and the peoples of the Congo basin in the nineteenth century and for the slave trade that ended under Belgian occupation; and for the adventurous journeys of Samuel Baker in 1856 and David Livingstone in 1866, who were searching for the source of the Nile, and the expedition of Stanley in early 1875, during his exploration of the great Congo. It was the subject of intense diplomacy during the scramble for Africa, which notably included agreements over the Nile water resources protection between the Congo Free State and the British government in 1894 and 1906; during the post-colonial period when agreements for cooperation were reached by the Nile basin countries; and also during recent conflicts involving the natural resources of the basin.

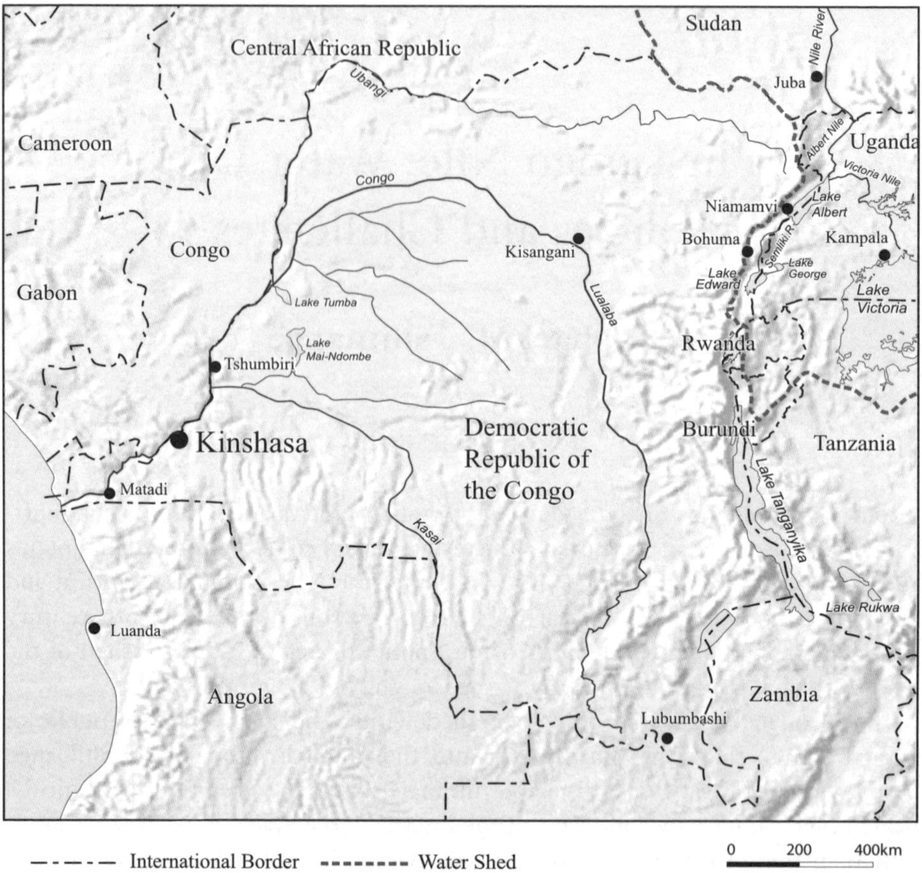


**Map 1:** Democratic Republic of Congo.

## THE GEOGRAPHICAL AND HYDROGRAPHICAL CONTEXT OF THE CONGO BASINS

The DRC extends from longitude 12.11 E to 31.21 E and from latitude 5.21 N to 13.25 S. It shares borders with nine countries: the Central African Republic, Sudan, Uganda, Rwanda, Burundi, Tanzania, Zambia, Angola and the Republic of Congo.

The hydrographical network of the DRC is the endowment of three main hydrological systems. The first is an isolated hydrological entity known as the Shiloango river, forming the political boundaries of the DRC, the Republic of Congo and Angola. The second hydrological system is the enormous Congo river of great international socio-economic and political significance, draining 90 per cent of the DRC's territory and involving no fewer than 11 riparian countries. The third river basin is the crest of the Albertine rift historically called the Congo Nile, which connects the DRC to the Nile system's other riparian states, and represents therefore opportunities for regional cooperation well beyond its borders.

### *The Shiloango basin*

The Shiloango river drains northwest Mayumbe and forms the political boundaries of Congo-Brazzaville, the DRC and Angola. The river has been considered relatively unimportant to the DRC as it remains poorly documented and outside of the national planning process.

### *The Congo basin*

The Congo basin rises most remotely from the tributaries of the Lualaba and Lufira rivers, south-east of Katanga Province, and drains about 90 per cent of the hydrological system of the country. The most distant of these rivers is the Chambezi, which, with its tributaries, rises on the southern slope of the plateau between Lake Albert and Lake Tanganyika.

The river issuing from the Katanga plateau at first flows north, then west and south, crossing the Equator twice in a great arc as it traverses a vast swamp over 4,375 km from east to west and up to 850 km from north to south. As it flows down from the Shaba plateau, the Lualaba runs for several hundred kilometers before reaching the 500 m contour at Kasongo (4°32'S/26°33'E), then rapids and a waterfall above Kindu (3°00'S/25°56'E). North of this area the river becomes much wider, 2–3 km in places, and passes through the seven cataracts of Stanley (Boyama) Falls (0°18'N/25°29'E), where it finally assumes the name of the Congo.

In its middle course the Congo varies in width between 3 and 15 km and loses only 115 m in altitude over a river distance of 1,740 km between Stanley Falls and Stanley (Malebo) Pool (4°11'S/15°35'E), a mean gradient of 1:15,000. Over 1,500 km of this stretch, between Kisangani and Tshumbiri (2°36'S/16°16'E), the river provides some 1,400,000 ha of very slow flowing open water with many of the characteristics of a lacustrine body. In this middle course, two major tributaries, the Ubangi and the Kasai, join the Congo towards the western side of the depression, providing a further 680,000 ha of open water and similar riparian environments. There are at least 12,000,000 ha of permanently or seasonally flooded forest in the Congo part of the basin, with another 6,500,000 ha on the west bank of the Ubangi, and it is quite possible that the former figure is a substantial underestimate.[2]

All the tributaries of the Congo are perennial, and flow in the river is fairly constant throughout the year. Mean annual flow rate at Kisangani is 7,000 $m^3$/sec, and at Kinshasa 41,300 $m^3$/sec. A maximum discharge rate of over 65,000 $m^3$/sec was recorded in 1908, and over 75,000 $m^3$/sec in 1961, while the record low discharge rate was 21,422 $m^3$/sec in the exceptionally dry year of 1905.[3]

Downstream at Tshumbiri the river enters a 220 km stretch fairway, where it is confined by low hills to a narrow channel, 900–1,600 m wide, and where swamp forest is largely absent. In this section the river deepens to 25–30 m and the current becomes swifter, and it is here that the Kasai enters on the south bank. The fairway ends in Malebo Pool, an almost circular lacustrine expanse of about 52,000 ha with a large central island.

In this section, the river extends into the neighboring Republic of Congo, taking in 65 million ha of very flat lowland between the 300 and 500 m contours in an ancient depression as it traverses the low rim on its way to the sea. It flows through a gorge some 360 km long and containing no fewer than 32 rapids between Kinshasa (4°18'S/15°18'E) and Matadi (5°50'S/15°52'E), descending from 307 m to 3 m. From the last rapids to the Atlantic Ocean is a distance of about 150 km, and the river is tidal for the last 95 km. Even below Matadi it flows through narrow gorges, and where it enters the ocean its bed occupies a deeply graven submarine canyon.

The potential of the Congo river for irrigation, domestic water use and industrial water supply, hydroelectric power, navigation and fisheries has been studied, and it is believed that the river is under-exploited and its flows remain consistent. Feasibility studies have been undertaken to evaluate opportunities for international interconnection, the results of which have foreseen a network between Egypt and the DRC involving the Central African Republic, Chad and the Sudan. The Inga Falls project included construction of transmission lines to Angola and Namibia, which are, in turn, connected to the network of South Africa. Visions for a southern African power grid have included western and eastern Africa, where some Nile basin countries were included. The African Development Bank funded pre-feasibility studies for this project in 1988, but the results have not yet been put to use.[4] Other opportunities for development based on the Congo river are increasingly discussed, as is the case for interbasin water transfers.[5]

As the Congo flows from the south-east, crossing the equator twice and collecting all the flows of the country's territory directed westwards to the sea, it gives rise to what may be called three lacustrine systems. The first system is part of the Great Lakes region and comprises Lakes Kivu and Tanganyika. Hard against the eastern border between those two lakes, the valley of the Ruzizi river contains extensive wetlands, as does the plain to the south of Lake Edward and the Semliki river valley draining Lake Edward into Lake Albert.

A second lacustrine system includes the lakes of the central basin. Two water bodies of major importance are Lakes Tumba (2,300 $km^2$) and Mai-Ndombe (7,000 $km^2$). Along the upper courses of the large rivers flowing northwards across the southern plateau in the Kasai Province there are innumerable large bogs and swamps, and wetlands comprise 20–45 per cent of the land surface over very large areas. Other seasonal wetlands, reed, sedge and grass swamps, flooded savannas, and riparian and riverine forests occur on the plateau in the north of the country between the Ubangi and Congo rivers.

In Bas-Congo, the province between Kinshasa and the Atlantic, the last lacustrine system encompasses the Malebo Pool and the lower course of the Congo river and several of its tributaries. The wetlands include reed and papyrus swamps around Malebo Pool and in the valleys of the Luina and Lunionzo rivers, peat bogs and grassy swamps in the valley of the Inkisi river, and periodically inundated savannas.

### *The Congo Nile basin*

The basin of the Nile is characterized by two mountainous plateaus rising some thousands of meters above mean sea level: the equatorial lake or lake plateau in the southern Nile basin; and the Ethiopian or Abyssinian plateau, which forms the eastern part of the basin.

The equatorial plateau, situated between two branches of the Great Rift, stands at a level of 1,000–2,000 m and has peaks of 5,109 m and 4,300 m. This plateau contains Lakes Victoria, George, Edward and Albert, which slope gently toward the north at an average rate of 1 m every 20–50 km. In the vicinity of this equatorial plateau rise the Rwenzori Mountains. Also famously called the Mountains of Moon, the Rwenzoris reach a height of 5,109 m near the border between Uganda and the DRC at 0°23'09 N, 29°52'18 E. The highest summits are permanently snow-capped. The Rwenzoris consist of six massifs separated by deep gorges: Mount Stanley (5,109 m), Mount Speke (4,890 m), Mount Baker (4,843 m), Mount Emin (4,798 m), Mount Gessi (4,715 m) and Mount Luigi di Savoia (4,627 m). Mount Stanley has several subsidiary summits, with Margherita Peak its highest point. The rock is metamorphic, and the mountains are believed to have been tilted and squeezed upwards by plate movement. They are in an extremely humid area, and frequently enveloped in clouds.

Over the last century, glaciers and their meltwater have formed headwaters of three main rivers in the Rwenzoris: the Mubuku flows in an easterly direction in Uganda, and the Butawu and Lusilube drain toward the west in the DRC. Under Lac Gris and Omurubaho glaciations during the Holocene, glaciers and their meltwater would also have contributed to the Ruanoli and Nyamugasani rivers. All of these rivers ultimately supply the Semliki, which discharges into Lake Albert. The contribution of the Mubuku to the Semliki occurs via a more circuitous route that includes Lake George, the Kazinga Channel and Lake Edward. The Nyamugasani discharges directly into Lake Edward. Of the remaining glaciers in the Central Rwenzori Massif, most of the largest (including the East Stanley, Speke, Vittorio Emanuele and Margherita) form headwaters of the Mubuku. The East Stanley, Margherita and Speke glaciers sustain meltwater streams that flow into Lake Bujuku, whereas meltwater from the Vittorio Emanuele glacier discharges to the east into a northern tributary (Bukurungu) of the Mubuku. Meltwater discharges from a smaller group of glaciers drain westward into headwater basins of the Semliki. From the Edward, Savoia and Elena glaciers, meltwater discharges into the

Kitandara Lakes that subsequently drain westward as part of the Butawu river basin. Meltwater discharges from the Moebius, West Elena, West Stanley and Alexandra glaciers also supply the Butawu. Meltwater discharges from the now fragmented Grant Glacier on Mount Speke form headwaters of the Lusilube river.[6]

As part of this complex hydrological system, the Congo Nile mainly comprises Lake Edward, the Semliki river and Lake Albert.

### *Lake Edward*

Lake Edward is 76 km long with a maximum width of 39 km. It is connected to Lake George, effectively a bay of Lake Edward, by the Kazinga Channel. Lake Edward reaches a maximum depth of 112 m, just 5 km from the western shore, above which the land rises precipitously to a high plateau over 2,000 m above sea level, with mountain peaks over 3,000 m. By contrast the lake floor slopes up gradually to the Uganda shore.

Lake Edward has numerous affluent streams, most importantly the Nyamugasani river from the Rwenzori Mountains, and the Ishasha, Rutshuru and Rwindi rivers from the Virunga volcanoes and the Rwanda highlands. Further inflow comes from Lake George, which also drains the Rwenzori range and to the east shares a watershed with Lake Victoria.

At the northern extremity of the lake, water spills over a rock shelf, which functions as a natural weir and discharges into the Semliki river, and thence to Lake Albert and the Nile. Apart from a database on aquatic biodiversity compiled by Hughes and Hughes[7] and other researchers, very little is known about the hydrological dynamics of Lake Edward, even though it is a feeder of the flow to the Nile.

### *The river Semliki*

The Semliki has a catchment area of 23,621 km$^2$, and is the net recipient of discharges from all rivers and lakes draining the Rwenzoris and the surrounding region.

The Semliki drains Lakes Edward and George and thus both sides of the Rwenzoris. Below Lake Edward most of its course is outside the DRC, but the main channel forms the border with Uganda before entering the southwestern end of Lake Albert through the Semliki flats. The Semliki loses 293 m of altitude over its 150 km course from Lake Edward to Lake Albert. The upper course is swift, rocky and boulder strewn, 30–40 m wide, and enclosed by the dense Ituri forest. The main descent occurs over two rapids in this part of the river. In places the river is confined to a narrow channel, only 10 m wide, between sheer rock walls. On emerging into the savanna of the lower Semliki plain, the river flows slowly, finally meandering to its delta towards Lake Albert. A chain of oxbow lakes to the west of the present lower course, and deltaic deposits in Lake Albert, indicate that the river bed has been moving eastwards over the lower plain. Swamps occupy an area 15 km wide along the southern shore of Lake Albert astride the Semliki delta, which projects a further 3 km into the lake. Upstream, seasonally inundated land extends back from these swamps in a belt 3–7 km wide for many kilometers, and a large area, centered upon the oxbows of the former river course, is poorly drained and comprises hydromorphic soils. Farther back upstream, smaller swamps occur along the Semliki at numerous isolated sites, mostly where tributaries enter the main stream. The net contribution of the Edward and George basins is combined in the outflows of the Semliki, and the flow is measured at Bweramule. The hydrological trend of the river from the discharge recorded from 1952 to 1978 shows a pattern of monthly flow variability, with low flow occurring in February and high flow reaching its peak in November.

The Anglo-Belgian agreement of 1894, modified in 1906, prohibited any work by the Congolese government that would modify the flow of the Semliki, an agreement that the successor state of the DRC has honored to this day. Taking into account population growth (about 4.6 million people in Ituri district), and the need for a safe and permanent water supply system and industrial and energy development along the Congo Nile, the agreement is likely to be revisited at some point.

### *Lake Albert*

Lake Albert lies between two parallel escarpments in the Western Rift Valley, at an altitude of 619 m, with an extreme length of 180 km and a maximum width of 43 km. Slightly less than 44 per cent of its surface is in the DRC. Its deepest point, at 56 m, lies 7 km off the western shore, from where the land rises steeply to a high plateau more than 2,000 m above sea level. The lake floor slopes gently upwards towards the east, but the eastern escarpment rises abruptly just a few kilometers from the east bank in Uganda.

The inflow to Lake Albert is generated from both the Semliki river and Lake Victoria. Approximately 100 km downstream from Lake Victoria, the so-called Victoria Nile flows into Lake Kyoga, which is a smaller, swampy lake whose mean depth is only a few meters. There are no other major inflows into Lake Kyoga, and local runoff is estimated to amount to less than 10 per cent of inflows from the Nile. The outflows from Lake Kyoga are unregulated and enter Lake Albert some 200 km farther downstream. There are no significant tributary inflows in either of the reaches between the outlet of Lake Victoria and the inlet to Lake Albert. For Lake Albert, a second main inflow comes from the Semliki, which originates at Lake George near Lake Victoria and passes through Lake Edward before flowing into Lake Albert. The outflows from Lake Albert are again unregulated and flow into the Bahr al-Jabal before reaching the *Sudd* swamps.[8]

Unlike some other water bodies of the Congo Nile, Lake Albert has been the subject of several studies that have helped to develop a water balance model.[9] The potential of Lake Albert as a reservoir has been considered for many years, because its relatively steep shores make it more suitable than Lake Victoria for overyear storage. These early studies led to detailed proposals[10] for the Jonglei Canal, which would bypass the *Sudd* as part of the Equatorial Nile Project. This project included a main reservoir to be located in Lake Albert, and a regulator at the Ripon Falls to control outflows from Lake Victoria in order to augment flows for irrigation below the *Sudd*. When extra water was required in the White Nile it could be controlled from the Albert dam and carried in a *Sudd* Diversion Canal in order to avoid losses from evaporation within the *Sudd*. In a later phase, there were plans to increase flows during periods of shortage on the main Nile at Aswan, using virtual Blue Nile storage in Lake Albert supplemented by Lake Victoria. This would require 'Century Storage' to equalize available water at Aswan over a period of 100 years and involve releasing higher flows down the White Nile in years of low Blue Nile flows. Storage in Lake Tana, on the Blue Nile, could form part of this project. The effect of flooding around Lake Albert was studied in 1956 to assist Uganda in possible negotiations, but the immediate need for overyear storage was reduced by construction of the High Dam at Aswan.[11]

Based on the rainfall regime of the Congo Nile, its seasonal pattern is bimodal and the area is considered among the zones of highest rainfall in the Nile basin, with an average annual rainfall of 1,214 mm. Precipitation in the Rwenzori Mountains occurs primarily during two pronounced seasons from March to May and August to November. The bimodal pattern results from the regional movement of air masses associated with the inter-tropical convergence zone. Apart from the seasonal control on precipitation exerted by movement of the inter-tropical convergence zone, there is a strong orographic effect on local precipitation.[12]

## HISTORICAL CONTEXT OF THE CONGO NILE

The establishment of the Belgians in the Congo, the Germans in Burundi, Rwanda and Tanganyika, the French north of the Ubangi, and the British in Uganda and the Sudan involved miscellaneous agreements relevant to the development and exploitation of the Nile waters. The one that most concerns the Congo Nile was signed on 19 May 1906 between Britain and the Congo Free State, amending the agreement signed in Brussels on 12 May 1894. The Anglo-Congolese agreement of 1906 aimed at delimiting the boundaries between the Congo and the Anglo-Egyptian Sudan, preventing modifications in the Nile flow, and prescribing measures for navigation and the transit of goods.

Although the agreement has been questioned by several riparian states as unfair, the attitude of successive Congolese governments since independence has been indifferent. Kinshasa has not been concerned with water scarcity in the region surrounding the Congo Nile, and in any case no pressing water demand in this area has as yet arisen.

## SOCIO-ECONOMIC DEVELOPMENT

The development status of a river basin depends on the socio-economic factors necessary for expansion. For our purposes these involve agriculture, fishing, navigation, power production and tourism, all of which are associated with local demographic trends.

### *Population*

The watershed drained by the Congo Nile lies in the North Kivu province and the Orientale province of the DRC. Statistics used in this chapter only involve the Ituri district of the Orientale province, which holds approximately 70 per cent of the watershed area. The Ituri comprises five territories: Djugu, Irumu, Mahagi, Aru and Mambesa, the first three of which are adjacent to the water bodies of the catchment. These territories remain under the administration of Orientale province, which also supervises five districts: Bas Uele, Haut Uele, Ituri, Kisangani and Thopo.

Ituri district is the most populous in Orientale province,[13] with (in 1994) some 2.2 million people or about 40.5 per cent of the total population of the province, the highest concentrations being in the territories of Mahagi, Djugu and Aru. Present estimates based on growth models are unreliable owing to the disastrous conditions that have prevailed in the region since 1998: civil war, epidemics, political instability and consequent displacement. According to estimates by humanitarian groups working in the district, the Ituri population is currently about 4.6 million. The main ethnic groups living in Ituri district are the Alur and Okedo (Mahagi); Lendu, Mabendi and Nyari (Djugu); Nyari, Bira, Hema and Pygmies (Irumu); Bira, Pygmies, Ndaka, Balese and Mbo (Mambasa); and Lugbara, Kaliko, Ndo-Vare, Madi Lugbara and Kakwa (Aru). The economy of the district is mainly based on agriculture, hunting and fishing.

### *Agriculture*

Statistics indicate that in the mid-1990s Orientale province, with 15 per cent of national agricultural production, was second only to Bandundu.[14] It was first in producing rice (27.4 per cent of the country's total) and bananas (30 per cent); second in cassava (13.82 per cent); third in beans (10.51 per cent); fourth in groundnuts (18.23 per cent); and fifth in maize production (7.69 per cent). The province was the country's leading producer of cotton, second in tobacco, rubber and cocoa, and third in palm oil, coffee and sugar cane. Ituri district in turn produced about 45 per cent of the province's maize, about a third of its groundnuts and cassava, and a quarter of its bananas. Ituri was even more important in animal production, accounting for almost 100 per cent of the province's cattle, over 90 per cent of its sheep, almost half its goats, and about a quarter of its pigs and poultry.

### *Fishing*

The Congo Nile watershed has a high range of aquatic biodiversity. The fishing industry centered in Lakes Albert, Edward and George is therefore an important activity contributing to the socio-economic development of the local population and to national income.

Statistics have fluctuated over the years, and many factors influence the general trend of the catch. In its 1990 report on the fishery management of Lakes Albert and Edward,[15] the FAO pointed out technical factors behind a decline in production. For example, some statistical data had grossly overestimated the rate of exploitation because fish were counted twice, once in Uganda when caught and again in the DRC when marketed. More seriously, illegal use of beach seines and heavy fishing in spawning areas were especially detrimental to the reproduction and growth of certain species (e.g. the tilapia). The report estimated the annual catch in Lake Albert at about 13,000 tonnes for the DRC and 10,000 tonnes for Uganda. Three major species constituted about 85 per cent of the total: hydrocynus (35 per cent), tilapia (25 per cent) and lates (25 per cent). The proportion of tilapia was higher in the DRC (about twice as much as in Uganda). The reverse was observed for hydrocynus. Fishing was concentrated in inshore zones close to the limited number of landing centers near commercial markets. There was no precise information on the exploitation of deepwater species.

As for Lake Edward, total production was estimated at 14–15,000 tonnes – 12,000 tonnes for the DRC and 2–3,000 tonnes for Uganda – and had already reached its maximum potential. Commercial exploitation in the DRC focused on five major species: tilapia (60 per cent of the total catch), baqrus (13 per cent), protopterus (12 per cent), clarias (3 per cent) and barbus (3 per cent). The relative importance of tilapia was less in Uganda (about 40 per cent).

The report pointed out signs of over-exploitation in both lakes, such as decreasing fish size and diminishing yield. Beach seines had been withdrawn in 1988 on the Ugandan side and in 1989 on the DRC side. In any case, the quality of statistics had been adversely affected by a lack of resources in the fisheries administrations and by illegal fishing.

### *Tourism*

The Congo Nile watershed is located in an ecological zone favorable to gorillas and other endangered species. Under a UNESCO initiative, a protected area of high value has been established in the border region of the DRC, Uganda and Rwanda. Known as the Transboundary Virunga Volcanoes Conservation Area, this consists of three national parks totaling approximately 434 km$^2$.

The largest part of the Conservation Area is located in the DRC, in the southern or Mikeno sector of Virunga National Park. (Mountain gorillas are located only in this sector.) Virunga National Park was created in 1925 as one of the first protected areas in Africa, and extends over an area of about 7,900 $km^2$. Since 1979 it has been on the World Heritage List because it offers a unique diversity of habitats and wildlife. The area occupies an important geopolitical position at the confluence of political borders and as part of the Nile basin.

Mountain Gorilla tourism in this area already generates considerable income. It is estimated that tourism and related activities could bring $20 million to the region each year.[16]

## INSTITUTIONS, POLICIES AND COOPERATION

It is convenient to view the evolution of water institutions in the DRC in two periods, with the dividing line at independence in 1960. From 1960 to 1965 the first Congolese Republic maintained the institutional framework inherited from the colonial regime. The regime that took power in 1965 had as its first aim the establishment of its authority over the entire country. An aspect of this was the concept of Zaïrianization, which advocated a return to traditional values. But its main achievement was nationalization of all property inherited from the colonial regime under the legal framework of the Bakajika Law. The Second Republic lasted 25 years until a transitional government was established to set up the mechanisms for a democratic regime. This was formally achieved in 2007 with the inauguration of the first democratically elected president.

Throughout these periods water institutions more or less retained the status they had acquired at independence. Modifications came about with the involvement of the DRC in international water law. The general institutional and legislative framework of water resource management in the DRC may be briefly summarized.

### *Post-colonial water institutions and policies*

At independence, the country's natural resources were subject to private property rights, land ownership and customary land systems. The colonial administration had established territorial limits to the traditional rural order, declaring all vacant lands property of the state. It also introduced a system of land registration and private ownership, in order to regulate commercial access to these lands. In 1973 the Bakajika Law gave the government full rights to – and total control over – all resources above and below ground throughout the whole country.

Responsibility for water resource management in the DRC is today divided among many ministries, with little coordination. These include the ministries of Environment, Nature Conservation, Forests and Water, Rural Development, Energy, Planning, Health and Public Works. To these ministries are attached in turn various specialized departments and units responsible for water-related issues. These include the Directorate of Water Resources and the National

Program of Sanitation under the Ministry of Environment; the National Rural Hydraulics Service under the Ministry of Rural Development; the national water supply company (REGIDESO) under the Ministry of Energy; the National Committee for Water Works under the Ministry of Energy; and the National Drainage Service under the Ministry of Public Works.

Water resource management is supported by a set of laws, most dating from the colonial period.[17] Legislation addresses, among other things, ownership of water resources (which is held by the state under the Bakajika law of 1973, modified and completed by law No. 80-008 of 18 July 1980), the legal nature of water rights, the effective and beneficial use of water, and the need to acknowledge and respect existing uses and customary entitlements. Legislation also deals with the quantity and quality of both surface and groundwater resources and with delivery of water services.

The legislative framework takes into consideration aspects of water resource management such as conservation, protection of water supplies and water quality, technology and efficiency requirements, creation of management areas, basin management principles, monitoring, information requirements, administrative rights, creation and enforcement of public rights, emergency measures, and registration and recording of uses and supplies.

Water resources belong in the public domain, reflecting the notion of water as a public good. Water-use rights are granted to private individuals or corporations under the provisions of national constitutions protected by the Bakajika Law of 1973, modified and completed by law No. 80-008 of 18 July 1980. Water rights are thus closely linked to land rights, and entitlement is held on the basis of productive capacity without the transfer of rights. In addition, the legislative framework sets out the duties and functions of water management institutions and water service providers.

The goals and objectives of water resource management on a national scale, however, and policies for regions, catchments and shared or trans-boundary water resources are not comprehensive. Matters of jurisdiction and delegation, the extent to which water management is decentralized or consolidated, the use of economic incentives, capacity building to meet institutional challenges, monitoring and control to reduce ecosystem degradation, measures requiring investment, and their costs and benefits are not fully legislated. An approach that would involve moving away from single sectoral water planning to multi-objective and integrated planning of land and water resources, recognizing wider social, economic and development goals and entailing cross-sectoral coordination is lacking. The priority of water use or the rank of use for allocation of water in times of scarcity or in case of competing applications is simply unaccounted for.

Water-quality protection includes both preventive and corrective measures. As of now, preventive measures are limited to effluent and discharge regulations, technical standards and requirements for treating polluted effluents, and control of point sources and non-point sources, but do not deal with economic instruments or quality standards set according to expected or existing water uses and services. Corrective measures are limited to an unclear polluter-pays principle that allocates responsibility for damages, but does not address the major issues of cease-and-desist orders, compensation for damage and economic losses, or alleviation and remediation requirements.

The policy of the DRC towards water resources is also shaped by the international conventions and treaties to which it has acceded.[18]

### *The DRC and the cooperative framework*

The movement towards establishing a Nile basin cooperative framework since the 1960s has witnessed four main milestones. The first was HYDROMET, a response to the sudden and unpredicted 20 per cent increase in rainfall that occurred between 1961 and 1964 on the lake plateau, producing extensive flooding in the lakes region and the disastrous inundation of the *Sudd* basin in the Sudan. The *Sudd* more than doubled in size from (13,100 $km^2$ to 29,800 $km^2$), costing many lives and incurring enormous economic loss.[19] The project lasted for 25 years, until 1992. It collected a considerable amount of hydro-meteorological data; conducted ground surveys and hydrographic surveys; and trained staff from participating governments.

In order to improve cooperation, in 1983 Egypt and the Sudan initiated Undugu ('brotherhood' in Swahili). The first Undugu forum was held in Khartoum, with founding support from the DRC, Uganda and the Central African Republic; Ethiopia, Kenya and Tanzania assumed observer status. The stated objective of the forum was cooperation in such fields as culture, the environment, telecommunication, electric power, trade and water-resource development. At a technical meeting held to evaluate UNDP sponsorship of the Undugu plan of action for the Nile basin, Ethiopia made an issue of Undugu's lack of legal standing and terms of reference and thus of its very competence to submit any plan at all.[20]

As HYDROMET approached its end, the water resources ministers of the DRC, Egypt, the Sudan, Tanzania, Uganda and Rwanda agreed during a meeting in Kampala in December 1992 to create TECCONILE, which was joined also by Burundi, Ethiopia, Kenya and Eritrea as observers. TECCONILE was superseded in 1999 by a Transitional Institutional Mechanism of the Nile Basin Initiative (NBI), which had a more flexible board and better organization than its predecessor.

The current organs of the NBI include the Council of Ministers of Water Affairs of the Nile Basin Countries (Nile-COM), which provides policy guidance and makes decisions on matters relating to the Nile; the Technical Advisory Committee (Nile-TAC), which gives technical advice and assistance to Nile-COM; and the Nile Basin Secretariat (Nile-SEC), which renders administrative services to Nile-COM and Nile-TAC. Nile-SEC oversees the Shared Vision Programme (SVP) and Strategic Action Programmes (SAPs).

### *Past development plans for the Congo Nile*

It has long been assumed that the DRC is indifferent towards the Nile waters. It is true that since independence the country has initiated no comprehensive water development plan, at least for that part of the Congo Nile made up of Lake Edward, the Semliki river and Lake Albert. But there is a long if fitful record of Congolese attention to the waters' potential. As early as the 1960s, following the death of Emery Patrice Lumumba, President Mobutu threatened to divert water from the White Nile if the Sudanese and Egyptian governments remained hostile towards his government.[21] In 1997, during the regime of President Laurent Desiré Kabila, the Societé d'Energie du Graben (SENERGRAB), under the Fédération des Entreprises du Congo (FEC), undertook an initiative to electrify Ituri and North Kivu districts. Funds were raised through a stock offering, but the project was still in its preliminary stages when political instability and war paralyzed the region in 1998; this project was believed to attract regional concern because it would affect the Semliki's flow. Meanwhile, President Kabila, in his election campaign, called for the construction of a hydroelectric dam on the Semliki to provide power for North Kivu and Ituri.[22] A feasibility study was done, but the results have yet to be made available.[23]

The only Congo-Nile development plan that was actually taken up involved fisheries. The DRC already had an important market for fish as compared to neighboring countries. Several steps to promote fishery development in the Congo Nile were initiated, which also improved relations with Uganda and supplied the fish market of the neighboring countries of East Africa. In 1990 a working group on Management of Fisheries of Lakes Edward and Albert, composed of experts from the DRC and Uganda and supported by the UNDP and FAO, compiled an inventory of fishing activities within the basin.[24] (It was at this time that the lakes' names were changed: Lake Edward became Lake Idi Amin Dada and Lake Albert was renamed Lake Mobutu Sese Seko. After the two presidents were overthrown the lakes recovered their former names.)

The working group noted that Lake Albert was exposed to a free-entry fishing regime in both Uganda and the DRC. There was no limit on the number of fish-landings and boats or the amount of gear on the lake. Minimum mesh size allowed for gillnets was 2.5" (60 mm) in both countries, but whereas Uganda had banned the use of beach seines, purse seines and trawl nets, the DRC had not. Ugandan regulations forbade fishing in lagoons and certain shallow bays where fish spawned and juvenile nursery grounds were located, and in the DRC spawning areas were delineated and fishing in them was prohibited. In Uganda, gillnets of not less than 5" (127 mm) and anchored baited hooks were the only fishing methods authorized, but in the DRC, although a similar minimum mesh size of gillnets was authorized, longlines were also allowed and beach seines were still in use despite having been prohibited in 1989. Only one fishing unit used a semi-industrial boat on an experimental basis. All other fishing methods were prohibited.

The working group found that exploitation of both lakes was severely constrained by the remoteness of the area. This affected fishermen at various levels: in the supply of fishing inputs, food and basic products; limited access to health, education and administrative services; and the difficulties and high costs of transporting and marketing the catch. Information from DRC and Uganda sources about the fisheries was inadequate and incompatible, thus diminishing its use for making decisions. Other constraints included inadequate financial resources for effective research; the location of research institutions – Makerere University and the Institute of Ecology at Mweya in Uganda, the Congolese Institut Zairois pour la Conservation de la Nature (IZCN) at Lulimbi, and the Service National pour le Developpement de la pecherie (SENADEP) at Kinshasa – far from the lakes, necessitating establishment and support of local sub-stations; and management issues associated with the bilateral nature of the project. Regular meetings were recommended. The two countries agreed to a sub-committee for bilateral collaboration, supported in part by external donors.

The working group found common ground for further management efforts. This included the need to manage for the sustainability of natural resources; realization that fish is a common resource to be jointly managed by riparian states if damage is to be avoided; and recognition that, because the two countries' fisheries are still artisanal, management strategy should be developed accordingly. The existence of the two national parks, the DRC's Virunga and Queen Elizabeth National Park in Uganda, also imposes management peculiarities.

The working group reviewed a number of project proposals prepared by the FAO. One of these emphasized a global and regional approach to fisheries development and management for the two lakes. Other projects focusing on the DRC emphasized institutional strengthening for the management of Lake Edward; development of semi-industrial fishing on Lake Albert; and the need for coordinated regional research for both lakes.

The working group stressed again the importance of regional collaboration in fisheries management and gave its full support for the creation of a joint sub-committee for the management of the fisheries of Lake Edward and Lake Albert. Noting that the creation of such a sub-committee had been agreed upon in principle by the permanent joint committee between Uganda and the DRC at its April 1990 meeting, the two sides decided that the mandate, functions and procedures of the sub-committee be finalized and agreed upon without delay.

### *Future Congo Nile development plans*

The DRC is currently undergoing major political, social and economic reforms with the main goal of poverty alleviation and economic growth. Important reforms in the water sector are under way. Strategies for managing cross-sectoral water resource issues are to be addressed at the institutional, policy and human-resources-development levels.

As stated already, the Congo-Nile hydrological system lies within the North Kivu province and the Orientale province of the DRC. The people in this north-eastern part of the DRC rely on agriculture, fishing, livestock and hunting for their livelihood. They have not made profitable use of the region's potential for development: its lakes with abundant fish biodiversity, the Semliki river's opportunities for irrigation and power production, fertile lands and mineral resources.

During the recent crises, under-development gave way to destruction and humanitarian disaster, and awareness of the region, its problems and potential has risen from the local to the international level. In Ituri itself, people have organized associations to defend their interests and to contribute to social development. One of the best known is 'Démocratie et Civisme pour le Développement Intégral de l'Ituri' (DECIDI).[25] In 2004 the organization undertook an assessment of opportunities for development, and this resulted in a document entitled 'Projet Agriculture, Travaux Publics et Energie pour le district de l'Ituri'.[26] The Congolese government, anxious to establish its authority in the area, saw an opportunity for confirming its sovereignty and gaining the trust of the local population. The project has since become part of President Kabila's reconstruction plans.[27] In an interview, he pointed out the inefficiency of bringing electricity from Inga to Butembo–Bunia, a distance of over 2,000 km, rather than buying it from nearby Uganda. But getting electricity from Uganda was a first step; the mid-term plan was to build a hydroelectric power plant on the Semliki.[28]

The Ituri project involves agriculture, public works and energy development. In the agricultural sector, including fisheries and livestock, the Ituri Development Plan (which we refer to here as the Congo Nile Development Plan) intends to make agriculture, fisheries and animal husbandry powerful levers for economic development, ensuring food security and poverty reduction by increasing food production, resuming the investment in cash crops, and reducing the current high mortality. These results are to be achieved through a variety of measures.

In the fields of agriculture, fisheries and livestock these include rehabilitation of the Ituri Project Management unit (Bureau de Projet de l'Ituri (BPI)), the Bunia slaughterhouse, and the cold storage facility at the Bunia airport to facilitate shipment of food; rehabilitation of the industrial fisheries of Kasenyi and Mahagi on Lake Albert; reconstruction of the Libi and Djugu dairy plants; construction of new dairies at Aru and Irumu; and development of rice cropping in Aru and Nzinzi, and of coffee, cotton, banana, maize, soybean and tobacco.

As for civil works and land use, the plan intends to restore the priority road network by improving the Ariwara, Engbokolo and Aba roads; rehabilitating feeder roads in farming areas; reconstructing the port facilities of Kasenyi and Mahagi on Lake Albert and establishing a ferry system between them; carrying out a feasibility study for a railway from Uganda to Kisangani; and rebuilding the roads linking Bunia, Djugu, Mahagi, Aru, Laybo and Aba, between Niania, Wamba, Mabunga and Isiro, and between Mambasa and Beni. Other measures include rebuilding the airports at Mambasa, Irumu, Mongbwalu, Mahagi and Aru, and construction of a bridge over the Ngezi river at Bunia.

In the energy sector, the plan intends to establish policies to promote rural electrification, public–private partnerships in energy production, increasing production at Inga; rehabilitation, construction and development of autonomous centers of electricity production and water supply and sanitation; and revitalizing oil exploration.

## CHALLENGES TO DEVELOPMENT

Conflict invariably complicates the development of resources. From a neo-Malthusian point of view, environmental problems result from scarcity, population pressure, population movements and resource distribution. These effects can be exacerbated by institutional and political weakness. Where environmental problems have political causes, the most frequent result is violence in one form or another. The Congo Nile region is no exception. It faces many challenges.

### *Demographic pressure*

Population density in the fertile Great Lakes region, including the eastern DRC, Burundi, Rwanda and, to a lesser extent, Uganda, is extremely high. Ituri is the most populous district of Orientale province, and the territories of Djugu, Mahagi and Aru are in turn the most heavily populated of Ituri, with densities of 86, 103 and 75 inhabitants/km$^2$ respectively. The case for relating population density and conflict is therefore easy to make. The first large-scale fighting in Ituri between the Northern Hema and Lendu ethnic groups, which broke out in Djugu in 1999, has been ascribed to conflict over land rights, fishing, agriculture and livestock.

### *Climate change and natural resource depletion*

The Rwenzori Mountains on the Uganda–DRC border contain glaciers and snow fields that nourish downstream water bodies and a range of fauna and flora, many of which are unique to the area. Glacial melt contributes to alpine river flow upon which the downstream BaKonzo and BaAmba communities rely for year-round water supplies and hydroelectric power generation. Glaciers are important reservoirs of fresh water that store seasonal precipitation and release melt waters during drier periods, thus serving the vital ecological function of regulating alpine stream flow and water levels in lakes and wetlands.

Studies show the impact of climate change in this region.[29] By 1990, glaciers on the Rwenzori Mountains had already receded to about 40 per cent of the extent recorded in 1955, and to less than one-quarter of that measured by the Duke of Abruzzi in 1906. Today the rate of recession is ~0.7 km$^2$ per decade, consistent with the linear trend observed over the last century. Assuming the current trend continues, the Rwenzori glaciers will disappear by 2030. This process of deglaciation is associated with rising air temperatures. From the available data, the same climatic variables (changes in humidity and radiative fluxes associated with rising air temperature) are expected to have contributed to observed glacial recession in the Rwenzori Mountains. However, considering the evaporation that characterizes the lakes, we assume, *mutatis mutandis*, that the increase in temperature will have a negative effect on the evaporation rate.

Based on sediment core archives of diatom and pollen, the lake ecology and flora of these Afroalpine areas do not appear to have undergone significant changes over the period of deglaciation. However, the aforementioned study detected evidence of a recent decline in epiphytic habitats and concomitant increase in algal productivity. Atmospherically deposited mercury, consistent with global trends and emissions, is also detected in lake sediment within alpine environments of the Rwenzoris. Trace-metal contamination via atmospheric deposits from more localized sources has been observed since the mid-1950s and coincides with the onset of copper mining downslope.

### *Oil and the Congo Nile*

The DRC is an oil-producing country. The 'Albert Graben', the portion of the Rift Valley around Lake Albert on the Ugandan–Congolese border, including Lake Albert and the Semliki river valley, is regarded as the most promising sector for oil exploration in the Rift Valley. Prospectors believe that oil reserves of up to 1 billion barrels may be found in the Albertine basin straddling a section of the border between the two neighbors.[30] In 1990 they signed an exploration and exploitation agreement calling for joint investment in infrastructure, revenue sharing and so forth. Officials of the current democratically elected Congolese government have since renounced the agreement, however, on the grounds that it was never ratified by both countries.

Following recent Ugandan petroleum exploration, a military confrontation occurred over Lake Albert. Under terms of the subsequent Arusha agreement of September 2007 the two sides undertook to strengthen efforts to settle their differences; to round up, demobilize, repatriate, resettle and reintegrate armed militants and rebels; to review border questions; to upgrade their diplomatic representation in the respective capitals; and jointly to explore for and exploit any petroleum in the border regions.

## CONCLUSION

A universal consensus is emerging that watersheds are both natural and logical units for planning and implementing sound policies for water resource management and development. The Congo Nile watershed holds important resources for the socio-economic development of the local population and the region, and brings together trans-border areas that share hydrological characteristics.

Although the Congo Nile has potential for hydropower, irrigation, navigation and fishery development, post-colonial policies have not included a development plan for efficient use of water resources in this region. The Congo Nile faces numerous challenges: population pressure, war and civil unrest, and global climate change, all of which jeopardize its development. A regional institutional structure could enhance efficient, equitable and sustainable use of the watershed's resources.

# 6

# The Nile Waters and the Socio-Economic Development of Western Kenya

**Mary C. Mwiandi**

The Kenyan Nile basin is defined as the drainage system of the entire area contributing water to Lake Victoria, and a section of the Kenya–Uganda borderlands that drain toward the Uganda Nile basin around Mt Elgon. This chapter defines that whole region as western Kenya, and examines the connection between the Nile waters and the economic activities they sustain and the extent to which available resources have been utilized.

Both the Nile Water Agreement of 1929 and the 1959 Nile Waters Agreement influenced the pace of economic development in the Kenyan Nile basin during the colonial period, but ceased to apply when Kenya adopted the 'Nyerere Doctrine' after independence. Before then, Kenya had to secure Egypt's approval before undertaking any irrigation or hydroelectric projects so that those projects did not reduce Egypt's water supply.[1] The colonial government's objections to the 1959 Nile Waters Agreement, and Britain's efforts before and after 1959 to persuade Egypt and the Sudan to rethink the situation, were largely ignored.

The situation has changed since the 1990s as Kenya, like all the other upstream states, has begun to object to the Nile Waters Agreement, demanding that she too has the right to use the Nile waters. In response to challenges facing the entire Nile basin – such as management of the resources for poverty reduction, resource protection and mutual benefit – the ten riparian countries began discussions on a long-term cooperative framework that culminated in the formation of the Nile Basin Initiative (NBI). Its broad vision was 'to achieve sustainable socio-economic development through equitable utilization of, and benefit from the common Nile Basin water resources', and thus to provide a framework by which to deal with Nile water issues.[2]

Since Kenya's Nile basin waters and other resources remained largely underutilized over the past half-century, the problem of seeking the approval of Egypt did not arise. Although western Kenya inherited and increased most of the economic development activities of the colonial state, which naturally increased the use of Kenyan Nile basin water, this did not alarm Egypt insofar as its share of the Nile water was concerned. Thus, the slow pace of socio-economic development in western Kenya since the 1980s cannot be blamed on observance of the

Nile agreements; rather, the problem is that under-development is a symptom of a more serious 'economic and political disease' afflicting Kenya in general and West Kenya in particular. The economic activities described in this chapter indicate that while the Kenyan Nile basin has high potential, most of this has gone to waste, under-utilized and mismanaged, leaving the area in great want.

## GEOGRAPHICAL AND HYDROLOGICAL BACKGROUND

Lake Victoria is the second largest freshwater lake in the world after Lake Superior in North America and the largest in Africa, with a surface area of 68,000 km$^2$. The lake is relatively shallow, with a maximum depth of 80–90 m and an average depth of 40 m, and has a volume of 2,760 km$^3$. It is shared by Kenya, Uganda and Tanzania; Kenya has the smallest portion of the lake, with only 4,113 km$^2$ (6 per cent), compared to Uganda's 31,001 km$^2$ (45 per cent) and Tanzania's 33,756 km$^2$ (49 per cent). In terms of shoreline, Kenya's is the shortest: 550 km compared to Tanzania's 1,150 and Uganda's 1,750 km.

The Lake Victoria basin, part of the larger Nile river basin area, has an area of 210,000 km$^2$. Tanzania has the largest area of lake basin, 79,570 km$^2$ (44 per cent), while those of Kenya, Uganda, Rwanda and Burundi are 38,913 km$^2$ (21.5 per cent), 28,857 km$^2$ (15.9 per cent), 20,550 km$^2$ (11.4 per cent) and 13,060 km$^2$ (7.2 per cent) respectively. The Kenyan Nile basin has an area of only 46,229 km$^2$ (1.5 per cent of the total Nile basin area), which is larger than her portion of the Lake Victoria basin; in the area around Mt Elgon, rivers drain into Uganda's Nile basin. Among these is the Lwakhakha-Malaba river on the Kenya–Uganda border, which flows into Uganda and the swamps of the southern part of Lake Kyoga. Other small rivers in Kenya are tributaries of the Lwakhakha-Malaba: the Ndakaru, Malakisi and Tisi join the Malaba before reaching Uganda, and that section is not included in the Victoria basin. For the purposes of this chapter, the terms Kenyan Nile basin and Kenyan Victoria basin are used interchangeably. The Kenya Nile basin constitutes about 8.5 per cent of the total area of Kenya.

The Kenyan Nile basin includes all the districts of Nyanza and Western provinces, and Kericho, Bomet and Nandi districts of the Rift Valley province. The districts in Nyanza province are Kisumu, Kisii, Migori, Nyamira, Homa Bay, Siaya and Kuria; those in Western province are Kakamega, Vihiga, Bungoma, Busia and Mt Elgon. (Districts are broken down further into divisions, locations and sub-locations.) The Luo community dominates the Nyanza province, Luhya the Western, and Masai and Kalenjin the Rift Valley. The area of the Kenyan Nile basin is therefore ethnically heterogeneous, but shares a common colonial and post-colonial experience.

Lake Victoria receives most of its water directly from rainfall. Still a substantial amount of water flows into the lake from rivers. These include the Mara, Kagera, Mirongo, Grumeti, Mbalageti, Simiyu and Mori in Tanzania. Kenya has two main

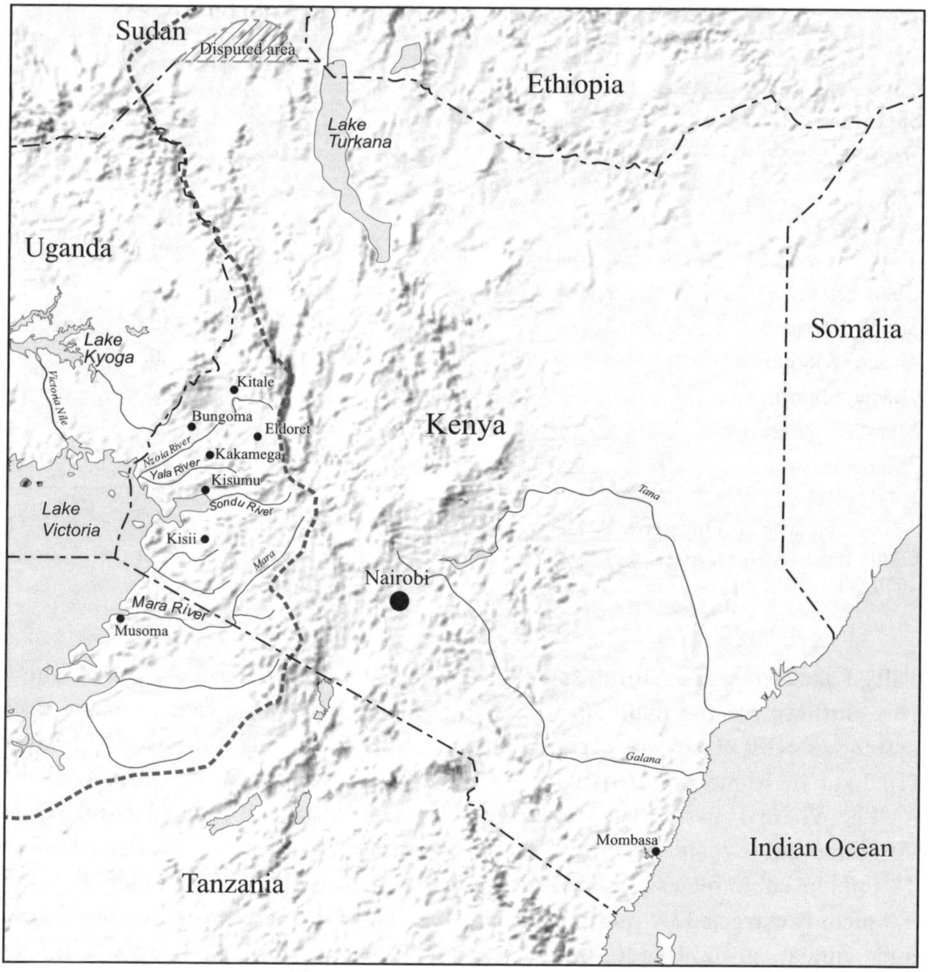


**Map 1:** Kenya and the Lake Victoria basin.

catchments: the North catchment, which comprises the Sio, Nzoia, Sosiani, Yala, Kibos, Nyando and Cheroget basins; and the South, of the Sondu-Miriu, Awach-Kiboun, Kuja-Migori, Awach-Tender (or Oluch). The Mara is shared by Kenya and Tanzania. In total, Kenya contributes 8.4 $km^3$ per year of water to Lake Victoria, which is a substantial proportion of a total flow of 20 $km^3$ per year from rivers in East Africa. The Kagera's total flow is 7.5 $km^3$; Tanzanian inflow into Lake Victoria is 3.2 $km^3$; and 1–2 $km^3$ come from the swamps in northwest Uganda.[3] The Kenyan side of the Victoria basin and its discharge are shown in Table 1.

Most water (85 per cent) from the lake is lost through evaporation, and the remaining 15 per cent largely flows out by way of the Victoria Nile, which leaves the lake near Jinja in Uganda at the rate of 600 cm/s, then flows via the Owen

**Table 1:** Surface runoff in the Kenyan Victoria basin.

| *Main rivers* | *Area (km²)* | *Runoff (million cubic m)* |
|---|---|---|
| Sio (Malakisi) | 1,080 | 287 |
| Nzoia | 11,900 | 1,777 |
| Yala | 2,650 | 1,114 |
| Kibos | 490 | 68 |
| Nyando | 2,650 | 247 |
| Cheroget | 560 | 23 |
| Sondu/Miriu | 3,220 | 845 |
| Awach/Kiboun | 610 | 180 |
| Gucha-Migori | 5,840 | 952 |
| Mara | 10,830 | 1,038 |
| Total | 39,830 | 6,531 |

*Source:* George S. Ongweny, 'Water resources of Lake Victoria basin in Kenya', in C.O. Okidi (ed.), *Natural Resources and the Development of Lake Victoria of Kenya*, Nairobi 1979, 71.

Falls, Lake Kyoga and Murchison Falls to join the outflow from Lake Albert: these two outflows are the main sources of the White Nile.[4] Besides the major river systems, satellite lakes are part of the lake basin. These include Lakes Kanyaboli (10.5 $km^2$), Nyamboyo (2 $km^2$) and Sare (5.2 $km^2$).

The Victoria basin has groundwater storage that is broadly estimated as 70 billion cubic meters (bcm). It has an annual groundwater discharge of about 18 million cubic meters (mcm), of which 7.3 mcm is discharged to streams and 4.9 mcm is extracted by pumping. Kenya's available groundwater is some 16 bcm, and annual groundwater discharge into Lake Victoria is about 6 mcm.[5] Groundwater is thus an important alternative water resource, which has not yet been exploited for socio-economic development.

Kenya's portion of the Nile basin is endowed with good climate and, for the most part, fertile soil capable of supporting a healthy and wealthy population. The climate of the catchments is mild, with small variations in monthly average air temperatures of 19–25°C year round. Rainfall in the catchments averages about 1,300 mm annually, varying from 2,000 mm in the highlands to 1,000 mm in the north, the south-west and the lowlands along the lakeshore. In general the rainfall pattern is bimodal, with long and short rainy seasons in March–June and October–December respectively. There are areas where rainfall is insufficient for the production of certain crops, hence the need for irrigation to supplement rainfall.

Western Kenya fits well into three ecological zones from the lowest point at the lakeshore to some of the highest points inland, north and south of the Gulf of Winam. The first zone runs along the lake margin with a width of about

**Table 2:** Long-term average discharge from river basins in the Lake Victoria catchment area.

| *Country* | *Basin* | *Discharge ($m^3/s$)* | *Percentage* |
|---|---|---|---|
| Kenya | Sio | 11.4 | 1.5 |
| | Nzoia | 115.3 | 14.8 |
| | Yala | 37.6 | 4.8 |
| | Nyando | 18.0 | 2.3 |
| | North Awach | 3.7 | 0.5 |
| | South Awach | 5.9 | 0.8 |
| | Sondu | 42.2 | 5.4 |
| | Gucha-Migori | 58.0 | 7.5 |
| | Sub-total | 292.1 | 37.6 |
| Tanzania | Mara | 37.5 | 4.8 |
| | Grumeti | 11.5 | 1.5 |
| | Mbalageti | 4.3 | 0.5 |
| | E. Shore stream | 18.6 | 2.4 |
| | Simiyu | 39.0 | 5.0 |
| | Magogo moane | 8.3 | 1.1 |
| | Nyashishi | 1.6 | 0.2 |
| | Issanga | 30.6 | 3.9 |
| | S. Shore stream | 25.6 | 3.3 |
| | Biharamulo | 17.8 | 2.3 |
| | W. Shore stream | 20.7 | 2.7 |
| | Kagera | 260.9 | 33.5 |
| | Sub-Total | 476.4 | 61.2 |
| Uganda | Bukora | 3.2 | 0.4 |
| | Katonga | 5.1 | 0.7 |
| | N. Shore stream | 1.5 | 0.2 |
| | Sub-total | 9.8 | 1.3 |
| | *Total Average inflow* | *778.3* | *100.0* |

*Source:* Sixtus Kayombo and Sven Erik Jorgensen, *Lake Victoria: experiences and lessons learned*, Brief, International Lake Environment Committee, 27 February 2006, p. 433.

15–45 km. Its landward limits lie approximately along the 1,400 m contour. In this zone, soils are poor and often poorly drained, with rainfall quite unreliable. The zone supports herding, and some pockets of fertility provide good cereal farming. The other two zones lie to the north and south of the gulf and are alike in many respects. Both have heavy and reliable rainfall and fertile soils. Both zones contain about one-quarter of Kenya's agricultural land.[6]

For the most part, the geographical and demographic boundaries of the Lake Victoria basin coincide. The basin embraces the plateau country sloping westwards from the highlands that border the central Rift geographical division.[7] The highlands of Kisii and Kericho and the slopes of Mt Elgon are fertile, with high potential. The lowlands have black cotton soil that, especially in the Bunyala lowlands and Kano plains, hinders agriculture. The ecological divisions of the lake basin are lakeshore; lakeshore savanna, which covers most of Kisumu Siaya, South Nyanza and Busia districts; then the high rainfall savanna zone, which lies between the first and second ecological zones covering Kakamega, Bungoma, Kisii, Kericho and Nandi districts. Translated into agricultural activities this succession involves cultivation of cotton, sugarcane, bananas, cereals and, in the higher parts, coffee, tea, pyrethrum and livestock farming.

## POPULATION DISTRIBUTION

The Lake Victoria basin is one of the most densely populated regions in the world. It supports over 30 million people and has an average annual growth rate of 3 per cent. The high population density is influenced by rainfall distribution. The Kenyan side of the basin harbors over 40 per cent of the country's population, mainly in Nyanza and Western provinces and a portion of the Rift Valley. According to the 2002 population and household census, Nyanza province had 4,392,196 people, Rift Valley 6,987,036 and Western 3,358,776, a total of 14,738,008 in 3,164,321 households.[8] More water is needed to support the growing population. This makes utilization of the Nile waters a more important and urgent matter than ever before. Continued adherence to the limitation of the colonial-era Nile water treaties impacts not only the population living in the basin but the entire country.

Population distribution in Kenya's Nile basin is closely linked to the environment and resources in a particular zone. The least endowed lakeshore areas are less densely settled than the higher and better-watered parts of the basin. The Nzoia–Yala river basin of about 15,143 km$^2$ has a population density of about 220 people per km$^2$; the Nyando basin area (3,517 km$^2$) about 175 people per km$^2$; the Sondu–Miriu river basin area (3,583 km$^2$) about 220 people per km$^2$; the Gucha basin area (6,612 km$^2$) about 225 people per km$^2$; and the Mara basin area (13,915 km$^2$), shared with Tanzania, a population density of only 50 people per km$^2$.[9] Land to support a growing population obviously needs better management, but of more immediate concern is the management and utilization of water resources to maximize production from the limited amount of arable land.

## BRITISH RULE IN WESTERN KENYA

For diplomatic and strategic reasons, Britain endeavored to control the region that supplied water to Lake Victoria, the source of the Nile. British control of

Kenya was not an end in itself, but only an aspect of that much larger goal. To secure the source of the Nile, Buganda was brought under a protectorate, and the British decided to put the whole area between Uganda and the Indian Ocean under their control.[10] Kenya's fate was sealed as a future British colony, not primarily because of any intrinsic value but because it was traversed by land routes to Lake Victoria.

In the diverse explanations for and interpretations of the *Pax Britannica* in Kenya in general and western Kenya in particular, the Nile issue has received little attention, even though it preoccupied strategists in the Foreign Office throughout the 1880s and 1890s. Construction of the so-called Uganda railway was undertaken to cement British control of the entire Nile watershed; economic use of the railway came much later. Even the British Imperial East African Company was chartered to occupy Uganda and checkmate explorers from other European countries. Most of the territory between Lake Victoria and the Indian Ocean was not critical to the British interest in controlling the Nile. Present-day western Kenya was an exception because it was within the Nile basin. Fredrick Lugard said that Uganda was important to Britain 'because Egypt was important, and Egypt needed security for Uganda and the sources of the Nile'.[11] For the same reason, then, western Kenya came under British control in 1894 when a protectorate was declared over Buganda.

In February 1895, C.W. Hobley was sent to take charge of the areas bounded by the River Sio, Mt Elgon and the Ravine. (This part of western Kenya, which extends from the Rift Valley to Lake Victoria, was transferred to the East African Protectorate – now Kenya – from the Uganda Protectorate in 1902.) Hobley had been posted in 1894 to the seat of Mumia, the ruler of Wanga, who for economic reasons was eager to establish an alliance. For their part, the company's remit included exploration of the territory and establishment of stations while cultivating friendly relations with the various 'tribes'. The Mumia–Hobley encounter set the pace for British rule in western Kenya.

Despite friendly relations between Mumia and the British, there was opposition from local communities to the newcomers. For instance, the neighboring Bukusu attacked a British mail party, taking away their rifles. Refusal to return the rifles led to a British expedition in 1895. Other communities that resisted faced a similar response. Hobley used various methods to conquer the area, ranging from direct attack to playing one section off against another. Some hostile 'tribes' – the Kitosh, Kakalewa, Nandi, Kakamega, Tiriki and Uyoma – must have known what Hobley was up to. In any event, there was minimal resistance thereafter, especially from the Luo, who honored an old prophesy of 'red strangers' emerging from the sea: 'the people were advised against showing any hostility to the intruders lest they incur the wrath of the ancestors'.[12] The area was thus effectively occupied, safeguarding the Nile basin for Britain.

Construction of the so-called Uganda Railway opened the interior of Kenya to British rule. To the British, the Mombasa hinterland was still essentially the road to Uganda. The railway reached the shores of Lake Victoria in 1901, linking together the British enclaves between Mombasa and the Kavirondo. Eastern Province of Uganda was transferred to the Kisumu and Naivasha provinces of the East Africa Protectorate (EAP), present-day Kenya. Thus, the Uganda railway and the lands it crossed became the responsibility of the EAP.

## THE 'WHITE HIGHLANDS' AND THE INTRODUCTION OF A CASH ECONOMY IN WEST KENYA

The process of introducing cash crops into Kenya's Nile basin is closely linked to the establishment of the 'White Highlands'.[13] Sir Harry Johnston[14] encouraged white farmers to come and settle in the Eastern Province of Uganda.[15] Later, Sir Charles Elliot, as the commissioner of the EAP, pressed for more Europeans to settle in the land he called 'empty of inhabitants'. The British government began systematically to recruit white farmers to occupy the land along the railway as a way of raising revenue to maintain the railway and the protectorate. Wealthy settlers came to Kenya to take advantage of the government's offer of land, labor and capital. They were offered the fertile and temperate highlands in the Central and Western districts, some 3.1 million hectares. Because some of these areas were far from 'empty', conflict was inevitable.

The fertile land controlled by the European settlers was put under intensive cash-crop cultivation using African labor. Poverty had forced many local people, especially the men, to move out to work for wages, thus creating a pool of voluntary African labor in the towns and for Europeans farms and factories. Laws were enacted to make sure the white settlers had an adequate labor supply. In 1920, *Kipande* was introduced; the poll and hut taxes and *matiti* tax were doubled,[16] compelling adult males to sell their labor in various European economic enterprises. At the same time, Africans were denied access to profitable cash-crop production and commercial credit. Since western Kenya experienced an earlier penetration by capitalism and western influence than most parts of Kenya, the region led the country in producing migrant labor during the colonial and post-colonial periods.

In western Kenya, the 'White Highlands' comprised huge tracts of land along the Uganda railway that the British appropriated in order to develop a large-scale export-oriented agricultural economy. Sugarcane was first introduced in Nyanza province in 1902, at Kibos near Kisumu by an Australian farmer for the production of jaggery (unrefined brown sugar). From that humble beginning, the growing of sugarcane in Nyanza on a commercial scale took root. Wheat, maize, tobacco, coffee and tea were extensively planted; in the lowlands near the lake, wetlands were cleared for sugar, rice and cotton. By 1960 the 'White Highlands' had about 4,000 European settler farms, producing 83 per cent of the total agricultural exports of the country. The foundations of the present-day agricultural economy of the region (30 per cent of GDP and 80 per cent of national employment) were thus laid in the early years of the twentieth century. Urban centers also developed on the lakeshore, and as population increased so did fishing pressure in the lake and farming pressure on the surrounding land.

Sugarcane cultivation under irrigation developed on a considerable scale in the Lake Victoria drainage area, and plans were underway to establish sugar factories, particularly in the Kano Plains, and the Colonial Development Corporation had shown interest in some of these projects. By 1960, the sugarcane investors had requested a total of 200 cusecs of water, which including some from Lake Victoria itself. This amount, according to the government, was within the 0.5 bcm notified to downstream states by Britain,[17] and these raised no objections.

Tea was introduced in Kenya in 1903, and by the 1920s it was one of the most important commercial crops in the colony. Tea was grown on large estates solely by Europeans until 1957, when legal production on smallholdings started. In 1961, the Special Crop Development Authority was established to promote cash crop production among the smallholders. This was renamed the Kenya Tea Development Authority (KTDA) in 1964 because it was by then dealing only with tea. In Kericho there emerged a large tea complex, which became an exclusive destination for rural migrants from western Kenya before the expansion of the sugar belt.[18] During the dry season, tea was irrigated using water from the rivers.

Cotton was introduced simultaneously in 1907 in Kenya and Uganda. Cotton growing was not enthusiastically embraced in Kenya because it had low cash returns and there was an alternative source of money – wage employment in the towns or on the European farms.[19] Cotton is a crop of the plains and that is why it was first grown on the Kano plains, from where it spread to other parts of the region. It requires less water than sugarcane and is grown on land below 1,400 m above sea level. In the 1950s and early 1960s, cotton was successfully grown in Belgut and Kericho District.[20] Other areas that grew cotton included Seme, Samia and Nyakach in the Lake Victoria basin. White settlers were discouraged by low prices to continue growing cotton but did not allow Africans to cultivate because that would disrupt the labor supply for their farms. Under the Cotton Ordinance of 1923, the governor was authorized to formulate rules controlling the quality of cotton. When no tangible results were realized, the government hired a British expert in 1925. But owing to competing labor demands, low prices, poor communication and lack of research facilities, the quality of cotton remained low.

Agricultural policy in the 1950s had an impact on practices in western Kenya, as indeed in other parts of the country. In 1954 the Swynnerton Plan aimed at accelerating agricultural development. It provided strategies for improving agriculture in various environmental conditions, hence raising productivity, intensifying land use and reducing livestock levels. In areas of high or medium potential, restrictions were lifted and African farmers, especially in western Kenya, began to engage in cash crop production. The plan introduced individual land ownership and consolidation. Most importantly, it advocated agricultural research, provision of credit for farmers, and marketing boards for farm produce. In the field of irrigation, the Swynnerton Plan also advocated irrigated agriculture in areas that did not receive adequate rainfall. A hydrological survey in the early 1960s showed that about 400,000 acres of land was potentially suited to large-scale irrigated agriculture.[21] In 1961 some of the former 'White Highlands' in western Kenya were opened up for land purchase and farming by Africans, a process facilitated by the Land Development and Settlement Board. A huge exodus of white settlers and civil servants ensued when it became clear that British rule was coming to an end.

In 1961 the Million Acre Scheme was established to settle thousands of families on 1.2 million acres of land bought by the colonial government from the white owners. The main aim of this project was to defuse rural unrest and promote the interests of the landed African class. Over half of the settler lands throughout Kenya were transferred almost intact by sale to wealthy Africans organized in partnerships or limited liability companies.[22] At independence, the former white settlers' farms continued to play a major role in Kenya's cash economy, depending on local labor at low wages. However, some Africans joined in the production of cash crops in response to a growing market economy, sometimes at the expense of food and health considerations. The growing of cash crops intensified after independence and had a huge impact on the use of available land and water in the Kenyan Nile basin.

## THE IMPORTANCE OF THE NILE WATER AGREEMENTS

Under international law, agreements signed by Britain on Kenya's behalf during the colonial period remained in force after independence, a restriction that Kenya refused to recognize. These divergent views already had a long history by 1963.

The major downstream state – Egypt – depended entirely on the Nile. To ensure that Egypt continued to have its 'fair share', the 1929 Nile Water Agreement was negotiated by Egypt and Britain (on behalf of all her East African colonies). That agreement was a tragedy for some parts of the Nile basin and a blessing for others. It consisted of an Exchange of Notes between Mohamed Mohmoud, the Egyptian prime minister, and Lord Lloyd, the British high commissioner in Cairo, on 7 May 1929, and provided a legal framework for water use and allocation in the whole basin.[23] Under this agreement, the Sudan was allocated 4 mcm/year, and 48 mcm/year was reserved for Egypt.[24] The need for water for irrigated agriculture in East Africa was hardly considered in the 1929 Nile Water Agreement when it stated in part:

> … Save with the previous agreement of the Egyptian Government, no irrigation or power work or measures are to be constructed or taken on the

> River Nile and its branches or on the lakes from which it flows, so far as all these are in the Sudan or in countries under British administration which would, in such manner as to entail any prejudice to the interests of Egypt, either reduce the quantity of the water arriving in Egypt or modify the date of its arrival or lower its level.[25]

The effect of this was that Kenya, like all the other riparian states, had to seek the consent of the Egyptian government to undertake irrigation and power works on rivers in the basin that passed through it. So far there have been no irrigation developments large enough to threaten the interests of Egypt in the Kenyan Nile basin.

Late in the colonial era, Britain was determined to make Egypt recognize the upstream countries' right to share the Nile waters. However, Egypt and the Sudan experienced strained relations in the mid-1950s, which made it difficult for Britain to introduce the Nile waters issue without worsening the situation.[26] Notwithstanding this, Britain pressed for an official recognition of East Africa's claims just a few weeks before the signing of the 1959 Nile Waters Agreement between Egypt and the Sudan. This came in the form of a Note dated 11 August 1959, which made the following observations. First, the East African governments reserved the right to review, at five-yearly intervals, their water requirements with other governments in the Nile basin. Second, the East African governments reserved the right to negotiate at any time, collectively or individually, for an increased quota of water required as the result of future development. Lastly, the note emphasized the East African governments' right to implement any scheme that might in future prove more immediately profitable or desirable than those upon which the current quota was determined, provided that the total of that quota was not exceeded.[27] Heedless of the British note, Egypt and the Sudan signed their agreement on 11 August 1959.

East African countries continued to press for their right to share the Nile basin waters. Based on surveys carried out by Sir Alexander Gibb and Partners, Britain estimated her dependencies' minimum water requirement for the next 25 years at 1.75 milliards of cubic meters, which was the equivalent of 2 per cent of the annual average flow of water at Aswan.[28] This amount was to be divided among the three East African territories as follows: Kenya to get 0.415 mcm/year, Tanganyika 0.647 mcm/year, and Uganda 0.642 mcm/year, for a total of 1.704 mcm.[29] Although Egypt and the Sudan ignored these estimates, the East African territories' share of the Nile water could not be wished away. But increased demand was not imminent, and the Foreign Office chose not to pursue the issue so as 'not to upset the prospect of the reestablishment of diplomatic relation between [the] United Kingdom and United Arab Republic'.[30]

The 1929 Nile Water Agreement remained in force throughout the colonial period and was reflected in the 1959 Nile Waters Agreement, which stated in part:

> If it became necessary to hold any negotiations concerning the Nile waters, with any riparian state, outside the boundaries of the two Republics (Egypt and Sudan), the Governments of the Sudan Republic and the United Arab Republic shall agree on a unified view after the subject is studied by the said Technical Commission. The said unified view shall be the basis of any negotiations by the Commission with the said states.[31]

Article V, paragraph 2 of the Agreement referred to the demands by other riparian states for water from the Nile system and made provision as follows:

> As riparian states, other than the two Republics, claim a share of the Nile waters, the two Republics have agreed that they shall jointly consider and reach one unified view regarding the said claims. And if the said consideration results in the acceptance of allotting an amount of the Nile water to one or the other of the said states, the accepted amount shall be deduced from the shares of the two Republics in equal parts, as calculated at Aswan. The technical commission mentioned in this agreement shall make the necessary arrangement with the states concerned, in order to ensure that their water consumption shall not exceed the amounts agreed upon.[32]

Such limitation did not prevent the colonial governments from undertaking economic activities around their parts of the Nile basin. Most farming in the 'White Highlands' was rain-fed, however, and the little irrigation did not threaten Egypt.

The British viewed the Nile as belonging to a 'community of users' once it dawned on them that the existing agreements would limit use of the water for their upstream interests. In the note referred to above, Britain called for settlement of the whole Nile waters question by proposing a conference of all riparian states, which would be sponsored by the International Bank for Reconstruction and Development (IBRD) and thus supposedly insulated from British influence.[33] But such a conference could succeed only if the Sudan and Egypt first settled their own differences.[34]

Since no objections were received from Egypt and the Sudan, the East African governments granted water permits to companies to ensure heavy investments, especially in sugar plantations, before the money meant for such endeavors was diverted elsewhere.[35] Water continued to be abstracted from the catchment area for irrigation and other economic purposes. Representatives of the riparian states agreed in a meeting at Khartoum in October 1961 that Kenya, Uganda and Tanganyika could continue abstracting water for their 'immediate irrigation requirement' so that development would not be retarded.[36] But the total amount of such water had to be reported to Egypt and the Sudan at periodic meetings of the technical representatives of the interested parties.

Problems arose over long-term projects. The East African territories argued that they needed about 5 milliards per year of water, measured in terms of the flow over the Owen Falls Dam, for their long-term projects. The Kenyan government's minister for agriculture argued that Kenya's long-term water need was 1.5 milliards while the immediate irrigation requirement was 0.26 milliards.[37] The latter figure was well below the 0.415 milliards requirement notified officially to Egypt and the Sudan in 1959.

It was becoming increasingly obvious that the East African territories' immediate and future needs for water in the basin could not be ignored because of a rapidly growing population and a rising political consciousness that was giving increasing expression to local opinions and aspirations through developing representative institutions.[38] When the East African governments' attempts to win recognition of rightful Nile shares through technical arguments failed, moreover, matters of sovereignty were bound to arise. In the early 1960s, East African ministers in charge of land settlement and water development held many meetings to press their claims.[39] A series of 'informal technical meetings' took place, with delegates from the East African territories and the Permanent Joint Technical Commission of Egypt and the Sudan, regarding use of water in the Lake Victoria catchment area. Political officials attended as observers.

The three East African states were at different stages of political development, but Britain was clearly losing its grip on all of them. Uganda and Tanganyika were independent by 1962 and Kenya in 1963. Why should Britain continue to be involved in Nile water matters? In a meeting of 'technical officers with no formal status' held on 21–28 March 1963, the then governor of Kenya, Malcolm MacDonald, urged the head of the Kenyan team to be careful because Tanganyika and Uganda were 'extremely susceptible to any suggestions of British dictation in matters of common concern' to all three territories and were 'particularly suspicious' that British influence might be 'exercised through Kenyan officials'.[40] Since the Nile water issue was not resolved by the downstream states and Britain before the East African states got their political independence, future discussions on the issue would undoubtedly face new expectations and difficulties.

Independence meant renouncing the Nile Waters Agreements. Instead, the new states subscribed to the so-called Nyerere Doctrine on Succession of Treaties, which meant acceptance of all obligations for a short period, followed by a declaration specifying which would continue to be accepted.[41] Kenya, for her part, declared:

> Insofar as bilateral treaties concluded or extended by former kingdom on behalf of the territory of Kenya or validly applied or extended by the former to the territory of the latter are concerned, the Government of Kenya is willing to be successor to them subject to the following conditions: (a) That such treaties shall continue in force for a period of two years from the dated of independence (which means until December 12, 1965); (b) That such

> treaties shall be applied on a basis of reciprocity; and (c) That such treaties may be abrogated or modified by mutual consent of the other contrasting party before December 12, 1965. At the expiry of the aforementioned period of two years, the government of Kenya will consider these treaties, which cannot be regarded as surviving according to the rules of customary international law as having terminated. The period of two years is intended to facilitate diplomatic negotiations to enable the interested parties to reach a satisfactory accord on the possibility of the continuance or modification or termination of the treaties.[42]

Thus Kenya continued to utilize the waters flowing into Lake Victoria and its part of the Nile basin with full confidence.

Lake Victoria continued to be the major unifying factor for East African countries, and its water a critical concern for downstream countries. To collect hydrometeorological data and develop hydrological models for river flow simulation in the basin, Egypt, Kenya, the Sudan, Tanzania and Uganda jointly launched the Hydromet project in 1967, with assistance from the UNDP and the World Meteorological Organization. Burundi and Rwanda joined later; Ethiopia never joined. This project was successful in providing valuable data until 1993, when it was superseded by the Technical Cooperation Committee for the Promotion of the Development and Environmental Protection of the Nile Basin (TECCONILE).[43] This in effect gave birth to a joint coordination forum to address common interests and objectives of the Nile basin countries. All riparian states participated in the founding meeting, either as full members or observers. TECCONILE established its secretariat at Entebbe.

The launching of the Nile Basin Initiative (NBI) was another milestone in attempts to resolve Nile water issues. In March 1998 the Council of Ministers of Water Affairs of the Nile Basin states reached an agreement at Arusha over sharing and managing Nile water. This led to the Nile Basin Initiative in February 1999, with all riparian states but Eritrea as members. The NBI secretariat was opened at Entebbe. The NBI was supported by a 'Nile Team' of multinational and bilateral donors. The NBI presented an important transitional arrangement until member countries were able to agree on a more lasting institutional framework for sustainable development of the entire Nile basin for the common good. The future lies in political goodwill to foster development of the entire Nile basin founded on an economic, ecological and political unit irrespective of political boundaries.

## ECONOMIC ACTIVITIES IN THE KENYAN NILE BASIN, 1963–2000

Political statements by government ministers in charge of water and irrigation create an impression that the poor state of socio-economic affairs in western Kenya is a result of the colonial Nile water agreements. The truth of the matter is that the

Kenyan Nile basin's resources are under-utilized, with water not only going to waste but also now killing people through floods. Reasons for the poor management of irrigation schemes, food shortages and the generally sorry state of socio-economic affairs in western Kenya will be found elsewhere.

The Lake Victoria basin in general is endowed with many natural resources that have remained under-exploited. Water is a basic resource that influences the quality, distribution and abundance of other natural resources, especially crops, forestry and soil. Water is also an important basic requirement for improvement and maintenance of human health, which in turn is part of, and a prerequisite for, development activities. The drainage basin is an important ecological unit where natural resources interact and influence one another.[44] The region is rich in biodiversity – fish, birds, reptiles, mammals and a host of plant species. The basin is well endowed with significant amounts of renewable energy resources, including biomass, hydropower, solar and wind. There is also great potential for hydropower in major rivers flowing into Lake Victoria that could be used for the socio-economic development of the region. As part of the larger Nile basin, Kenya shares Lake Victoria and its catchments. Management of water provides the basis for examining economic activities in the basin.

People of the Victoria basin are involved in several socio-economic activities in support of their livelihoods, including fishing, farming, bee-keeping, trading, quarrying and sand mining. Exploitation of natural resources is embedded in local cultures. In Kenya, rights to agricultural land are vested in individual ownership with title deeds. There are still parcels of land that fall informally under customary law: these are places of cultural significance involving historic episodes or mythical incidents and rituals. There is no restriction to the use of water from lakes, rivers and streams, but cultural norms and belief systems also affect natural resource management.

The basin is one of Kenya's economic regions with the capacity not merely to sustain but also to accelerate socio-economic development. Its catchment is endowed with natural resources not found anywhere else in the country, yet it is the home of the poorest Kenyans. Over 40 per cent of Kenya's population depends directly or otherwise on the resources of the lake basin for their livelihoods. In addition, the region has an enormous amount of water – Lake Victoria, numerous rivers, deep underground water, and rainfall for domestic, industrial and agricultural use, as well as for the production of hydropower. Lake Victoria even has potential as a source of tidal energy.

The wetlands are sources for goods and services and contribute to climate stability, and are a source for water and food production in dry seasons. Papyrus is harvested in wetlands and is used for thatching houses and for making mats, furniture, boats, rafts and so forth, while both shrubs and papyrus are used for fuel. Wetlands also are breeding grounds for certain types of fish.[45] In short, the lake basin has high ecological potential that has remained largely under-utilized.

**Table 3:** Irrigation and swamp reclamation possibilities.

| *Province* | *Area* | *Rivers* | *Immediate irrigation possibilities* | *Areas requiring hydraulic and soil surveys* |
|---|---|---|---|---|
| Nyanza | Kano Plain | Kibos | | |
| | | Nyando | | |
| | | Miriu/ | | |
| | | Sondu | 30,000 acres | 240,000 acres |
| | Nzoia Plain | Nzoia | 3,000 acres | 10,000 acres |
| | Yala | Yala | | 3,000 acres |
| | S. Nyanza | Kuja-Migori | | 2,000 acres |
| | Mara | Mara | | 3,000 acres |
| | SWAMPS | Yala | | |
| | | Nzoia | | |
| | | Small Swamps | | |
| | | N. Nyanza | | |
| | | Miruka | | |
| | | Kuja | 4,250 acres | 50,000 acres |
| Total | | | 37,250 acres | 308,000 acres |

*Source:* R.J.M. Swynnerton, *A Plan to Intensify the Development of African Agriculture in Kenya*, Nairobi 1954, 40.

The lower Sondu river has potential for hydropower generation and irrigated agriculture covering an area of 85,000 ha. It has been estimated that a flow of 30 $m^3/s$ could be regulated to provide hydropower generation. The Nzoia offers a great opportunity for irrigating 50,000–100,000 ha if the river is controlled to a steady discharge of 60 $m^3/s$. Dams could provide a hydropower supply with a capacity of 33 MW, in addition to facilitating flood control and providing water for domestic and livestock development in the area. Similarly, the Yala catchment has the potential for successful irrigation schemes; the Gucha and its principal tributary, the Migori, with assistance of some storage facilities, have the capacity to support irrigated farming. The Mara catchment, which is shared by Kenya and Tanzania, has a suitable dam and reservoir site near the Nyangores river for 300 mcm of water, which could be used during the dry seasons both for irrigation and livestock. (The river also is the source of water for many wild animals, and thus a tourist attraction for Kenya and Tanzania.)

Kenya's great potential for irrigation could solve the food problems and poverty that have engulfed the country. Estimates provided in 1954, by Swynnerton, on possible irrigation areas in western Kenya give a clear scenario of what Kenya could have done since independence to boost agricultural production through irrigation. A total of 60,000 acres had irrigation possibilities and another

727,250 acres could have been irrigated after hydraulic and soil surveys were done. In the Nyanza province, which encompasses the Kenyan Nile basin, a total of 37,250 acres could have been irrigated immediately, while 308,000 acres could have been irrigated after further surveys.[46] Out of all this only 67,000 ha is irrigated today. Table 3 illustrates this in a more detailed manner.

Groundwater potential in the Kenyan Nile basin is reasonably high. The Victoria basin has underground water in the order of 70 bcm. Only 18 mcm of the underground water is discharge – 7.8 mcm into streams and 4.9 mcm extracted by pumping.[47] It is estimated that annual underground water flow into the lake is about 6 mcm. Roughly 16 mcm of underground water is readily available for the economic development of the region.[48]

Limited quantities of minerals have been found in the lake basin. These include gold, silver, copper, iron, tin and valuable stone including limestone and soapstone.[49] Additionally, the basin has varied soil textures and qualities capable of supporting multiple economic activities after careful planning and scientific research have determined the best crops and practices for maximum yields.[50] Current land uses include mining, tree-growing (both natural and planted), rain-fed and irrigated agriculture, and livestock keeping; some land is used for national parks and game reserves, and of course for human settlement and all kinds of physical infrastructure.

## KENYA'S DEVELOPMENT POLICY AT INDEPENDENCE

In Kenya in general, and western Kenya in particular, the informal economy occupies a central role in the general economic framework. Literacy levels and informality are directly linked; most of those engaged in the sector have low levels of education. The informal economy involves unregulated activities in both urban and rural areas. Agriculture, informal in nature, offers employment for almost 90 per cent of the population while contributing to food security and the export sector. Women in the informal sector have peculiar disadvantages, including much lower educational levels.

Kenya inherited the colonial infrastructure and policies, and an economy dominated by foreign capital. The new government adopted policies aimed at ensuring that Kenyans had the greatest share of subsequent development. The main principles and strategies for economic and social development were spelt out in a 1965 document entitled 'African socialism and its implication to planning in Kenya'. The Kenyatta government seemed to reject both western capitalism and eastern communism. African socialism guaranteed every citizen full and equal political rights, control of the national wealth for society as a whole, and freedom from want, disease and exploitation.[51] These guarantees remained on paper only, because 44 years later most Kenyans, especially in western Kenya, live with want, disease and their resources exploited.

The presence of private investors in the Kenyan Nile basin should not come as a surprise. Part 17 of the African Socialism document encouraged private enterprise and joint private–public investment. In this regard, the Kenyan government has lived up to its pledge. Nationalization of foreign-owned companies and enterprises could be undertaken only if national security was threatened and resources were clearly misused. What this means is that, right from the beginning, the Kenyan leadership opted for clear strategies of economic growth based on keeping existing ties with the major western industrial nations so as to attract aid and investment, build overseas markets for local products, and expand tourism. Kenyan economic policies were biased towards earning foreign exchange, rather than towards economic growth and equitable distribution of resources to her people. The role of multinational businesses in Kenya and their influence on prices and markets has undeniably increased since independence.

Attempts were made during the first years of independence to get some Africans into trade. The government implemented trade licensing, which aimed at excluding non-citizens from trading in rural and non-urban areas. Rural areas lacked capital and infrastructure for any meaningful trade, so this provision benefitted a small number of Africans. The government also put in place some credit transfer corporations, which could extend credit to Africans. These included the Agricultural Finance Corporation, Kenya National Trade Corporation, Kenya National Housing Corporation, and the Industrial and Commercial Development Corporation.

## AGRICULTURAL ACTIVITIES IN THE KENYAN NILE BASIN

About 80 per cent of the people in the basin practice agriculture: subsistence farming, livestock keeping and agro-pastoralism. In western Kenya rain-fed agriculture is supplemented with irrigation. Subsistence agriculture and cash crop cultivation form the backbone of the local economy. Full exploitation of water and land for enhancement of agricultural activities has fallen short of expectations owing to poor governance (corruption and bad policy), a lack of emphasis on research geared towards improved agricultural methods and practices, and limited involvement of women in decision-making that directly deals with agriculture despite their being central to the activity. Sadly, educational and research institutions in Kenya are hardly incorporated into economic processes meant to improve livelihoods.

Agricultural improvement was a priority in the development agenda at independence. Towards this end, the government established development authorities with well-defined agricultural roles and clear managerial terms and powers sufficiently wide to permit strong executive action.[52] Among these were the Kenya Tea Development Authority (KTDA), Coffee Development Authority (CDA) and Kenya Sugar Authority (KSA). In 1979 the government established the Lake Basin

Development Authority (LBDA) to regulate and control the land-use system of the basin and recommend specific programs for Western and Nyanza provinces. Each of the authorities mentioned had a role to play in the Kenyan Nile basin.

A smallholder system of agricultural production predominates in the region, consisting mainly of mixed farming on relatively small plots. Areas with adequate rainfall (101–152 cm) have more developed farms than those with limited and unreliable rainfall. The latter suffer from alternate droughts and flooding. The drier areas support indigenous livestock because they are better able to cope with harsh conditions, and some indigenous crops are cultivated for the same reason. In areas with adequate rainfall, farmers engage in cash-crop cultivation, for instance of rice, sugar, coffee, tea, cotton and pyrethrum. They also grow maize, bananas, pineapples and other fruit for domestic consumption and cash. Some farmers keep improved dairy cattle, some wool sheep and poultry.

Smallholder farms are characterized by low investment in inputs, small size, reliance on family labor, production of food crops for domestic consumption, and only peripheral links to agricultural markets. More important, during the off-season, production is very low because of poor irrigation facilities, especially in the lowlands. Loss of traditional seed banking for farmers who cannot afford certified seeds and lack of access to emerging technological innovation have worked against maximum exploitation of the available resources in the basin.[53]

On the other hand, the basin has a long history of connection with the export marketing of agricultural products, mainly from large farms. Some parts of the basin are endowed with rich and fertile soils that support a variety of crops. The black cotton soils in Migori, Homa Bay, Suba, Kisumu and Siaya districts in Nyanza province have proved to be a boon to cotton growing, while in other districts the soils support high rice yields, sugarcane farming, sorghum and other crops in the Kano plains.[54]

Inhabitants of the lake basin grow a variety of food crops sold for cash. Among the common crops grown are maize, sorghum, millet and rice. Sorghum is mainly cultivated in areas with limited rainfall, while maize is best grown in areas with reliable rain. Smallholders cultivate cassava, yams, potatoes and groundnuts. However, with persistent reliance on traditional agricultural practices, farmers can hardly satisfy their own family requirements.

## IRRIGATION ACTIVITIES IN WEST KENYA

Although the Kenyan Nile basin is endowed with numerous rivers that would provide water for large irrigation schemes, very little acreage is under irrigated agriculture. Kenya's lakeshore and river basins constitute about 387,000 ha of land that can easily support irrigated agriculture. Additionally, western Kenya has about 200,000 ha that can support irrigated agriculture: about 65,000 ha along the middle and lower areas of the Nzoia river; about 15,000 ha in the Yala swamps;

60,000 ha in the Kano Plain; over 25,000 ha along the lower reaches of the Kuja-Migori; and 20,000 ha along the upper Mara. Other parts of the lake basin (constituting 15,000 ha) are irrigable. The lakeshore itself adds another 157,000 ha of land good for irrigation.[55] Despite such high potential, irrigation activities undertaken in the region are negligible. By 1985, for instance, only 3,000 ha was irrigated, a mere 1.5 per cent of the total potential irrigable land in the Victoria basin. Of that total, about 1,700 ha was under the National Irrigation Board and the balance in small-scale farmers' irrigation schemes.[56]

### *Rice irrigation*

The area under irrigation has been very small, but it has contributed significantly to socio-economic activities in Kenya, providing food crops, cash crops and employment opportunities. In addition to private irrigation for cash crops such as vegetables and coffee, the National Irrigation Board (NIB), under the ministry of agriculture, was established in 1966 to be responsible for irrigation schemes other than their research and training components. Irrigation development, whether large-scale or small-scale, is part of the 'technological improvement' of African agriculture linked to external inputs ranging from technical advice and the guidance of experts to the installation of large-scale engineering works and pumping facilities.

The NIB was basically a project-oriented organization and as such was not involved in large-scale investigation and planning issues. Its main functions included formulation, design, construction, operation and maintenance, agricultural management and irrigation agronomy research on lands specifically listed by the government for the purpose.[57] In all NIB-run schemes, farmers were tenants on plots of 1.2–1.6 ha. Tenants were expected to provide the necessary labor. They also sold their crops to the NIB at a fixed price, with all management and overhead costs deducted at that point. The NIB had insufficient trained staff to handle irrigation matters. For this and many other reasons, after some initial successes its schemes fell upon hard times.[58]

Among the several projects under the NIB, the Bunyala Irrigation scheme in the Busia district of West Kano may serve as an example. The Bunyala project, fed from the Nzoia river, was inaugurated in 1968. Its initial 400 ha of paddy received 5 mcm of water per year, with drained water returned to the river.[59] Irrigation began on a piece of land that was not initially bought by the NIB, unlike all the other NIB irrigation schemes in the country. In this case, farmers set aside a portion of their land that they wanted to irrigate. At the beginning, there were only 131 farmers, each with an average four-acre plot, with a few with larger plots comprising a total of 530 acres.[60] The project began with rice irrigation, two crops per year. Pressure from the NIB to do double cropping killed the spirit and morale of farmers. Many complained of too much work, and that long hours in the water were not good for health. Production dropped from 27 bags per acre in 1969 to less than half of that in 1972. Double cropping was dropped in 1978 and annual production rose to record levels. However, the production declined in the subsequent years due to poor management and low morale of farmers. The scheme collapsed in 1999, but was revived in 2004 by a farmers' initiative with government funding.

In the Kano Plains near Kisumu several rice irrigation schemes were established. Land in Ahero and West Kano was expropriated and plots were divided and distributed among previous owners and smallholders from neighboring areas. Each household was allocated 1.6 ha of land. The two schemes covered 4,800 ha, out of which 840 ha were cultivated by 519 tenants in Ahero, while 553 tenants cultivated 880 ha in West Kano in the 1980s.[61] Tenants also grew rain-fed crops, usually on small areas near their houses, and cultivated plots outside the scheme on their own land or rented farms. The scheme's village soon became crowded and unhealthy, prompting people to establish homes elsewhere while retaining the scheme plots. Depending on where tenants lived in relation to the plot, they became either 'resident tenants' or 'non-resident tenants'. This categorization is important in that 'resident tenants' usually had small plots and lived in the crowded scheme village, while the 'non-resident tenants' had much larger plots and also farmed sizable plots of land elsewhere. Yet another category of farmers in the Kano Plain were non-rice growers and others who grew rice as individuals, and combined the resources of farms and small plots of rice farming.

In West Kano, irrigated farmland was divided between paddy and sugarcane production. The West Kano Scheme on the shores of Winam Gulf began in 1975 with approximately 900 ha of sugarcane cultivation. The water requirement for this scheme was estimated at about 15 mcm/year; dikes protected the scheme from flooding, and excess water was drained back to the lake, but at such a cost as to render the effort unprofitable. Farmers selected and settled as tenants on land acquired by the NIB got little income from the irrigation. Like most of the irrigation schemes, the Kano projects are in disuse and the water flowing into the plains poses a threat to people and livestock. The Ahero Scheme, started in 1966 on the Kano plains near Kisumu, similarly failed.

The Provincial Irrigation Unit, established in 1978, supported a number of large-scale irrigation schemes in the Kenyan Nile basin. The PIU's main task was to promote irrigation, drainage, reclamation and flood protection. It was not mandated to initiate any project, but to support a local initiative by smallholders' groups and to some degree public institutions if they asked for assistance. PIU was partially assisted by external donors, especially financially and technically. For instance, the Dutch assisted Nyanza, Central, Coast and Rift Valley provinces, while the Germans supported irrigation schemes in Eastern province and the Danish government in Northeastern province. In the absence of local experts, these external groups perpetuated their presence and domination; dependence on foreign assistance has not helped the people in developing their own initiative and ownership of projects.

### *Sugar production*

Kenyan sugar is produced largely in western Kenya by both large- and small-scale producers, most of it under irrigation. The major sugarcane growing districts in Lake Victoria basin are: Kisumu, Bugoma, Trans-Nzoia, Nandi, South Nyanza, and to a small extent, Kericho, Siaya and Kisii Districts. By 1969, there were huge farms growing with sugarcane in the northern part of the Kano plains and the adjacent foothills which used overhead irrigation. In addition, there was about 5,000 acres (2,000 ha) of smallholder cane in the area. Other large sugarcane estates includes the Mumias with 3,238 ha; Chemilil, 2,833 ha; Muhoroni, 2,024 ha; Miwani 3,238 ha; and Ramisi, 4,856 ha. The area of sugarcane acreage increased over the years.

The small-scale farmers contribute a lot to the sugarcane industry. Small-scale farmers are in all the sugar-growing districts of western Kenya in which farmers cultivate below 40 ha. Like the large-scale farmers, they supply the sugarcane factories with the cane. This group accounts for 71.4 per cent of total cane supply in Mumias factory, 66.7 per cent in Muhoroni factory, 28.2 per cent in Chemilil factory and about 16 per cent in Miwani factory. So to have better bargaining power, small farmers formed cooperative societies.[62] Farmers in the so-called Million-Acre Settlement Scheme cultivate sugarcane on land between 1.01 ha and 3.03 ha. Yet, with such production, the people in the region live in abject poverty. Smallholder producers are under the company rules to produce sugar for the company at a price determined by the government. Inputs such as fertilizers, seed and hired labor are provided by the government as needed for a fee. The total of all inputs are deducted from the gross sales and the farmer gets the net. Sugarcane production was and still is an alternative source of employment or complementary to other economic activities. The sale of the sugarcane provides income, which in turn produces positive effects on households in terms of food supply and other amenities.

Other parts of western Kenya began to grow sugarcane for cash. Bungoma District was among them and in 1976 a sugarcane factory was started at Nzoia, located 25 km east of Bugoma town. The factory's objective at inception was to provide income and enhance the socio-economic status of its farmers' livelihoods.[63]

Industries dealing with sugar processing in western Kenya use hydroelectricity and petroleum. In addition, sugarcane industries in Kenya use co-generation, which is a combination of heat and electricity to generate power. This is done by using bagasse – a waste product from sugarcane processing – as the primary fuel for the bulk of domestic sugar factories. All the seven sugar companies in the region have been involved in one way or another in co-generation, but only Mumias Sugar Company is self-sufficiently supplied with electricity from the use of bagasse. Opportunities are there to co-generate in all the industries not only for Kenya, but other riparian states.

Lake Victoria basin is also a major producer of cotton, accounting for between 50 and 80 per cent of national output. Kisumu and Siaya have over 432 ha of land that is potentially good for cotton growing, while South Nyanza, Bugoma, Busia and Kakamega have 566 ha, 253 ha, 163 ha and 325 ha of potential land respectively. Production has remained low due to poor application of modern technology and agronomic practices.[64] Irrigation of cotton has been tried at the Ahero Pilot Irrigation Scheme. Following this lead, small-scale farming in cotton was encouraged, but – due to a lack of technical know-how and financial constraints – the yields were low and most farmers abandoned it. In addition, competition for labor as farmers engaged in other agricultural enterprises further contributed to the poor yields and interest.[65] Cotton production led to the emergence of cotton mills all around the lake ports; among them, Kisumu Cotton Mills (KIKOMI).

In addition to sugarcane, rice and fish industries, several other industries were established primarily because there was water nearby. For instance, the PanAfrican Paper Mill at Webuye on the banks of the Nzoia river, which drains into Lake Victoria, has greatly altered the local environment. Discharges into the river killed many aquatic organisms.[66] Vast areas in the neighborhood of Webuye have become wasteland. The river is continually polluted by chemicals produced during pulping, thus denying clean water to both people and livestock. Dumping of the mill's solid waste in the fields has led to a decline in local agricultural production. Demand for wood to feed the mill has also led to deforestation. Simply put, both people and animals have been exposed to real health hazards. Webuye is termed a 'sick town' by many observers. Information from local hospitals indicates that 60 per cent of the children aged 1–5 had been treated for breathing problems during a recent 20-year period.[67] The private-sector interest has been protected at the expense of Kenyans' health, in full view of the political elite, the World Bank and the government over the years.

### *Lake Victoria fisheries*

Fishing is one of the oldest economic activities associated with Lake Victoria and the main rivers of the Kenyan Nile basin, especially among the Luo community. The lakeshore provided the Luo with an opportunity to practice a mixed economy – fishing, farming, cattle herding and trading. They developed a variety of techniques and fished both for their own consumption and for trade. Tilapia was the fish of choice because it was easy to catch and the locals liked its taste.

Fishing in Lake Victoria was an activity closely integrated into the local culture. The lake communities had traditional and territorial rules and regulations, ensuring that the fishery was exploited in a sustainable manner. These limited access to a specific group of people and to certain stock areas. The traditional technology also tended to reduce the danger of over-fishing.[68] During pre-colonial times and much of the colonial era, fishing was limited to areas near the shore, in the wetlands, and at the river mouths. Rights were vested in clan elders, whose jurisdiction encompassed not only the land but also the water near shore. Among the lakeshore Luo in Kenya, rights extended up to what may be termed the '*kira* line'. Waters within the *kira* line were referred to as *puodho*, a term normally used to denote owned land, implying that access to it was restricted.

Clan elders regulated the use of the lake water and fisheries thereof. They had the responsibility of ensuring that the rules and regulations for fishing were adhered to. These rules included strict observance of closed seasons during periods of fish breeding and spawning. During the closed periods, when people were forbidden to fish, they could eat fish that had been sun dried and stored. Restrictions on canoe ownership and on fishing gear meant that only two fishermen per beach were allowed to own and operate seine nets. These two were expected to have high moral values, meaning in this context that they would share their catches. Fishermen had to be at least 20 years old and married. Outsiders were compelled to seek permission to set up their equipment in *kira* waters. They had to adhere to the rules and regulations of the locals, and had to report to the clan elders before fishing. They had to give details of where they had come from and declare how much and what kind of gear they had. It was only after all these conditions were met that outsiders were allowed to settle within the clan and to fish. Similar control measures have been reported among the hundreds of other riparian communities. In some instances authority was vested in 'beach leaders' and in others in 'chief fishermen'. In Uganda, 70 per cent of beach landings still fall under the control of an *agabunga* (head fisherman).

Commercialization of fisheries, especially the tilapia, locally called *ngege*, and later the Nile perch, was closely associated with the development of the colonial economy. Demand for fish came especially from the European population of the highlands and major towns, and from the emerging wage-labor sector in need of protein in their diet. To meet this demand, more efficient fishing nets were devised, which in effect commenced the 'industrial fishing' of Lake Victoria; the five-inch flax net was first introduced by a Scandinavian fisherman named Aarup in 1905.[69] To facilitate fishing, bales of netting twine were imported from Ireland so that Africans could make their own nets. Prisons became net-making factories; a long-term prisoner was moved from the Kisii jail to train prisoners elsewhere to make nets. The provincial administration supplied the Local Native Council with motor boats as well, for which the council supplied maintenance and crews.[70] Armed with new nets, local fishermen were joined by Europeans and Asians. European-managed boats put out 10–20 imported nets totaling about 1,000–2,000 yards.

By the 1920s the Kenyan fisheries' yields were in decline. Lake Victoria, in one official's words, now suffered from an 'indiscriminate slaughter',[71] which, if not checked, would result in the disappearance of fish from the lake. By the mid-1920s, both the local fisher-folk and the European and Asian fishermen were already forced to go far out from shore to get fish.

The decline of fish stocks prompted the colonial government to enact laws to govern the fisheries, and these displaced traditional authority. Under the 'Fish Protection Ordinance' of 1908 (Cap 163 Laws of Kenya) the governor of the colony was empowered to regulate fishing, impose fees, register boats, issue licenses and determine the times of fishing. It became a crime to be found in possession of immature tilapia.[72] Local communities were forced into this new fishing arrangement, not so much for their benefit but to give way to sustainable commercialization of fisheries for the colonial state. A new development was the introduction of the Nile perch, which changed the lake fisheries forever.

The three East African states sharing the lake also aggressively shared the exploitation of the lake's fisheries. (Kenya's portion of the lake, although small, has long had the highest fish production. This is attributed to its greater biological productivity and aquatic media.[73]) For harmonious utilization of the lake's fisheries, cooperation was established in 1927 under the East African Governor's Conference (EAGC). The three governors began to meet once or twice a year to discuss the fisheries (among other subjects of mutual interest). A Fisheries Development Service covering the whole of East Africa was founded in 1927 as a response to the decline in the catch. In 1948 the EAGC became the East African High Commission (EAHC), the umbrella under which were placed many activities of an inter-territorial character, among them the Lake Victoria fisheries.[74]

The decline of fish in Lake Victoria alarmed the colonial governments of the three territories. To reverse the trend they embarked on scientific research aimed at stepping up fish production. In 1927, for example, they collaborated to fund a fishing survey of the Victoria Nyanza. Among the first scientists to conduct research of the fisheries of the East African lakes were Michael Graham and E.B. Worthington.[75] Graham was a British naturalist whose major contribution was to inquire into the fisheries of the lakes and the means for preservation and proper exploitation.[76] Between the 1920s and early 1940s similar scientific research was conducted in other East African lakes, but achieved little due to limited personnel and finances. In 1945, however, the British Colonial Development and Welfare Act provided a substantial amount of money for the development of fisheries in Lake Victoria.

The governments and scientific foundations supported scientific and anthropological research on the trends in fisheries in Kenya. Worthington presented his findings in a lecture on 25 January 1932, which provided a scientific analysis of the prevailing situation in fishing and aroused much interest in Britain. Among those present during the presentation was a former Deputy Governor, G.A.S. Northcote, who commented:

> The fact of the matter is that about six years ago (1927) in Nairobi, where I lived comfortably – certainly not in such surroundings as have been depicted to-night – we heard a dreadful rumour that our breakfast tables were in danger of being deprived of their daily fish, which was principally derived from Lake Victoria. I had the honour, or audacity, then to suggest – money was easier to get [in] those days – that we might have an expert out to see if our breakfast tables could be maintained.[77]

In 1930, Worthington and Graham led an expedition to Lake Nabugabo in Uganda to determine the consequences of introducing Nile perch. Lake Nabugabo was once part of Lake Victoria, but had been cut off from it by the growth of a sand bar from the Budu coast. If Nile perch could establish themselves there, one of two things was likely to happen – either they would multiply enormously and eventually exhaust their food supply and die out, or they would strike a natural balance with the other species and an economic fishery for Nile perch could be established.[78] Since Lake Nabugabo contained a complete sample of Lake Victoria fauna and flora, it made sense to begin the experiment there. Although little information is readily available to ascertain the success of this experiment, it can be assumed that the result was good enough to deduce that Lake Victoria could support the Nile perch in future.

In the same way that land was tamed for economic gain, the colonial government saw the possibility of 'farming' Lake Victoria. The government recommended introduction of the Nile perch provided that measures were taken to ensure that the new species did not escape to other waters.[79] Although the introduction of exotic species was to be deprecated from a scientific point of view, the Kenyan government was keen to introduce the fish anyway; the economic development of the colonies rendered interference with the native flora and fauna inevitable.[80]

In the 1930s, Lake Victoria became a major source of fish consumed locally and in towns beyond western Kenya. The recorded catch in Kavirondo Gulf between July and December 1933 was over 580,000 tilapia, 14,250 catfish, 17,200 snoutfish and 11,500 lungfish, fetching about 6,176 shillings.[81] About a third of this fresh fish was sold at Kisumu market then shipped by rail to Nairobi; some fresh and smoked fish was transported by road to Kakamega, Eldoret and Kitale for European consumption. The remainder was locally sold for African consumption. During January-March 1933, 70 per cent of the fish caught by local fishermen was smoked and shipped to Uganda by dhow in readiness for the opening of the cotton season; some went also to the North Kavirondo Reserve.[82] But the process of stocking Lake Victoria with the Nile perch began in earnest only in 1960.[83]

### *The fishing industry since independence*

The second half of the twentieth century witnessed dramatic changes in the fisheries of Lake Victoria. Since the 1960s, the collapse of traditional regulatory structures and of remedial actions taken during the colonial period have paved the way for unprecedented exploitation, and the extinction of some fish species from the Lake Victoria and satellite lakes and rivers in the basin. Commercial fishing involves three species: the Nile tilapia, *omena* and Nile perch. The last two accounted for 3.5 per cent of the total landing in 1970, but 95.5 per cent 20 years later. The situation remained the same into the late 1990s. A substantial export trade developed. The World Bank and International Monetary Fund encouraged this trend as a way of offsetting the big loans owed by Kenya. However, the cost of promoting fresh-water fish export in terms of food security, malnutrition and poor health, and changes in the local economy and gender relations, are impossible to quantify and have largely been ignored.

Development of the fishing industry was a way of alleviating poverty. It increased economic growth, incomes and employment opportunities, as well as increasing the output and quality of government services. Kenya has recorded a higher rate of fish production than Uganda and Tanzania, although she has only a 6 per cent share of the lake. This is because the biological productivity on the Kenyan side of the lake is greater than in the other parts. The most exploited – and over-fished – species is the tilapia.

The introduction of exotic species increased the total catch. In 1960 the Nile perch was introduced into Lake Victoria, and its population soon exploded. Between 1979 and 1989 the annual total catch of all species rose from about 100,000 tonnes to over 500,000 tonnes; while the Nile perch catch from Lake Victoria increased from about 1,000 tonnes in 1981 to 363,000 tonnes in 1993, with Kenya accounting for 29 per cent.[84]

There has been a dramatic transformation of fisheries since the mid-1990s, from locally-based fishing with little external intervention and capital investment to domination by national and international capital. The reason for this change is the great international demand for the Nile perch. Nile perch export is a major foreign exchange earner for Kenya. In 1998, estimated income from Nile perch exported from the three countries amounted to a total of $219.4 million. Of this, Kenya accounted for $34.8 million.[85] Had this income reached local fishers, their welfare would have appreciably improved.

Almost every part of the Nile perch has some economic use. The skin is tanned for belts, purses and shoes sold in faraway locations. The Kenya Industrial Research and Development Institute (KIRDI) uses perch hides to make leather products. The internal organs that help fish rise and sink in the water are sent to England for filtering beer and wine and to the Far East for thickening soup. Only the smallest Nile perch and leftovers – heads, tails and 'frames' – are left for the people of the lake.[86]

To meet external demand, fish processing factories have been established along the shores of Lake Victoria. Samaki, which started in Nairobi in 1965, opened another branch at Kisumu in 1986. Many more factories have since opened in the area, most owned by Kenyan Asians, but providing employment and income for many Kenyans. These are financed by international and local institutions, which have in a way legitimized the processes of exploitation of the Lake Victoria fisheries.

The Kenyan fish processing factories include Samaki Industries, East African Sea Food, Prinsal, Peche Food Ltd, Tropical Food Product International, Modern Fishing Ltd, Afro Meat Co. Ltd, Kendag, Capital Fish, W.E. Tilley (M) Ltd,

Cold Storage (Foods) Ltd, Lake Victoria Fish, Star Fisheries and Eagle Fisheries.[87] With processing factories in place, and fishing technologies improved, Lake Victoria has been transferred from the local and regional market economy to the global economy.

Since the late 1990s fishmeal-based animal feed industries have been established in the area. The main raw material for this industry is R. *argentea* (*omena*, in the local language) and Nile perch skeletons. This industrial venture targeted regional and international commercialization. The market is not designed for local food consumption, but rather export for animal feeds. Associated with the fish processing factories, especially the filleting ones, is Nile perch frames processing: the *mgongo wazi* (the fish skeleton), which used to be discarded as waste or consumed by local communities. This changed when the animal feed manufacturing industry began to use it for making fishmeal.

Fishmeal factories sponsored by the Dutch government were introduced, notably at Mwanza in Tanzania, in the late 1970s. In addition to offal from slaughterhouses and crushed animal bones, these used the *dagaa*, a small sardine-like fish dried and consumed locally, as a main ingredient for fishmeal. This led to over-fishing. In the 1990s fish frames became a raw material for the industry, causing stiff competition between local communities and the factories. Factories and human beings now compete for fish frames, used as food by local people and by others for fishmeal. Thus, the three most important fish species of Lake Victoria, which together made up as much as 98 per cent of the catch, became integrated into the global market.

Most of the factories are owned by foreigners and/or local Asians. Their main objective has been to secure sufficient supplies of raw material in order to exploit as much as possible the installed technical capacity of the factories. There is larger technical capacity in the region than the available raw material, hence the competition for fish among the factories. The factories are reported to have entered into long-term agreements with some of the major supermarket chains in Europe and the USA. Although the factory owners are convinced that the demand for Nile perch is 'unlimited', many of the factories around the lake have problems in delivering quality fish, which has adversely affected earnings. Fish factories in other parts of the world, including Canada and the USA, have closed down for lack of raw material and the Lake Victoria fisheries are likely eventually to follow suit.

### *Small-scale fish farming*

During most of the twentieth century the lake was a multi-species fishery with a diverse ecosystem, in which native food fishes were targeted. The lake ended the century with a much more productive fishery, but one in which three species – two of them introduced – made up the vast majority of the catch. Although many fish stocks in Lake Victoria had declined before the expansion of the Nile perch population, a dramatic increase in the population of Nile perch in the 1980s roughly coincided with the drastic decline or disappearance of many indigenous species. Now, two decades after the rise of the Nile perch in Lake Victoria, this species too has shown signs of being over-fished, and some of the native species that were in retreat – or even thought extinct – are re-emerging. Data on the resurgence of the indigenous species suggest that heavy fishing of Nile perch may enhance biodiversity, and this has spawned renewed interest in management options that promote both fishery sustainability and biodiversity conservation.[88]

The Lake Victoria basin has huge potential for fish farming because there is plenty of water from the lake itself and the rivers, under climate regimes ideal for both warm and cold aquaculture development. Fish farming constitutes part of the economic activities in the basin in which residents have been actively involved, although at the artisanal level. The Kenyan government encouraged the formation of cooperatives as a way of improving fisheries and the livelihoods of the fishermen. Cooperatives are essential for providing markets for the fish catch. The number of fishermen cooperatives has steadily increased. By 1994 there were 3,188 farmers owning 4,382 fish ponds in the region. The majority of these were individually owned, though women's groups, youth groups, self-help groups or institutions owned a few. Most produced tilapia; in 1993 some projects introduced *Clarias gariepinus* (African catfish).

Women occupy a central place in the fishing sector. They predominate in the Lake Victoria fisheries, representing 70–80 per cent of workers involved in this activity, especially in the artisanal fish trade. The fishery sector (as traders or processors) is indeed characterized by high participation of single, divorced and widowed or separated women. Women have joined the sector because fish are easily accessible and stored, profits can be divided without difficulty, and the initial capital requirement is low. Fishing is a ready job opportunity and can pay relatively well.[89]

Women sell *dagaa* because the fisheries are closer at hand than those of the Nile perch and *dagaa* are easily stored. Some say fishing is in their lineage: they were introduced to fishing by parents, friends and husbands upon marriage. The lake makes fishing easy, and fish is the most readily available resource with immediate demand as a commodity for sale. Women can also barter fish with other communities for food during dry spells when food is scarce. The lake is literally the 'mother' of most families on the lakeshore.[90] Traditional Luo regulatory mechanisms were usually harsh on women: women did inshore fishing, and men did the offshore fishing, at night. Women were involved in a type of fishing called *kitenga*, in which they drove fish into basket traps, usually in shallow waterways in the lake's wetlands. Women now do more serious fishing, and some hire boats to venture offshore. Others have formed groups for fish farming; 84 per cent of the marketing sector is in women's hands.[91]

With the formation of the Lake Basin Development Authority (LBDA) in 1978, the country began to treat the resources of the Victoria basin more comprehensively. The LBDA was established as an economic development agency used by Kenya primarily to coordinate the development with Uganda and Tanzania.[92] In 1994 a convention for the establishment of the Lake Victoria Fisheries Organization (LVFO) was signed by the three countries. More assistance came from the Food and Agriculture Organization of the United Nations (FAO), the EU through the Lake Victoria Fisheries Research Project (LVFRP), and the World Bank and Global Environment Facility (GEF) through funding of the Lake Victoria Environment Management Project (LVEMP). The LVFO is specifically responsible for promoting proper management and optimum utilization of the fishery resources of Lake Victoria. The main objectives of LVFO are to foster co-operation among the contracting parties in matters regarding Lake Victoria, as well as to harmonize national measures for the sustainable utilization of the living resources of the lake.

LVEMP is a comprehensive environmental program for the conservation of Lake Victoria. It was developed according to a tripartite agreement signed in 1994 by the governments of Kenya, Tanzania and Uganda. Planned as a five-year project, lasting from 1997 to 2002, it was funded by the International Development Association (IDA) and the Global Environment Facility (GEF). The government of Norway through the World Bank has provided funds for development of a Shared Vision and Development Strategy to be used as a basis for a second phase. The riparian states agreed to two regional task forces: one dealing with fisheries management and control of water hyacinth and other weeds; and the second dealing with water quality and land use, including wetlands.

In 1995 the International Union for Conservation of Nature (IUCN) approached the Ministry of Foreign Affairs in Oslo to obtain support for the planning and implementation of a project related to the Lake Victoria fisheries. IUCN's research project, 'The socio-economics of the Nile perch fishery on Lake Victoria', investigated the effects of globalization on the Lake Victoria fisheries. This project was funded by the Norwegian Agency for Development Cooperation (NORAD). A second project under the IUCN, the 'Network of Environmental Journalists for Lake Victoria', was intended to improve information exchange and raise knowledge levels on environmental issues. This project was funded by the Swedish International Development Agency (SIDA). The two projects aimed to improve access to information about Lake Victoria fisheries, encourage stronger participation of lake-dependent communities in fisheries management, and develop regional policy options for the sustainable use of fisheries resources.

'Friends of Lake Victoria' (OSIENALA, as abbreviated in the local Luo language) was established in 1992 at Kisumu as a non-governmental organization by a group of professors, researchers and community leaders, to represent the interests of the fisher-folk and express their views to decision-making bodies. In order to ease the demand affecting the size and diversity of Lake Victoria's fish stocks,

OSIENALA works closely with local groups to educate the people who are dependent on the lake about the advantages of fish farming. For instance, OSIENALA made its mark in highlighting health and pollution issues. It has begun to study the virtues of waste treatment in industries in the basin, especially the Muhoroni Sugar Factory and its molasses-producing counterpart, Agrochemical and Foods Ltd. In this case OSIENALA was faced with two options: exposing the two industries and eventually getting them closed and thereby causing thousands to lose their jobs; or working with them to steadily reduce pollution levels. OSIENALA chose the latter. They began to work with the District Development Committee and managed to get the government of Kenya to offer tax breaks. This exercise was not cheap for the companies, which installed new treatment works to reduce the toxicity of their effluents.[93] Diplomacy and threats worked in this case, as indeed did the Kenyan judicial system.

### *Hydropower within the Kenyan Nile basin*

Kenya has hydropower potential of 30,000 MW, but only a total of 611 MW is installed. In 1987, Kenya Power and Lighting estimated that the Nzoia river has gross capacity of 60 MW, while the Yala, Sondu and Kuja rivers have 60 MW, 120 MW and 18 MW respectively.[94] This potential remained untapped until 1999 when, at the Sondu/Muriu Dam, the government-owned Kenya Electricity Generating Company (KENGEN) began to produce 60 MW of electricity using 40 mcm/s of water. Unlike other major hydropower projects in Kenya, the Sondu/Miriu project relies on the flow of the river, with only a small storage capacity at the intake. Some water is diverted to the power station via the intake tunnel, while the rest is left to flow downstream into the original course.

The problems associated with the project are not so much related to the Nile because the water eventually goes back to the river. The issues are environmental, economic, cultural and even religious. Local people responded angrily to the destruction of forests, especially the Koguta forest, for the tunnel, and this deforestation is likely to affect the microclimate of the area, with implications for food security. Moreover, the electricity produced was clearly not intended for local benefit: step-downs were not installed. Apart from environmental concerns, the construction of this hydropower station has interfered with the cultural and religious shrine at Odino Falls because, according to the local communities, it is God's abode. This project will also disrupt the whole hydrological basin, especially wildlife.

## CONCLUSION

Political rhetoric that blames the two Nile agreements for the sorry state of affairs in the Kenyan Nile basin displays ignorance of socio-economic and political changes that demand a new look at the availability, utilization and management of water in the basin and elsewhere in Kenya. Other factors – population growth, consumption practices and patterns, diversion of water resources, climatic and environmental conditions, and rainfall patterns – have changed so much since the late 1950s that areas that once had adequate rainfall no longer do so. The Kenyan population living within the basin depends on land and water resources; western Kenya has great potential, but is ironically the home of Kenya's poorest population. The Egyptian–Sudanese Nile agreements do not stand in the way of such development: the water and land resources in the region are under-used and mismanaged.

# 7
# Independent Uganda and the Nile: Hydroelectric Projects and Plans

James Mulira

## INTRODUCTION

Mainstream Ugandan history writing has basically overlooked a very important fact: Uganda is geo-strategically located in the Nile basin and this has influenced her modern history in fundamental ways. This chapter takes as its starting point the fact that almost all Uganda is located within the Lake Victoria basin, and that from here, at Jinja, the Nile eventually flows into the Sudan and down to Egypt. Uganda's development has moreover been complicated by the physical fact that she is both an upstream and a downstream country with regard to the Nile, and that its water resources are almost wholly trans-boundary in nature.

This dual position of Uganda in the Nile basin sometimes puts the country in a challenging situation with respect to negotiating for its equitable share of the Nile water resources. Being an upstream country implies that Uganda's negotiating position vis-à-vis Egypt and the Sudan could roughly coincide with that of the other upstream countries such as Kenya, Tanzania, Rwanda, Burundi, the Congo and Ethiopia. On the other hand, being a downstream country implies that Uganda's legal interests when it comes to principles for sharing the Nile waters could coincide with those of the other downstream countries such as Egypt and the Sudan. Uganda therefore faces the challenge to reconcile the demands of both positions at the same time as they offer her more negotiating options and therefore potentially greater influence in the basin. Uganda's location in the Nile basin is therefore crucial for her socio-economic and socio-political development.

This study analyses the historical development of Uganda in a Nile perspective; the importance of Uganda's strategic location in the Nile basin vis-à-vis the 'source' of the River Nile; the implications of British colonization of the Nile upstream of Egypt in the late nineteenth century to secure her interests in Egypt; independent Uganda's participation in, and role in the development and sustenance of, the sub-basin and basin-wide projects and programmes; and Uganda's position on current water agreements. But first and foremost this chapter examines the extent to which the Nile and Uganda's position in the basin have affected and are still affecting the course of Uganda's socio-economic development.

The empirical focus is on hydropower development since this has by far been the most important Nile-related activity in the country. Irrigation, fisheries, tourism and oil are also considered, but in more general terms.

## GEOGRAPHICAL BACKGROUND: HYDROLOGICAL AND CLIMATIC CONTEXT

The geographical and economic importance of Nile waters in Uganda is so obvious that its neglect in many social and historical analyses of Uganda's development points to these traditions' 'water blindness'. Uganda covers an area of 236,000 km$^2$ of which 15 per cent is water – despite it having no sea coastline – and most of it being part of the Nile system.

There are several large freshwater lakes (Victoria, Albert and Edward), which are shared with neighbouring countries and make up the basin for the Upper White Nile. The main river into Lake Victoria is the Kagera, which rises in the highlands

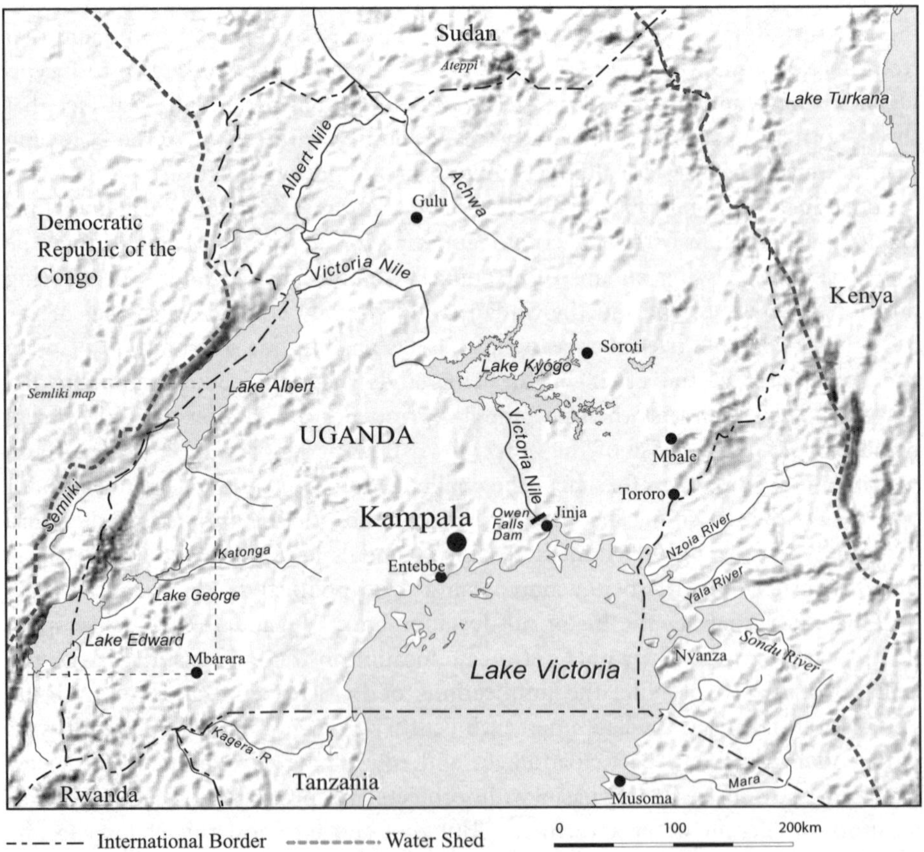


**Map 1a:** Uganda.

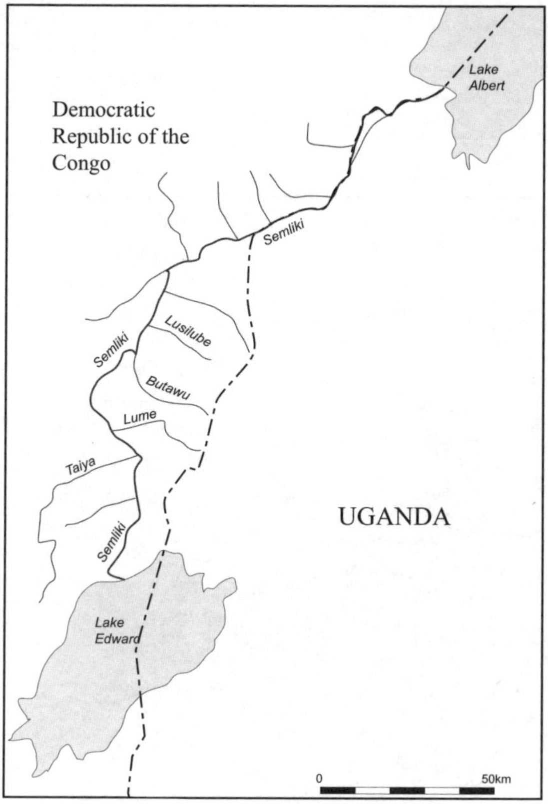


**Map 1b:** Upper White Nile basin.

of Rwanda and Burundi.[1] Lake Victoria catchments (see Map 2) are constituted by five countries – Kenya, Tanzania, Uganda, Burundi and Rwanda – and are drained by a number of large rivers – Kagera, Nzoia, Gucha-Migori, Sondu Miriu, Mara, Yala, Issanga, Nyando and Biharamulo – plus many small rivers and streams (see Table 1). Lake Victoria is one of the most important shared natural resources of the original East African countries – Uganda, Tanzania and Kenya – it serves several common interests of the three countries and is a resource of great socio-economic potential. Covering an area of 69,000 $km^2$, Lake Victoria is the second largest freshwater body in the world after Lake Superior in the USA. Uganda's share of the lake is 45 per cent, Tanzania's is 49 per cent and Kenya's is 6 per cent.

The main rivers that flow into Lake Victoria and the long-term average discharge, based on 1950–2000, from the river basin are summarized in Table 2.

The mass balance of water in the lake based on inflow and outflow is shown in Table 3. The outflow in the White Nile is correspondently larger than the outflow from the catchment.[2] The White Nile provides a small but steady flow of water and accounts for 14 per cent of the Nile's yearly flow. Lake Victoria has both an

**Map 2:** The Lake Victoria basin.

*Source:* United Nations Environment Program. 1997, World Atlas of Desertification, 2nd edition. London, UK: UNEP.

**Table 1:** Long-term average discharge from river basins in the Lake Victoria catchment area.

| *Country* | *Basin* | *Discharge ($m^3/s$)* | *Percentage (%)* |
|---|---|---|---|
| Kenya | Sio | 11.4 | 1.5 |
| | Nzoia | 115.3 | 14.8 |
| | Yala | 37.6 | 4.8 |
| | Nyando | 18.0 | 2.3 |
| | North Awach | 3.7 | 0.5 |
| | South Awach | 5.9 | 0.8 |
| | Sondu | 42.2 | 5.4 |
| | Gucha-Migori | 58.0 | 7.5 |
| | Sub-total | 292.1 | 37.6 |
| Tanzania | Mara | 37.5 | 4.8 |
| | Grumeti | 11.5 | 1.5 |
| | Mbalageti | 4.3 | 0.5 |
| | E. shore stream | 18.6 | 2.4 |
| | Simiyu | 39.0 | 5.0 |
| | Magogo moame | 8.3 | 1.1 |
| | Nyashishi | 1.6 | 0.2 |
| | Issanga | 30.6 | 3.9 |
| | S. shore stream | 25.6 | 3.3 |
| | Biharamulo | 17.8 | 2.3 |
| | W. shore stream | 20.7 | 2.7 |
| | Kagera | 260.9 | 33.5 |
| | Sub-total | 476.40 | 61.2 |
| Uganda | Bukora | 3.2 | 0.4 |
| | Katonga | 5.1 | 0.7 |
| | N. shore stream | 1.5 | 0.2 |
| | Sub-total | 9.8 | 1.3 |
| Total average inflow | | 778.3 | 100.0 |

*Source:* COWI, Consulting Engineers, *Integrated Water Quality/Limology Study for Lake Victoria Environmental Management Project, Part 2 Technical*, 2002, retrieved 17 October 2007 from www.ilec.or.jp/eg/lbmi/index.htm.

international and a regional dimension. The lake is the main source of the White Nile, which is vital for the socio-economic interests of many riparian countries.

The Lake Victoria basin is one of Africa's largest trans-boundary water catchment areas, covering an area of about 194,200 km$^2$. The Lake Victoria catchment is shared among five states in the following proportions: Tanzania 44 per cent (85,448 km$^2$), Kenya 22 per cent (42,724 km$^2$), Uganda 16 per cent (31,072 km$^2$), Rwanda 11 per cent (21,362 km$^2$) and Burundi 7 per cent (13,594 km$^2$).

**Table 2:** Average inflows and outflows from Lake Victoria.

| *Type of flow* | *Flows* ($m^3/s$) | *Percentage (%)* |
|---|---|---|
| Inflows | | |
| Rain over lake | 3,631 | 82 |
| Basin discharge | 778 | 18 |
| Outflows | | |
| Evaporation from lake | –3,330 | 76 |
| Nile River | –1,046 | 24 |
| Balance | +33 | |

*Source:* COWI, Consulting Engineers, *Integrated Water Quality/Limology Study for Lake Victoria Environmental Management Project, Part 2 Technical*, 2002, retrieved 17 October 2007 from www.ilec.or.jp/eg/lbmi/index.htm.

The lake and its catchment provides food (fish), hydropower generation, transport and communication, tourism and water for domestic, agricultural and industrial use; it also allows wastewater disposal and is a source of recreation. The lake is also vital for weather and climate modulation.

It is not only that the Lake Victoria basin is a water body shared among the three East African Community states; its geographical location has international legal implications, especially for the Sudan and Egypt, within the River Nile basin, and for Rwanda and Burundi, owing to its connection with the Kagera river basin.

## THE COLONIAL NILE HISTORY – A SHORT SUMMARY

The 'discovery' of the source of the Nile by John Speke had far-reaching implications for European imperialism in the Sudan, Egypt and East Africa. Since then Uganda has taken centre stage in the international political affairs of the Nile basin.

At the Congress of Berlin of 1885 it was agreed that the whole of East Africa was to be German and British spheres of influence. But the 1886 Anglo-German Agreement left the question of who was to take over Uganda unanswered. The British dilemma was how to take up Uganda in the face of German interests in the East African region. For instance, before the treaty of 1890, the United Kingdom had claimed an area south of Lake Victoria extending to Lake Tanganyika, and Germany had claimed a boundary that would have extended north-westwards from the eastern shore of Lake Victoria, bisecting the Kingdom of Buganda, currently a part of Uganda. The latter claim would have placed western Buganda and an access route to the Nile valley within the German sphere of influence.

And yet Uganda's strategic location in the Nile basin, in relation to Egypt and beyond, remained of paramount interest to Britain in her empire building. Britain felt that in order to preserve her interests in Egypt she had to secure the entire Nile valley not only against European rivals, but also in order to develop the irrigation economy and cotton production in Egypt, and the latter was dependent upon waters from the White Nile. The British strategists also regarded power over the headwaters of the White Nile as an important strategic asset vis-à-vis political forces in Egypt that wanted to get rid of British control of the country. It is essentially in this context that Uganda was to become a British colony, and British interests in East Africa centred on securing the area that today is called Uganda.

Uganda was necessary to serve British interests in the Sudan and Egypt.[3] The men in charge of Egypt's water said in 1894, the year Uganda was put under London's control, that the White Nile was the tributary that contributed most to the total flow of the Nile, during the dry summer season, when cotton was being grown, and the waters of the White Nile were consequently described as being 'as valuable as gold'.[4]

It was noted that the presence of the source of the Nile gave Uganda strategic importance both in relation to Egypt and the Suez Canal through which ships sailed to the jewel in the imperial crown – India.[5] Hence Uganda served the geopolitical and economic interests of Britain in Egypt, the Sudan and East Africa.

## HYDROPOWER DEVELOPMENT IN UGANDA BEFORE INDEPENDENCE

Regulation of Nile waters for Egypt was a concern of Britain as early as 1894, as exemplified by a report written by William Willcocks, one of the influential British water engineers in Egypt. The report states that 'what the Italian Lakes are to the plains of Lombardy, Lake Albert is to the land of Egypt'. By damming the lakes, he asserted, a constant and plentiful supply of water to the Nile valley during the summer months could be ensured. Another senior water planner in the Egyptian administration, Justin C. Ross, had speculated in 1893 that raising the water level of Lake Victoria by only 1 m would produce a water flow 30 times what Egypt would ever need.[6] Britain's main interest in setting up dams in Uganda was to facilitate irrigation in downstream countries.

Plans for hydropower development in Uganda started as early as 1904, when the Uganda Company suggested Ripon Falls as a potential site for hydroelectric power. Then, in 1907, Winston Churchill, at the time Under Secretary of State for Colonies, on his tour down the Nile from Lake Victoria to Egypt, pointed out the potential of damming Lake Victoria at Ripon Falls.[7] Churchill thought it was possible to establish a dam on the Nile at Owen Falls for the modernization of central

Africa; Uganda, he said, should be the regional powerhouse. However, although the colonial authorities were interested in hydropower development in Uganda, the Foreign Office in London was against it because they did not wish to jeopardize their relationship with Egypt.

The next survey was carried out by Uganda Electricity Survey in 1935. In the 1940s the plans for a hydropower plant were again mooted, this time by Governor Hall. The construction of the Owen Falls dam has had developmental implications for Uganda in particular and for the basin-wide development of the Nile in general.

## OWEN FALLS DAM (NOW NAMED THE NALUBALE DAM)

It was believed that:

> The construction of the Dam at Owen Falls will serve two main purposes: It will provide Uganda with a source of hydro electric power and at the same time it will assist irrigation projects in the lower Nile valley by enabling the flow of water from Lake Victoria to be regulated and for the Lake itself to be used for water storage.[8]

One of the driving forces behind the setting up of a hydropower project for Uganda's development was the British Governor, Sir John Hall. He likened the role of electric power in developing Uganda's sectors to that played by coal in Britain during the Industrial Revolution in Europe.[9] Apart from serving the socio-economic interests of Uganda, the Owen Falls Dam was to turn Lake Victoria from a natural lake into a large reservoir that controlled the lake's flow to the White Nile on its journey to the Sudan and Egypt. The dam was also set up to supply electricity to Uganda and to Kenya.

### *Dam design and its implications for the Nile basin*

The plan involved damming up to the level of Lake Victoria, and submerging Ripon Falls to produce a head of 18 m on which the turbines would operate. This required a constant discharge in the Nile of 632 $m^3/s$ of water. The total installation capacity of Owen Falls Dam was 150 MW. The construction of the dam resulted in notable changes to the natural setting of Lake Victoria. For example, the dam transformed Lake Victoria from a natural lake to a reservoir that controlled the lake's outflow to the Victoria Nile. Originally the outflow of Lake Victoria was controlled 'hydraulically' by Ripon Falls, which acted as a natural weir and constriction that allowed the flow of water to exit the lake depending on the level of the water in the lake. Nalubale Dam, therefore, submerged Ripon Falls, which was also excavated for the construction of the dam. The dam from then on assumed hydraulic control of Lake Victoria. The level of the lake is maintained partly by the Owens Falls Dam on the river, which raises the level of the lake about 0.9 to 1.2 m.

### *'Agreed curve'*

Based on the agreements of 1949 and 1953 between Uganda and Egypt, it was resolved that the operation of the Owen Falls Dam would be according to the 'agreed curve' (to mimic natural flow). This operating rule was developed for the operation of the Owen Falls Dam to dictate how much water should be released from Lake Victoria based on the water level in the lake. It was intended to retain the original (natural) pre-Nalubale-Dam relationship between lake level and outflow. In order to ensure that the 'agreed curve' was maintained, engineers from Egypt have been stationed at the dam since it became operational.[10] Egypt's interest in the management of the dam, dating from before the construction of the Aswan Dam, was due to her dire need to use Lake Victoria as a seasonal storage site, so that 'timely' water could be released during the period when the main Nile river runs low. This necessitated raising the dam 1 m higher than was needed for power generation. For this modification, Egypt paid Uganda a one-time compensation of 980,000 British pounds for the construction of the extra metre.[11] Egyptian interests in Uganda were further guaranteed by the Anglo-Egyptian Agreement of 1950 for hydrological and meteorological surveys.[12]

The Owen Falls Dam, constructed at a cost of 16 million pounds, was opened by Queen Elizabeth II in April 1954. The importance of this dam in the overall socio-economic development of Uganda and Egypt has been underscored by one leading scholar on the Nile thus: 'Seen from Uganda, it was as if the life of the country had started anew ... the hydroelectric project had transformed Lake Victoria into the largest reservoir in the world. This was the beginning of Uganda.'[13] In a note to the Uganda Electricity Board (UEB) in 1949 the British government guaranteed Egyptian interests in the dam. The Owen Falls Dam since then has become a major point of interest among some riparian countries because of its location at the only outflow of the Victoria Nile, in Jinja, Uganda.

### *Uganda and the Suez Canal crisis*

Long before the Suez Canal crisis, Britain had anticipated the control of the Nile upstream as an instrument against Egypt. In 1893, Sir Samuel Baker urged Britain to raise the level of Lake Albert in order to control the water supply of Egypt. Cromer understood the potential political power of this leverage, since he knew that the Egyptian leaders feared control of the upstream by foreign powers.[14]

At a glance it would appear that Uganda was far removed from the conflict between Egypt and Britain over the control of the Suez Canal. Still, the blocking or reducing of the Nile flow from Jinja, Uganda to Egypt was considered by the British government a possible option against then president Gamal Nasser of Egypt.

The nationalization of the Suez Canal by Egypt naturally angered Britain because this action would affect British economic and strategic interests in the region and beyond. Britain took a series of measures against President Nasser in the hope that these measures would pressurize him into changing his new policy on the Suez Canal.[15] One of the measures was the use of the Owen Falls Dam in Uganda to reduce water levels in the White Nile that flowed to Egypt. This plan was confirmed by a Foreign Office minute that stated: 'Ministers have recently shown much interest in the possibility that by stopping or reducing the flow of the White Nile at Owen Falls, severe damage may be caused.'[16] This plan was partly based on a report provided by a consultant hydrologist, G. Hawes, based in Uganda. He argued that a reduction of 30 per cent flow in the East African component of the Nile would somewhat have adverse effects on Egypt. The scheme to cut or reduce the Nile water flow to Egypt was, however, found to be unfeasible because, among other things, the impact on the Aswan Dam would be minimal and there was likely to be flooding in the Lake Victoria region. Despite British realization of the unworkability of the scheme, some British officials believed in using the Nile flow as a psychological weapon: 'Any suggestion of restrictions on the flow of the Nile would have a strong psychological effect; it might be possible to spread the word to the more illiterate Egyptians that unless Nasser climbs down Britain will cut off the Nile.'[17]

## HYDROPOWER DEVELOPMENT IN UGANDA SINCE INDEPENDENCE

Uganda's main interest in the utilization of the Nile resources, unlike most other riparian countries, has always focused on hydropower development for her socio-economic development. Other sectors, such as irrigation, tourism and water communication, which have been of concern to some other riparian countries, are for now of secondary consideration to Uganda. For example, in the Lake Victoria basin, whereas Kenya and Tanzania are interested in inter-basin water transfers to meet their domestic, industrial and irrigation water needs in the water-scarce parts of their countries, Uganda, on the other hand, is interested in the uninterrupted flow of water into Lake Victoria to support its hydropower production at Owen Falls and Kiira Dams, as well as other future hydropower dams.

It is also worth noting that currently virtually all of Uganda's hydropower potential is on the River Nile – the White Nile. It is estimated that the hydropower potential on the Nile within Uganda's borders is about 2,750 MW. Of this only 9.6 per cent has been developed, at two sites. Uganda's electrification rate is very low, with a grid access of only 5 per cent for the whole country and less than 2 per cent in rural areas. This means that only 200,000 customers are connected to the grid, with an annual growth rate estimated at between 5.5 and 7.5 per cent.[18] Overall, hydropower is relatively little developed in Uganda, but its demand is increasing rapidly.

### *Power sector reforms in Uganda*

In light of increased economic activities and subsequent increased power demand and consumption in Uganda, it was considered necessary to reform the power sector. The Uganda power sector was a monopoly of UEB until 1999. It was responsible for generation, transmission and distribution, as well as being the regulator. The power sector has been operating under the legal framework of the Electricity Act (1964).

The Uganda economy has been liberalized, with privatization taking root in many sectors. The electrical power sector, as part of the general utility reform process, has undergone the same changes in policy. The most notable move was the repealing of the Electricity Act (1964) and the enactment of the electricity law (The Electricity Act (1999): Electricity Regulatory Authority (ERA)). There is substantial potential for increased development of reliable, low-cost power, for example through expansion of hydropower production as planned in Table 3. This table indicates the hydroelectric dams under operation and those planned for the

**Table 3:** Major sites on the River Nile for hydropower development in Uganda.

| *No.* | *Site* | *Hydro-capacity (MW)* | *River* | *Remarks* |
|---|---|---|---|---|
| 1 | Nalubale (Owen Falls) | 180 | Nile | In service |
| 2 | Kiira (Owen Falls extension) | 200 | Nile | 80 MW in service; 40 MW expected by June 2002 |
| 3 | Bujagali | 250 | Nile | Construction started July 2007 |
| 4 | Kalagala | 225 | Nile | Not developed |
| 5 | Busowoko | 230 | Nile | Not developed |
| 6 | Karuma (Kamdini) | 300–50 | Nile | The next to be constructed after Bujagali |
| 7 | Ayago (North) | 310–400 | Nile | Not developed |
| 8 | Ayago (South) | 230–50 | Nile | Not developed |
| 9 | Murchison Falls | 450–50 | Nile | Not developed |
| | TOTAL (max.) | 2,635 | | |

*Source:* National Water Development Report 2005. http://www.unesco.org/water/wwap/wwdr/wwdr2/case_studies/uganda/pdf/8_water_energy.pdf retrieved 24/7/2007.

future in view of increased demand for urban and rural electrification through the public and private sectors. Apart from Bujagali, the other planned dams face a grim future due to current lack of funding.[19]

### *Uganda's competitive advantage in the energy sector*

Uganda is endowed with water resources in various forms, such as in the form of direct precipitation, ground water, runoff and surface water. Three of the lakes in the country are shared with other riparian states, and each of the eight major rivers has an estimated length in excess of 100 km with varying discharge rates. Uganda has tremendous hydroelectric potential, particularly along the Victoria and Albert Niles. Uganda is also strategically located to export power to its neighbours, as indeed it is already doing to Kenya, Bukoba in Tanzania, Rwanda and Burundi, and has been selected by ESKOM of South Africa, an energy body, as its regional headquarters.

Efforts are underway to refurbish, extend, improve and maintain the existing power supply to different parts of Uganda by Uganda Electricity Transmission Company Limited.

### *Kiira hydropower dam*

The ever-increasing demand for electricity by different sectors, especially the industrial sector in Uganda, necessitated building another dam to supplement the existing Owen Falls (Nalubale) Dam. By 1986 the Nalubale Dam, with an installed capacity of 150 MW, was only producing 60 MW. Yet the demand for electricity supply increased from 200 MW in 1994 to 350 MW in 2004.

Kiira Dam was the second hydropower project in Uganda and the first in independent Uganda. It was built to supplement the deteriorating Nalubale (Owen Falls) Dam. Kiira Dam was built as an extension of Nalubale Dam and was located only 1 km away. It was intended to increase the badly needed power supply in Uganda, and designed to utilize 'excessive' water that was being spilled by the sluices of Nalubale Dam, thus generating more electricity. A 1.3-km canal above Nalubale Dam diverts water to Kiira, in such a way that the two dams, now in combination, control the Lake Victoria water level and flow. The power capacity of Kiira Dam was planned for 200 MW. Work started in 1993 and major construction was completed in 1999. The choice of the Kiira site was influenced by many factors, among which were: a) it was relatively cheaper to construct and it was a donor-funded project; b) the infrastructure in the neighbourhood was already developed; c) it was easier and cheaper to expand the existing switch yard at nearby Nalubale Dam to connect a shorter transmission line (132 kV) from the station; d) it was found necessary to utilize the spare hydro-capacity that was being spilled by the sluices; and e) political and social pressures. Some government officials claim that there were plans to decommission Nalubale Dam in the near future and then keep Kiira as the main dam.[20] But Kiira Dam failed to perform as expected and the result was more power shortage.

### *The performance of Nalubale and Kiira Dams*

The Nalubale and Kiira Dams gradually and steadily failed to supply the urgently needed electricity in Uganda and the performance of the two dams came into question. This development has been closely linked to the fact that for years there was a noticeable drop in the water level of Lake Victoria. It started to drop significantly in 2003. Since then, the Lake Victoria water level has dropped over 1.1 m from its ten-year average. Consequently, as of December 2005, the lake level was approximately 10.69 m, reaching its lowest since 1951. It was reported that the Lake Victoria water level had receded by more than 2 m between 2000 and 2005, causing water and electricity shortages in Uganda, and it was anticipated to drop further. By the same date the lake lost 75 $km^3$ of water, about 3 per cent of its volume. And the lake level in Uganda has dropped by 2 m in the last three years. Figure 1 shows that the water level has changed over time. In the early 1960s and early 1970s the water level was higher compared to 2003–05. It was reported, however that there was a rise in the water level of the lake due to heavy rains in 2007.[21]

Some reasons have been advanced to explain the cause of the drop in the water level of Lake Victoria, and the issue has generated intense and controversial debates among hydrologists, engineers, the governments of Uganda, Kenya and Tanzania, technocrats and environmentalists. The decreasing water levels of Lake Victoria also have consequences for other basin countries.

A UN report accused Uganda of flouting a 50-year-old international agreement designed to protect the lake's waters. It accused Uganda of draining Lake Victoria to maintain its electricity supplies, despite an impending environmental catastrophe as water levels in Africa's largest lake dropped to their lowest in 80 years.[22] Dramatic stories have been told: The Lake Victoria water is 3 m below its normal level, leaving the jetties where pleasure boats moor and the landing sites where fishermen sell their catch high above the water. The falling water level is affecting 30 million people in Uganda, Tanzania and Kenya whose livelihoods depend on the lake. As the waters recede, silt and vegetation are encroaching on the lake and goats nibble the green shoots where fish once swam in the shallows. The Sudan and Egypt, both of which rely on the River Nile, which runs out of Lake Victoria, for their water supply and for agricultural irrigation, will also be affected. In October 2006 the UN warned that African lakes were the worst affected by climatic changes.

One position regarding the decreasing water levels of the lake has been espoused by Danniel Kull, a hydrologist engineer and UN consultant based in Nairobi, Kenya. According to Kull's findings, recent severe drops in Lake Victoria (2004–05) were approximately 45 per cent due to changes in rainfall patterns or drought and 55 per cent due to over-release from Nalubale and Kiira Dams.[23]

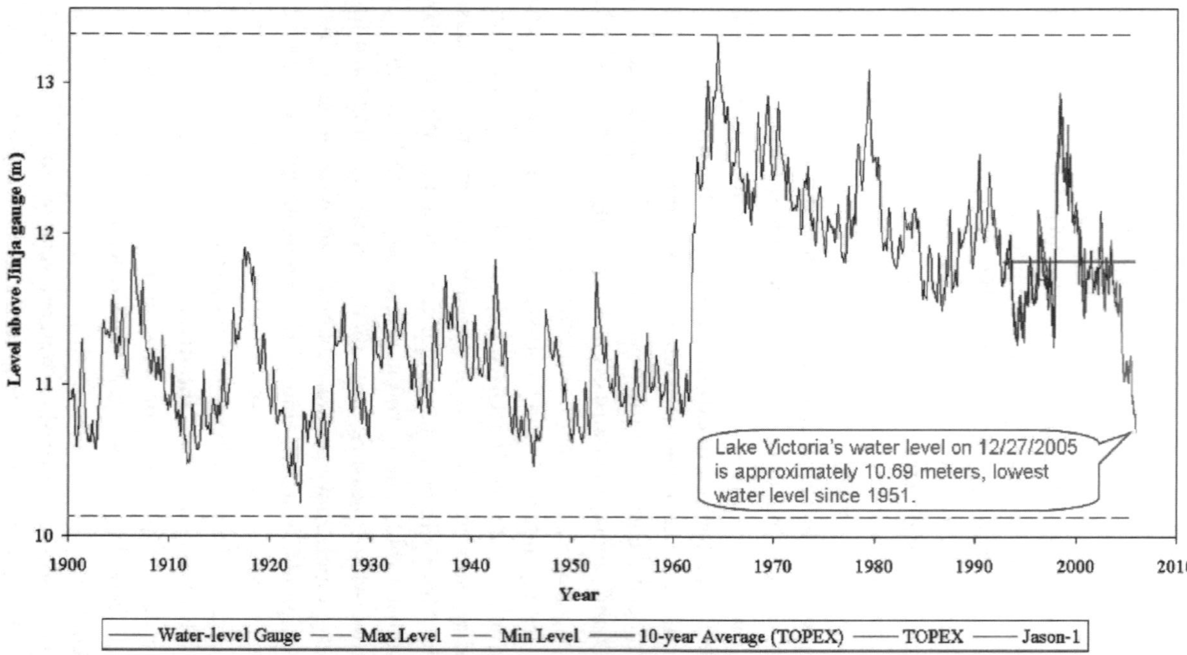


Data Source:
Historical water level gauge data from Jinja, Uganda (near Lake Victoria's outlet).
Satellite radar altimeter data from USDS/NASA/UMD at:
http://www.pecad.fas.usda.gov/cropexplorer/global_reservoir/

U.S. Department of Agricultural (USDA)
Foreign Agricultural Service (FAS)
Production Estimates & Crop
Assessment Division (PECAD)

**Figure 1:** The water levels of Lake Victoria, 1900–2005.

Engineers and environmentalists blame the World Bank and the Government of Uganda for ignoring their advice against the construction of Kiira Dam. For example, M.H. Onek, a Member of Parliament and a specialist water engineer, warned as early as 1996 that Lake Victoria would not be able to handle the new dam. Kull and Onek attribute the drop in the water levels of the lake to essentially poor design and management of the two dams, and they argue that the construction of Kiira Dam made matters worse. Design-wise, the turbines of Kiira Dam are a few metres lower than those of Nalubale and they therefore utilize the same 'head' from Lake Victoria. A 1.3-km canal diverts water to Kiira in such a way that the two dams in combination control the Lake Victoria level and outflow. The construction of Kiira parallel to Nalubale had the effect of sucking in over twice the amount of water that was being pumped out of Lake Victoria.

Krans Nuri Moeller, a consultant in the Ministry of Works, Uganda, also warned that there was simply much more water running out than coming into the dam and that the water that has been let out was obviously not coming back.[24] Consequently, the hydropower supplied by the two dams has drastically declined. Kiira is currently producing only 40 MW instead of 120 MW because there is just not enough water to produce electricity. The government and most technocrats argue that the cause of the drop in the water level of the lake must be attributed almost exclusively to continuous droughts.[25]

While the Ugandan government is keen to point the finger of blame for the lake's falling water levels at the lack of rain, not everyone agrees. Sewagudde, a hydrologist at the Directory for Water Development in Uganda, argues that outflows were increasing because of excess water releases at the Jinja dams and adds that the amount of water coming into the lake had decreased dramatically – from 21.8 billion cubic metres (bcm) in 2003 to 12 bcm in 2005.[26]

The drop in the water level of Lake Victoria has the potential to affect the upstream countries of Egypt and Sudan and the downstream countries of Rwanda and Burundi, if the situation is not rectified in time. The Lake Victoria basin is not just a water body shared among the three East African Community states: its geographical location has international legal implications, especially for the Sudan and Egypt, within the River Nile basin, and for Rwanda and Burundi, owing to its connection with the Kagera river basin. The mechanisms governing the utilization of water and living aquatic resources shared by two or more states raises many concerns, partly because of the value of the resource for national policy and partly because states often invoke legislative or diplomatic interest. As for the East African region, Lake Victoria is a vital valuable resource providing, among others, potable water, hydroelectric power and transport. It supports many industries, such as agriculture, trade, tourism, wildlife and fisheries. The lake basin serves as a major source of employment for an estimated 30 million people. Approximately 3 million people are engaged directly or indirectly in subsistence and commercial fishing and more than 80 per cent of the population is engaged in agricultural production, as small-scale farmers and in livestock keeping. So the importance of this lake basin cannot be overemphasized, thus the falling levels of Lake Victoria are having a negative socio-economic impact on the sub-basin region of East Africa.

In view of the above, some riparian countries are concerned about the decreasing water levels of Lake Victoria. For example, recently the East African Legislative Council Members blamed Uganda's hydroelectric dams for heavy receding of the Lake Victoria waters and claimed that the reduction in water levels is the result of a 45 per cent water outflow to Uganda dams.[27] The Secretary General of the East African Community warned that the drop in water levels would affect the East Africa countries. He noted with concern: 'Trade has been made expensive, traders are now importing and exporting fewer goods due to low level of water that could not hold ships.'[28] Kenya's Permanent Secretary warned of adverse effects on the fishing industry in the region due to shrinkage of the lake.

Uganda has, however, continued to maintain that the decline in the lake level is essentially due to drought rather than over-release from the lake resulting from the construction of Nalubale and Kiira Dams. Accordingly, the hydraulic statements from the power generation company – Eskom Uganda Limited – and the Ministry of Energy, say that the lake level has been increasing steadily from 10.90 m in January 2007 to the current high of 11.30 m above sea level in September 2007. Lake Victoria has been affected by one of the most devastating droughts in history since 2000. The same source added that the lake level dropped by 152 cm between 2000 and 2005; that excessive release of water accounted for 35 per cent, while drought accounted for 65 per cent.[29]

### *Bujagali 1 hydropower project*

When the government commissioned the Kiira power station in mid-2000, it was hoped that the era of load-shedding of the mid- and late-1990s would finally be over. The commissioning was greeted with so much optimism that President Museveni announced that Uganda would eventually raise its electricity exports to Tanzania, Rwanda and Kenya, and that electricity was expected to form the basis of Uganda's export earnings in future. However, the government's intentions to turn Uganda into a regional powerhouse did not go according to plan. Persistent droughts and a drastic loss of water level in Lake Victoria caused a fall in the level of water by about 2 m, the result of which is power shortages. Uganda needs (2007) about 370 MW, but the country's two major power stations, Kiira and Nalubale, are generating only about 140 MW, despite the fact that Uganda's hydropower potential is more than 2,700 MW. The demand for electricity is growing rapidly at the rate of 27 per cent per year. By 1999, Uganda began to experience a power shortage of 80 MW per day due to increased demand and the poor performance of Nalubale and Kiira dams. It was under these pressing circumstances that the Bujagali scheme was conceived.

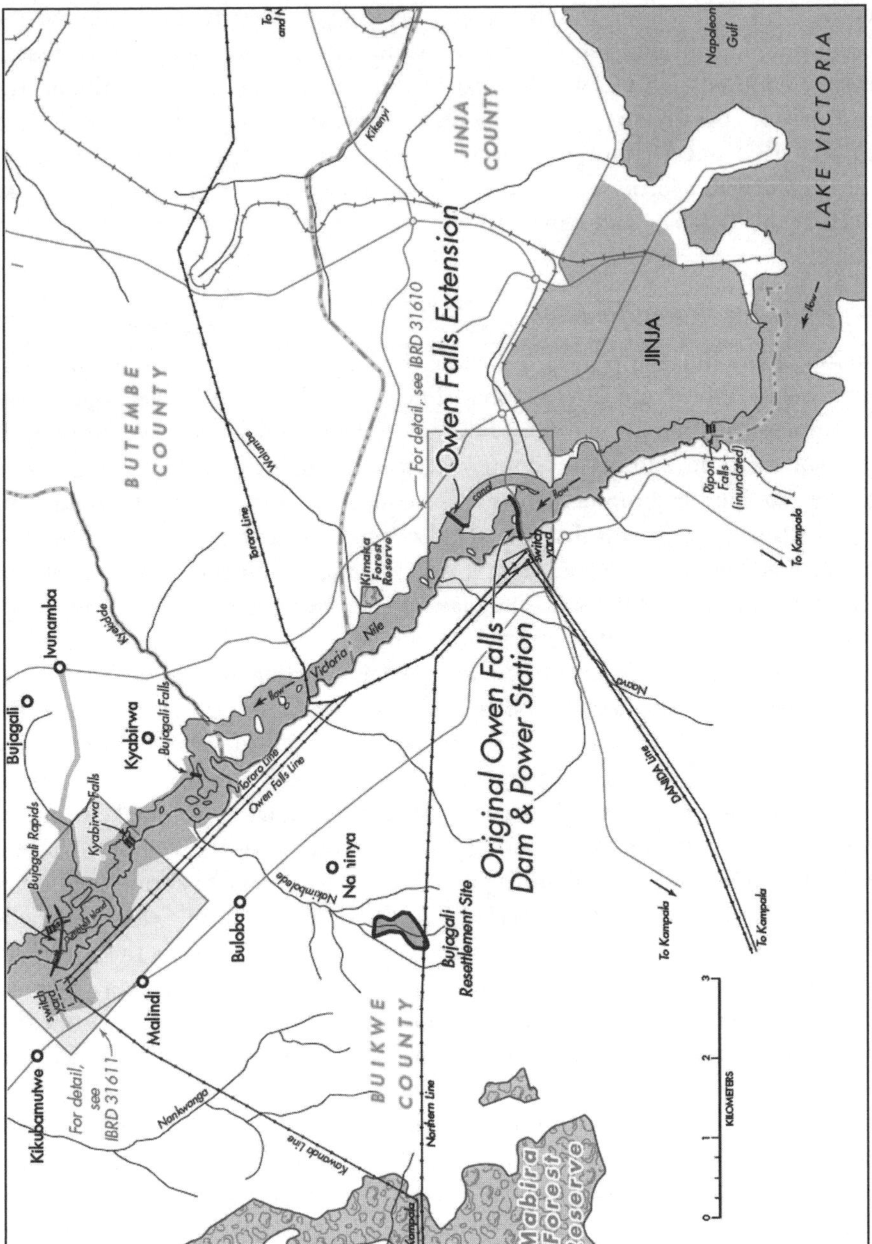


**Map 3:** The proposed Bujagali hydropower project.

The 30-m Bujagali dam site is located 10-km north of Nalubale and Kiira dams, at Dumbbell Island. The 250 MW power plant was planned to be linked to the existing Nalubale plant. A USA-based company, AES, was selected to be a sponsor; AES Nile Power was to develop, construct, operate and maintain the Bujagali power plant and sell electricity to the Uganda Electricity Transmission Company Limited (UETCL). The cost of the project was US$530 million, the largest private investment, to date, in East Africa. Funding was by World Bank (IDA and IFC) and African Development Bank, among others. Water flow arriving at the Bujagali hydropower facility will be controlled by discharges from Lake Victoria at the Nalubale and Kiira Dams.

### *Criticism of the Bujagali 1 project*

Criticism of the Bujagali project has come from different organizations and individuals, both national and international. They were all united in their opposition to the implementation of the project because of their various concerns for Uganda in particular and the sub-basin region of East Africa in general. The critics' views were diverse. The environmentalists were concerned that the dam would submerge the Bujagali falls and that the island's vegetation and biodiversity would be lost, and that the project would have a negative impact on the environment and the people living around the site. The highlands in the vicinity of the Bujagali falls provided breeding sites and a refuge for a number of bird species, and, according to the environmental impact assessment, the inundation of the islands would result in loss of habitat for the birds and other fauna as well as flora and the extinction of some fish species. The advocates of the tourism industry were opposed to the project because Bujagali falls were one of the world's attractions for both local and international tourists and that Bujagali's rapids make it one of the most exciting white-water rafting rivers in the world. Cultural advocates opposed the project because the Bujagali site serves as a home to shrines and other cultural regalia of the Basoga, especially the Baise Mwena clan, whose religious foundation and annual ceremony are performed at the Bujagali falls; the construction of the dam would flood the area, including the shrines.[30] From a technical point of view, the construction of Bujagali Dam raised the fear that the cumulative environmental effects of having three dams close together could be fatal. The possible negative consequences include flooding, disruption of biodiversity and the water level downstream, as well as a reduction in the total amount of power produced. Economically, according to The National Association of Professional Environmentalists (NAPE), electricity produced at the proposed dams would be too costly for ordinary Ugandans. For instance, the power tariff rate proposed by AES under the Power Purchasing Agreement (PPA) was US 10.5 cents per kilowatt-hour as opposed to the current rate of US 6.5 cents.[31] Other major criticisms of the Bujagali project were its high cost of construction and lack of transparency on the part of the contractors. For example, a study commissioned by an NGO

International River Network (IRN), and carried out by Prays Energy Group Consultants of India, revealed a lot of financial flaws in the Bujagali project. The following were some of the firm's findings:

- The capital cost of the project was found to be excessively high and that a comparable project with similar features in India would have twice the generating capacity of Bujagali, but would cost less than the AES's project in Uganda.
- The Power Purchase Agreement (PPA) with AES was not transparent: for example, the Ugandan government would have to make yearly payments of up to US$132 million for the project and not US$111 million as in the World Bank's project appraisal document.
- If the PPA requirements and the capital costs were more in line with international standards, Uganda's yearly payment obligations for Bujagali would be reduced by US$40 million initially and by an average of US$20 million over the lifespan of the project. This would result in savings for Uganda with a net value of US$280 million; this therefore means that the Bujagali project was over-priced by US$280 million.

The report finally recommended that the Uganda government should invest in the energy sector under its control and supervision rather than allow private investors and developers to do it under excessive terms as those under the Bujagali PPA. In addition, the World Bank was accused of publishing misleading or incorrect information on important issues.[32]

Based on the above report, the NAPE and IRN made the following conclusions:

- The Bujagali project, including the project's PPP, is fundamentally flawed and not in the best interests of the Ugandan people and the Uganda government, and the Uganda Electricity Board should therefore cancel its funding of the project.
- A balanced and participatory process should be immediately launched to assess all available options to bridge the gap between Uganda's energy needs and supply, including the potential of thermal power and the Karuma hydropower project.[33]

Apparently, as a result of these findings and recommendations, AES was investigated by the US Justice Department and the Uganda Inspectorate of Government for corruption. The Uganda High Court finally forced PPA to disclose the terms that were concluded with AES. In June 2002 the Inspection Panel (IP) unit of the World Bank found that the Bujagali project violated five operational policies of the bank. In July 2002 the World Bank discovered that AES bribed the Minister of Energy in Uganda, Sydah Bumba, who then was an alternate executive director for African countries on the board of directors of the World Bank.[34]

In its quarterly filings to the Securities and Exchange Commission, AES, possibly in a face-saving move, said the company would write off its investment in Bujagali, amounting to approximately US$75 million, and pull out of the project. The following reasons were given by AES:

- The project would produce less profit for the company with a greater degree of risk than originally forecast.
- Its withdrawal from the project was not connected with the corruption investigations.[35]

### *Bujagali 2 project*

The demand for electricity in Uganda continued to grow sharply, resulting in massive power shortages and hence constant load-shedding during peak periods. Since 2003, Uganda has experienced even more acute power shortages, especially due to the failure of the first Bujagali project to take off.

Despite intense criticism of the Bujagali 2 project, the Uganda government was committed to implementing it. President Museveni was unequivocal about this. The World Bank fully endorsed Bujagali 2. As was reported: 'The World Bank Group and the government of Uganda believe that Bujagali remains the long-term least-cost electricity supply option of Uganda and that it is in the interest of the country to develop the project at the earliest opportunity.'[36]

Given the continued electricity shortage and the resultant negative impact on the economy, the Uganda government was under intense pressure to implement the Bujagali 2 project. The project consists of the development, construction and maintenance of a power plant with a capacity of up to 250 MW at Dumbbell Island, 8 km north of the existing Nalubale and Kiira power plants. The project will be an Independent Power Project (IPP) and will sell electricity to UETCL under a 30-year Power Purchase Agreement (PPA), and will be guaranteed by the government of Uganda.[37]

In early 2004, the Government of Uganda, with help from the World Bank Group, embarked on an international, transparent and competitive bidding process (and was determined to be transparent for Bujagali 2, as opposed to the bidding system in Bujagali 1). In April 2005 the government selected the preferred sponsor, who finally entered into relevant agreements with the government of Uganda and UETCL in December 2005. The selected sponsor is a joint venture between Industrial Promotion Services (IPS) (Kenya) and the US-based Sithe Global Power LLC. IPS (K) is the industrial development arm of the Aghakan Fund for Economic Development (AKFED) in Kenya. The two sponsors have extensive track records of developing and operating power plants in other countries.

The dam's cost has been revised upwards – from US$550 million to almost US$800 million. On completion it is hoped that the current capacity of 380 MW will be increased, allowing exports to Kenya, Rwanda and Tanzania. The project has acquired all the money needed for its implementation from major world lenders. These include International Finance Corporation (US$130 million), European Investment Bank (Euro100 million), African Development Bank (US$110 million), Barclays Bank and Standard Chartered Bank (US$115 million), Dutch Development Agency (US$73 million), French Agency for Development (US$73 million) and German Development Bank (US$45 million).[38]

When commissioned, the Bujagali project is expected to have a major socio-economic developmental impact on Uganda in particular and on the East African sub-basin in general. The Bujagali project, therefore, when it becomes operational, is expected to:

- displace at least an estimated 100 MW of expensive emergency thermal power;
- relieve residual load shedding; and
- increase electricity supply for both domestic and commercial consumption.

But far more important is the commitment of the new Bujagali sponsors to preserve the water level of Lake Victoria by 'recycling the same water that is already used to generate power at Nalubale and Kiira'; thus the project will reduce the pressure to over-release water from Lake Victoria, thereby helping to preserve lake levels and facilitate the government of Uganda's compliance with the 'agreed curve'.[39]

It is expected that Uganda will increase its power supply to the Bukoba region of Tanzania, northern Rwanda and possibly Kenya.

Unlike in the Bujagali 1 project, the Bujagali 2 project managers are attempting to address publicly and transparently the environmental and social issues relating to implementation of the project. For example, the likely negative impact on the environment and resettlement of affected people at the Bujagali site is addressed through a body called Bujagali Emergency Ltd (BEL), which has been conducting public consultation with affected communities.[40] The environmental and social impact assessments of Bujagali have been made public in accordance with the requirements of the World Bank Group and Environmental and Social Standards.

### *Criticism of the Bujagali 2 project*

Despite the efforts of the government of Uganda, the sponsors and the World Bank to appear transparent and genuine in all ways to justify the Bujagali 2 project, old and new criticism of the project continued unabated. Environmentalists, NGOs and representatives of civil society organizations wrote a letter to the World Bank President, protesting against the revival of the Bujagali project. They stressed the importance of stakeholders' participation in an open and systematic assessment of options in project planning and policy formulation. They argued that commitment to resurrect Bujagali before undertaking a comprehensive options assessment would be premature, that transparency and accountability and PPA should be put to public scrutiny.[41]

On technical grounds, the site of Bujagali was criticized by M.H. Onek, an MP and water engineer, who said that Bujagali, located about 6 km away from Nalubale and Kiira Dams, would depend on the water released from the two installations and that the short distance between Bujagali and Nalubale meant that the two installations would share the same structural risks. He further noted that there was already cracking in the powerhouse, which could also be happening in the dam structure. Should Nalubale structurally fail, Bujagali would absorb the load and could be washed down. The impact of Bujagali on the water levels of Lake Victoria is also a matter of concern to some hydrologists, based on the following arguments: the over-release upstream at Nalubale and Kiira poses problems for Bujagali as well. Should the levels continue to drop, discharge from the lake will decrease and power generation will drop with it. All these developments will negatively affect not only Uganda, but also the Lake Victoria basin countries and possibly beyond.

The cost of the Bujagali project came under constant and intensive criticism, more so in light of the unexplained increased total cost of the project. Originally, the figure was US$430 million, then US$550 million, then US$580 million and the current cost (2007) is approximately US$800 million.[42]

Ugandan energy experts explain this rise as resulting from 'rising construction costs due to the high oil, metal and cement prices in the world market'. Pursuant to blocking the Bujagali 2 project, environmental activists and other critics sent a letter to the World Bank country director in Uganda in which they opposed the proposed project because of the continuing decline of Lake Victoria water levels, and the need for a Sectoral Environmental Assessment (SEA) to analyse the cumulative impact analysis (CIA) on the Nile from multiple dams.[43] In responding to the concerns raised above, the World Bank country representative in Uganda confirmed that the Bank is committed to dealing with the Bujagali project in a transparent and professional manner.[44]

### *Bujagali 2 project approved*

After a protracted struggle between the government of Uganda, hydrologists and environmentalists over the Bujagali project, the National Environmental Management Authority (NEMA) Uganda cleared the construction of the dam: 'We have approved the construction of the dam with conditions and a wide range of mitigation measures. I hope this settles the dust concerning the environmental considerations of the dam.'[45]

The World Bank approved the Bujagali project on 27 April 2007. Construction work started in July 2007 and the dam will be commissioned in 2011.[46] Costs and terms were finalized. The loans for this project have a 16-year maturity period, with a grace period of 50 months.

For the supporters of the Bujagali project it was regarded as a deserving victory and they re-emphasized the importance of the project in the socio-economic development of Uganda. 'This is the biggest investment in sub-Saharan Africa … the Government's medium-term strategy in meeting the power shortages has materialised', said Kabagambe Kalisa, Permanent Secretary Ministry of Energy, Uganda.[47] '[T]he WB considered the dam an integral component of Uganda's strategy to close an energy supply gap that seriously constrains social and economic development', stated Judy O'Connor, the World Bank Country Director for Uganda. In pursuit of this objective, the World Bank granted to Uganda a credit line of US$300 million for the development of the power sector, including a government guarantee for the Bujagali Dam project.

However, the environmentalists were still sceptical about the viability of the project: '…the bitter truth is that the dam is over-priced and will not provide the expected output … Already the World Bank … approved Kiira power station which has drained Lake Victoria. Where will they get water required to generate the 250MW?'[48] asked Frank Muramuzi of NAPE (National Association of Professional Environmentalists – Uganda).

## SHORTAGE OF HYDROPOWER SUPPLY AND UGANDA'S SOCIO-ECONOMIC DEVELOPMENT

The intermittent power shortages in Uganda since 1986 have had a major negative impact on almost all aspects of Ugandan society. The reversal of this trend is therefore imperative, if Uganda's socio-economic plans, policies and projects are to be achieved. Although Uganda's hydropower potential is above 2,700 MW, only a fraction of it has been tapped and even that in use has not been operating at optimal capacity owing to a number of factors, some of which are already mentioned in this study.

The current debate about electricity shortages is dominated by different groups: hydrologists – for example, M.H. Onek, Kull and others – have maintained that the power shortages are largely due to the drop in the water level of Lake Victoria, and therefore that Nalubale and Kiira Dams cannot generate power optimally. Conversely, the Uganda government and its technocrats continue to argue that the power shortages are mainly caused by prolonged droughts. While environmentalists (NAPE) attribute the power shortages to poor government planning and lack of a power policy since 1999, and accuse the government of lack of transparency in the privatization of the power sector, President Museveni has dismissed the environmentalists' arguments and blames them for the power crisis, claiming that if they had not blocked the Bujagali project Uganda would not have faced the current electricity crisis.[49] Other factors responsible for the power crisis may include environmental degradation and global warming.

Uganda's electricity supply has decreased substantially while the demand has gone up. At least 340 MW are now required and yet Nalubale and Kiira Dams currently can only generate 120 MW. The impact of electricity shortages has been felt almost in all facets of Ugandan society. For example, power tariffs doubled in 2006; Uganda loses 90 days' worth of work per year; industrial production has dropped by 30 per cent; electricity bills have gone up by 25 per cent; and incessant load shading is the order of the day in Uganda.

The recent poor performance of the economy shows the extent of the impact of power shortages on the country. For instance, the Bank of Uganda announced in August 2006 that the power shortages were contributing adversely to the economy because the GDP dropped from 6.5 per cent to 4.5 per cent in 2006, partly due to power shortage. Information from the Ministry of Energy indicated that 'the government loses about $254m annually due to the power crisis'.[50] The Uganda Manufacturers Association (UMA) showed that industrial production was cut by 30 per cent due to power shortages. This has resulted in the high cost of production and the eventual loss of jobs.[51] Consumer prices have gone up on most products. The new power tariffs are a further burden to already an over-stretched business community and domestic consumers. The reaction of the manufacturers and domestic consumers has been anger and frustration. The chairman of UMA, James Mulwana, demanded that the government should subsidize the fuel that the business community uses to power generators and that the government should also expedite use of thermal power as a stop-gap measure. It was reported recently that some manufacturers are shifting their businesses to Kenya owing to the high cost of production caused by the use of generators whose fuel costs are high.

## GOVERNMENT'S MITIGATION AND FUTURE PROSPECTS

The government of Uganda has taken a series of measures in its short- and long-term planning to address the energy crisis in the country. The government has put in place policies and laws favouring and encouraging private investment and capital flow in the energy sector. In the 2006/07 budget, the government allocated US$123 million for subsidization of electricity costs, imported 800,000 energy-saving bulbs worth US$1.2 million for distribution, imported 20 MW of electricity power from Kenya, and has installed a 100 MW thermal generator.[52] In the medium term the Bujagali hydropower project started construction in July 2007 and is to be commissioned in 2011.The government has decided to lend US$100 million to the developers of the Bujagali hydropower project to kick-start construction work.[53]

As a long-term strategy, the government has proposed 14 sites for hydropower generation. Some of the sites named were: Bujagali, 250 MW; and Karuma,

300–59 MW. It is planned that 2,752 MW will be generated by 2025 to satisfy the demand of the 42 million people.[54] The excess power can be exported to the sub-basin countries of Rwanda, Burundi and Bukoba region of Tanzania, which are potential markets for electricity supply from Uganda. Currently, northern Rwanda receives 5 MW and Bukoba region receives 9 MW, while Kenya receives 30 MW from Uganda.[55]

Uganda's hydroelectric development along the Nile has some implications for the Nile basin. First, it boosts socio-economic development in Uganda. Second, it contributes to the socio-economic development of the sub-region countries of East Africa through current and increased future exports of electricity from Uganda. Uganda's hydropower potential can be harnessed to supply more energy to the surrounding countries. The hydropower potential of Uganda exceeds 2,700 MW, the largest in the region. Plans are under way to exploit this resource. Third, Uganda's dual position of being both upstream and downstream, with regard to the Nile, may raise suspicion among other riparian countries that Uganda's dams may interfere with the smooth flow of the Nile, if not checked.

## UGANDA'S NON-HYDROELECTRIC PROJECTS, PROGRAMMES AND PLANS IN THE NILE BASIN

Uganda's water richness and position in the Nile basin have also been of some importance to other economic activities in the country. Although this study focuses on Uganda's hydropower development and related activities in the Nile basin, Uganda's other important socio-economic activities, in the same basin, are briefly discussed below.

Lake Victoria offers potential opportunities for investment in fisheries, tourism, transport and communications, water, energy, agriculture, trade and industry. For the whole basin, the potential is further extended to cover the abundant natural resources in wildlife, forestry, minerals and fertile soils. Hence, some of the non-hydropower Nile-related activities in Uganda – irrigation, fisheries, tourism and oil – are discussed below.

### *Irrigation*

Using the Victoria-Nile river and its tributaries, or the larger inland lakes, for irrigation purposes would have serious consequences for irrigated farming and the supply of water for countries to the north, especially the Sudan and Egypt. But there is enough water generated within Ugandan territory that could be utilized for supplementary irrigation without impinging upon the water rights of other riparian countries in the Nile basin.

In general, Uganda is fertile and benefits from adequate rainfall, offering favourable natural conditions for agricultural production, with the exception of the far north and north-east Karamoja, where the rainfall is confined to one rainy season. In the south, rainfall conditions permit more or less continuous cropping. In view of this, irrigation has not been considered to be very important either by the government or the individual farmers. For example, in 1987, the total water-managed area was estimated to be only 9,120 ha, or about 0.2 per cent of the cultivated area. The full or partial control-equipped area was 5,550 ha and the equipped wetland, 3,570 ha. On a pilot scale (80 ha), sprinkler irrigation was practised. Rice, sugar cane, citrus fruits and vegetables are the main crops grown under irrigation.

According to African Development Fund (ADF) estimates, Uganda has potential for expansion of its irrigated agriculture, with the amount of land which could potentially be cultivated under irrigation estimated at 202,000 ha. Uganda is well-endowed with freshwater resources, comprising small and large lakes, rivers and an extensive network of wetlands. Currently these resources are under-utilized. Uganda's agriculture is predominantly rain-fed. It remains a low-intensity, low-input and low-output system. A number of factors contribute to this situation, of which rainfall variability and unpredictability are important aspects. Since the late 1980s, however, the yields of most crops were stagnant or have declined. There has also been more demand for home and export markets, so there arose a need to use irrigation on a small scale.

The government of Uganda has formulated a Medium-Term Plan (MTP) for the Modernisation of Agriculture to produce sufficient food to ensure food security, increase and diversify the production of export commodities, and create sufficient employment and income generating opportunities in the agricultural sector. The MTP, among others, proposes to minimize over-reliance on rain-fed agriculture and promotes irrigation technology for high-value crops such as flowers, vegetables and fruits (horticulture), especially for specialized export markets in Europe, the USA and Asia.[56] In pursuit of this plan, Egypt sent a team of five senior irrigation experts to Uganda to undertake a water project as a preliminary to creating a production strategy and investment plan. Specific objectives of the irrigation mission included the preparation of detailed feasibility studies for the rehabilitation and new development of irrigation schemes for public- and private-sector financing. It established irrigation demonstration facilities for the practical training of irrigation extension staff and farmers in water-supply and water-use management techniques. It also prepared institution capacity development proposals in terms of equipment and logistical facilities to enable the relevant central and local government institutions to effectively undertake irrigation development, with a special focus on small-scale irrigation.[57]

### *Fishery industry*

Lake Victoria is the largest inland water fishery sanctuary in East Africa. The fishery resources from the lake are a major source of food, of revenue to governments and of employment for many people. It supports the livelihoods of 3 million people who are directly involved in the fishery industry. It is estimated that the annual fish catch from Lake Victoria is about 750,000 metric tonnes, generating more than US$400 million per year, of which US$250 million comes from exports.[58]

Fishery is an important resource of Lake Victoria for the people of the Nile sub-basin of East Africa. It is one of the most productive inland fisheries in Africa. There are over 500 endemic fish species in the lake. One of the most important species in the lake is the Nile perch (*Lates niloticus*), locally called *emputa*. It can grow up to 2 m in length and weigh up to 200 kg. The Nile perch first appeared in Lake Victoria in the late 1950s. Ecologically, the Nile perch is blamed for causing the extinction or near-extinction of several hundred native species. Since the late 1990s, the Nile perch stocks have decreased owing to commercial fishing, and some of the native species are returning.

Uganda, like other riparian countries of the Nile sub-basin – Kenya, Tanzania, Rwanda and Burundi – has benefited enormously from the fish resource. The fisheries industry has since the late 1980s become one of the major economic activities in Uganda both for domestic consumption and for export. Because of the importance of this sector the government of Uganda has taken a series of measures to improve it. The government of Uganda's long-term vision for the fish sector is articulated in the National Fisheries Policy Draft, and embraces, among other things, stable fish production of over 330,000 metric tonnes per annum, improved domestic fish consumption and improved fish exports.[59]

The fishery industry is one of the largest employers in the country. For instance, more than 350,000 people are directly employed by the fisheries industry in Uganda and over 1.2 million are indirectly employed. Uganda has over 20 fish factories exporting over 30,000 tonnes of fish worth US$150 million annually. Uganda's fish is exported to the EU, Asia-Japan, Australia, South America, Cuba, the USA, Canada, South Africa, Egypt, and the Common Market for the eastern and southern Africa region, among others.[60]

### *Tourism*

There are many tourist attractions in the Lake Victoria basin and along the River Nile in Uganda. They include: viewing falls such as Murchison, Bujagali and Kabalega; birdwatching of rare species along rivers and lakes; visits to the Owen Falls and the 'source' of the Nile; rafting; sport fishing in the lakes and along the Nile; watching animals such as crocodiles and hippos along the shores of the lakes and rivers; and bungee jumping at the falls.

As noted earlier, the potential of Lake Victoria also lies in tourism attractions throughout the White Nile basin. The five East African countries unveiled an initiative that could see the region marketed as a single tourist destination. The East African Tourism and Wildlife Coordination Agency (EATWCA), will, among other things, implement a regional strategy for tourism promotion whereby individual efforts are reinforced by regional action, and this will be in conformity with the spirit of the East African Community.[61]

The tourism industry is one of the major foreign-exchange earners for Uganda. The government objective for the tourism sector is to contribute to the growth, sustainable development and diversification of the Ugandan economy. A strengthened tourism sector will also bring benefits to rural communities surrounding protected areas, as well as diversify the community tourism initiatives already operating. Enhanced tourism revenue will also benefit the biodiversity conservation aims of the government.[62] There are several and varied tourist sites in Uganda. In the Lake Victoria part of Uganda there are many islands that are home to a variety of wildlife. Upon the shores of Lake Victoria are beautiful, undiscovered tropical areas. Lake Victoria is popular for fishing, boating and exciting canoe trips. The Ugandan shore is characterized by a number of small islands and many peninsulas. Lake Victoria is an extensive expanse of water that attracts tourists to the Ugandan shores year after year. Uganda has some of Africa's major attractions. For example, it is regarded as the 'source' of the Nile at Jinja; it is home to more than 1,000 species of birds, making Uganda the richest birding destination in Africa; and it has the highest concentration of primates on the continent. Uganda is the only country where the endangered mountain gorilla can be seen in its natural habitat. In addition, sport fishing has been introduced, and many tourists engage in this sport in Uganda each year. Uganda is a country of diversity and it has been termed 'the country of lakes and rivers' because almost a third of the country is covered by water. Endowed with numerous rapids and waterfalls, the River Nile has become a famous white-water rafting destination.

Tourism is the second largest foreign-exchange earner for Uganda. For instance, it contributed US$316.6 million in 2004. The Uganda tourism sector, with its continuous growth, was expected to earn the country over US$500 million in 2009.[63]

### *Oil sector*

Oil discoveries will significantly contribute to the region's economic fortunes. Results of recent geological surveys suggest that East Africa may soon become one of the world's hottest oil exploration zones, based on data analysed in 2008 by Jebco Seismic, a UK-based geophysical contractor, which showed major oil deposits off the coasts of Kenya, Tanzania, Mozambique and Madagascar. This prediction is further confirmed by the fact that the same rock formations now yielding large quantities of oil in the Sudan are known to stretch into Kenya, Uganda and Tanzania. It is further argued that the source of all these potential deposits is the East African Rift Valley and adjoining formations. While the interior sections of the rift zone consist mainly of volcanic rock, the system has been

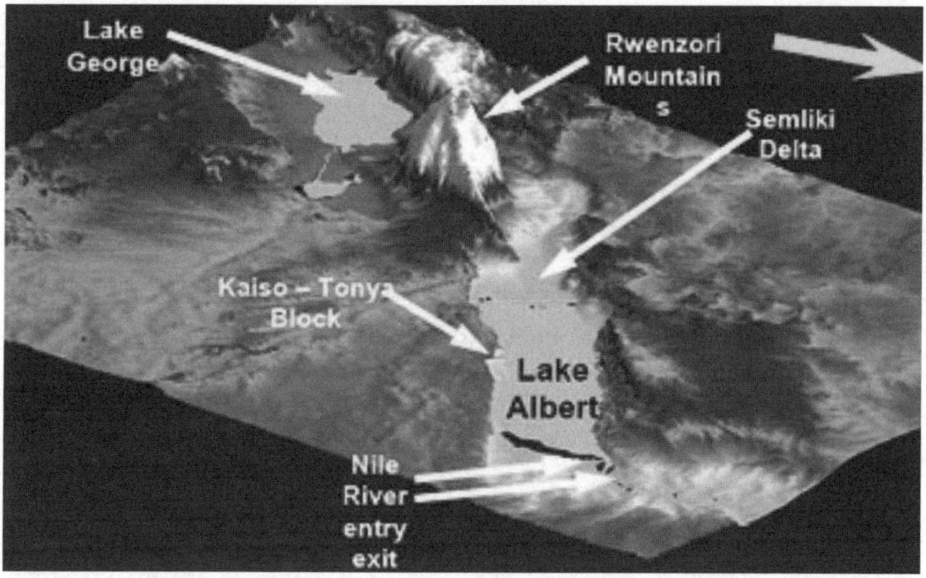


*Source:* Emmy Olaki, 'Sources of oil and gas in Uganda. Uganda Finds Natural Gas', *The New Vision*, 12 July 2007.

**Figure 2:** Oil resources in Uganda.

found to contain several sedimentary basins as well. It is from those onshore and offshore basins that oil would be extracted.[64]

The 'Albert Graben', as the portion of the Rift Valley around Lake Albert on the Ugandan–Congolese border is known, including Lake Albert and the Semliki river valley, is regarded as the most promising sector for oil exploration in the Rift Valley.

At the time of writing, oil companies are involved in drilling for oil in Lake Albert, Uganda.[65] The announcement for the discovery of oil around Lake Albert in Uganda was first made in 2006. In total, when the discoveries of Heritage and Tullow Oil are combined, Uganda can now output 27,000 barrels of oil per day and prospecting for more oil is continuing in other areas. The three fields in western Uganda where the oil has been discovered are estimated to have reserves of between 100 million and 300 million barrels, with 30 million barrels ready for extraction at just over 12,000 barrels a day.

More explorations turned up more deposits of oil. For instance, it was reported that oil reserves from the Kingfisher well discovery could be ten times greater than previously believed, a multibillion-barrel oil potential, estimated at US$7 billion.[66]

Another important recent discovery for Uganda in the energy sector is the discovery of natural gas. Oil explorers in western Uganda have unexpectedly chanced upon natural gas reserves. Gas found in Hoima flows at a rate of 420,000 cubic metres per day. At the current world market prices of US$7 per 30 cubic metre, this amounts to about US$98,000 (Ush162 million) a day from the one well where gas has been discovered.

## UGANDA'S PARTICIPATION IN THE SUB-BASIN AND BASIN-WIDE PROGRAMMES AND PROJECTS

Uganda, like some other upstream riparian countries, is critical of the two colonial Water Agreements of 1929 and 1959, and has demanded their replacement with a new treaty for equitable utilization of the Nile resources by all riparian countries. Uganda's effort to introduce a new treaty is regarded as one of her major participations in the basin-wide activities for reforms.

### *The 1929 Water Agreement*

In the early 1900s there was a relative shortage of cotton on the world market. This put pressure on the Sudan and Egypt, then under the British condominium, to produce cotton. However, this required use of modern methods of irrigation using the Nile waters. British support for this scheme was in addition to her strategic and economic interests in Egypt, which could not be sustained without supporting Egypt's interest in controlling the Nile waters. It was essentially under these circumstances that Britain became a signatory, on behalf of her East African colonies including Uganda, to the 1929 Water Agreement.[67] To date this agreement remains the centre of controversy and debate among the upstream riparian countries, including Uganda. This agreement gave Egypt more control of the River Nile than any other riparian country, including countries from which the river flowed.[68] Britain favoured Egypt at the expense of her other colonies.[69] But Sir William Gowers, the then Governor of Uganda, criticized the agreement in a dispatch to the Secretary of State.[70]

### *The 1959 Water Agreement*

Uganda, as an upstream riparian country, is critical of the 1959 Water Agreement that favoured only Egypt and the Sudan over the use of the Nile waters.

It is noteworthy that the 1959 Nile Water Agreement only involved Egypt and the Sudan. Neither the colonial riparian states and their colonizers nor sovereign Ethiopia were invited to participate in the negotiations. In the 1950s, Egypt was planning the Aswan High Dam project to collect the entire annual flow of the Nile waters. The main objective of the 1959 Agreement was to gain full control and utilization of the annual Nile flow. Britain, this time around, defended the interests of her East African colonies, including Uganda, against those of Egypt in the 1959 Water Agreement. Britain now advocated the principle of 'Community of Users'.

To this end Britain declared: 'The territories of British East Africa will need for their development more water than they at present use and would wish their claims for more water to be recognised by the other states concerned.'[71] At this point Britain thought that her East African colonies would need about 2 bcm to meet their irrigation and power needs.

### *Uganda's position on the two water agreements since independence*

Upon obtaining independence in 1962, the Uganda government announced its policy on treaties concluded before she became a sovereign state. They included the Nile water treaties of 1929 and 1959. In a letter to the Secretary General of the United Nations dated 12 February 1962, the Uganda government's position was stated as follows: 'In response of all treaties validly concluded by the United Kingdom on behalf of Uganda protectorate … before October 9th, 1962 the Government of Uganda will continue on the basis of reciprocity to apply the terms of such treaties from the time of its independence until December 31st, 1963. It is the earnest hope of the Government of Uganda that during the aforementioned period the normal process of diplomatic relations will enable it to reach satisfactory accord with states concerned upon the possibility of continuance or modification of such treaties…'.[72] The 1929 and the 1959 Water Agreements inescapably fell within this new Uganda policy on colonial treaties. Since 1963 there have been more explicit condemnations of the two water agreements by the government of Uganda, Members of Parliament, academia, the press and the public. For instance, a parliamentary committee once declared that the Nile Treaty imposed on Uganda at independence by the British was not a binding treaty, and it also mentioned that damage caused by the rise of water levels following the construction of the Owen Falls Dam in 1950s was never paid for by Egypt. The parliamentary committee concluded that new negotiations over the use of the Nile waters would be the only way forward, and that the states of East Africa, with tributaries towards Lake Victoria and Albert, must be receiving compensation for their waters as well have sovereign decision over their use for agricultural irrigation and as sources of potable water for the populations in these countries. Amon Muzoora, a Member of the Uganda Parliament, specifically proposed that Egypt should pay US$1.2 million per annum to Ugandans as compensation for using Nile water.[73]

President Museveni called upon Egypt to halt what he called the 'egoistic' approach over the use of the Nile waters. He complained that the treaty favoured Egypt and he called for a new one.[74] The Minister of Water Resources in Uganda added: 'They were taking care of people who knew what they wanted – Egyptians. We were not consulted. We are now independent. We can sit at a table and negotiate.' The *3 March 2007 Uganda Foreign Policy Release* further confirms Uganda's official criticism of the water agreements in their current form. The Ministry statement lamented: 'While Uganda has great endowment of fresh waters which among others, are critical for implementation of the Programme of Modernisation of Agriculture, these resources are not yet harnessed in any significant manner. Even if the capacity was in place to harness these vital resources, Uganda is still unfairly constrained by the agreements entered into between the British colonial authority and Egypt in 1929 and 1959, which agreements continue to bind Uganda to date. The unfairness in these agreements lies in the unfortunate requirement that for Uganda to use her water resource for survival, prior concurrence of Egypt must be secured. However, there is no reverse requirement upon Egypt in her use of the same water resource.'[75]

It is doubtful that it is appropriate, in view of the criticisms quoted above, based on official pronouncements on the water agreements, to call Uganda 'Egypt's Unwilling Ally', as John Waterbury does.[76] While both Uganda and Egypt have some shared interests and have had some cooperation in the Nile basin activities, Uganda is not in support of the status quo, as Waterbury argues.

Criticism of the treaties has also come from academic circles. According to John Ntambirweki, of the Makerere University Faculty of Law, 'One of the characteristics of the 1929 Agreement, is its unilateral character, especially in the rights it purports to confer on Egypt. The East African territories on the other hand obtained no rights in Egypt. It is argued that a state cannot succeed to obligations created by a treaty in which it has no *quid pro quo* and where the *quid pro quo* belongs to other states.'[77] In view of these arguments, Ntambirweki made the following observations about Uganda's position regarding the Nile Water Agreements:

- that Uganda successfully repudiated the water treaties after independence in accordance with international law;
- that the repudiation of the Owen Falls Agreements by Uganda was further affirmed by the breach of those treaties by Egypt, since the latter did not pay compensation for damaged lakeside properties as agreed; and
- that the Nile Waters Agreement was neither a disparities treaty nor did it create servitude in favour of Egypt regarding the River Nile in the territories of the East African states.

He concluded that the 1929 Nile Waters Agreement is a one-sided agreement, and that Uganda should subscribe to the principle of equitable utilisation of shared water resources and the concomitant obligations and rights of all co-riparian states.[78] In another related study, Ntambirweki and Mujaju argued that the 1949 agreement specifically between Uganda and Egypt over the Owen Falls Dam was imposed on Uganda, and that it called for review and revision 20 years after storage at Owen Falls had begun. Since this was not done, the 1949 agreement is no longer binding on Uganda.[79]

A variety of views have been published in the press, basically expressing what can be described as outrage over the Nile Water Agreements in their present form. A sample of these views is given here. In a press release, Charles Onyango-Obbo declared: 'Egypt cannot enjoy the benefits of having access to the sea, while blocking a landlocked country like Uganda from profiting from the fact that it sits on the source of the Nile.'[80] G. Kateihwe retorted: 'God makes no mistakes. If the Egyptians wanted more water than any other countries through which the Nile flows, why didn't he make another river from Mediterranean Sea to Egypt? This means that all countries should be given a chance to use this water freely according to their needs. After all, the problems of 1929 are far different from those of 2004.'[81] Another critic added: 'I find the Egyptians' remarks very annoying that the altering of the Nile basin treaty would be regarded as war … if Egyptians think that their livelihood depends on the Nile waters what about the other people in nations in which the river originates and flows…?'[82] One East African parliamentarian, Yonasani Kanyomozi, commented: 'What bothers me is that when Uganda developed a scheme to divert the Nile to Karamoja, the plan was opposed by Egypt, yet for them they can do anything with the Nile waters.'[83]

### *Uganda and the proposed new Nile water treaty*

Uganda is a strong exponent of a new water treaty that will serve the interests of all riparian countries equitably. After many years of bitter complaints by upstream countries about the two Water Agreements of 1929 and 1959, which favoured lower-stream states, a new treaty is being proposed: The Nile Commission. In 2001 the East African countries, as a Nile sub-basin group, Uganda, Kenya and Tanzania queried the two colonial agreements, which favoured Egypt and the Sudan over the Nile against the rest of the riparian states.[84] Negotiations have been going on under the auspices of the Nile Basin Initiative (NBI). The new treaty will result in the formation of the Nile Basin Commission (NBC) that will oversee its implementation. Negotiations have concentrated on developing 'win–win' projects benefiting multiple nations. In a joint statement, water ministers said efforts were under way to draw up a legal framework aimed at what they called 'reversing the old path of blind confrontation, missed opportunities and wasted time from which our people have suffered'. The Uganda water minister added: 'If we agree, no country should use water without informing other basin countries …'; and the Egyptian water minister, Mahmud Abu Zeid, responded: 'We view the Nile waters as source of cooperation and not of confrontation.' Important articles in the proposed treaty which generated heated debate, and in previous negotiations that needed either to be repealed or revised, were addressed. The new treaty was expected to be signed in January 2007.[85] When the Nile Commission is signed, it will replace the NBI.

### *Uganda's cooperation in the Nile sub-basin of East Africa*

The East African states of Uganda, Kenya and Tanzania share many common interests, among which are culture, history, geography, socio-political and socio-economic interests. These interests have made it almost inevitable for these countries to cooperate in many sub-basin activities and sometimes take up a common position on wide-basin issues. There are a number of factors which dictate or compel East African states to take up a common position on Nile issues. First, the commonality of interests expressed in the form of the binding commitment of Egypt and the Sudan to the 1959 Water Agreement dictates that other countries with common interests, such as the East African Community, should also take up a common position on the Nile waters.[86] Hence, it may be appropriate for the East African Community of Kenya, Tanzania, Uganda, Rwanda and Burundi to have a common position over the Nile waters.[87] Second, the upstream/downstream relationship between the two groups of riparian states dictates that East African countries should develop a position that is common amongst themselves, but different from that of the lower riparian states. Third, the programme for regional integration of the East African states has made it necessary for the states involved to take up a common stand on the Nile basin issues. There have been cooperations among the East African states which have resulted in the current treaty of the East African Community, comprising the states of Uganda, Kenya and Tanzania. In June 2007, Rwanda and Burundi joined this body. Fourth, among the things the partner states commit themselves to undertake is the adoption of common regulations for the protection of aquatic resources in the Nile waters and Lake Victoria. Last, in the absence of a basin-wide agreement on the utilization of the Nile waters, a number of sub-basin initiatives have been developed. In the East African region, there are three sub-basin initiatives that are most relevant to the utilization of Nile waters:

1. The Kagera Basin Organisation (KBO) 1977, whose members are Uganda, Kenya, Tanzania, Rwanda and Burundi. The objectives of KBO are cooperation among member states in hydropower development, water, fisheries and the environment.
2. The Lake Victoria Environmental Management Programme (LVEMP), established in 1994. Its objectives are management of water resources, quality and use.
3. The Lake Victoria Fisheries Organisation (LVFO), established in 1994. Among its objectives are the management and optimum utilization of the lake.

All these sub-basin initiatives have tended to bring the East African Community countries together and enable them to take up a common stand on basin-wide matters.

### *Uganda in the basin-wide cooperation: projects and programmes*

Uganda's strategic, economic and other interests in the Nile basin are imperatives that make her participate fully in Nile basin projects, activities and programmes.

Indeed, Uganda has participated and continues to take an active part in all Nile basin cooperative activities. It may be because of Uganda's active involvement, and her strategic location, in the Nile basin that she was chosen to house the NBI secretariat, at Entebbe.

Uganda was an active member of the Hydromet project, which was formed by UNDP and the World Meteorological Organisation in 1967. The purpose of the project was to evaluate the water balance of the Lake Victoria catchment area, in order to assist in regulating the water levels as well as the water flow of the Nile.[88] This project was important for Uganda in so far as it would prevent potential flooding in future. Uganda was also a member of the Undugu group (*undugu* is a Swahili word for 'brotherhood'), which was formed in 1977. Uganda was a beneficiary of some of the activities of this organization, which included pollution control and water management in the upstream states. Uganda actively participated in the Technical Cooperation Committee for the Promotion of the Development and Environmental Protection of the Nile Basin (TECCONILE). It conducted a series of training sessions for staff members of water resource agencies and also organized a series of annual Nile 2000 conferences, in which Uganda was fully involved. Uganda is a founder member of the Nile Basin Initiative (NBI), which was launched in February 1999. The NBI was founded on the principles of sustainable development and management of Nile waters in order to promote basin-wide cooperation in integrated water resource management. The Nile basin states accepted the NBI in its transitional form until a permanent legal framework was put in place. Uganda is currently involved in all NBI projects and programmes. Under the Subsidiary Action Programme of the NBI, Uganda belongs to the Nile Sub-basin Group (Equatorial Nile Sub-basin) along with Egypt, the Sudan, Kenya, Rwanda, Burundi, Tanzania and the DRC. Uganda was honoured with hosting the NBI secretariat. Finally, Uganda has participated fully in the formulation of the proposed Nile Commission (NC), which is expected to replace the NBI.

## UGANDA ON THE NILE: CONCLUSIONS

This study has shown that the Nile basin over the years has contributed to the course of Uganda's modern history and development in various fields, and in ways that until now were either inadvertently ignored or unresearched. For example, Uganda has benefited from various Nile basin projects on the Nile and in Lake Victoria, especially in the hydropower sector. The study reveals further that Uganda's main interest in the Nile basin since 1948 has been in hydroelectrical power planning and development. For example, whereas Kenya and Tanzania are more interested in inter-basin water transfers to meet their domestic, industrial and irrigation water needs in the parts of their countries where water is scarce, Uganda is more interested in the uninterrupted flow of water into Lake Victoria to support her hydropower production at Owen Falls, Kiira and other future hydropower dams. The chapter refers the discussion about the Nalubale and Kiira Dams and to what extent they are responsible for the drop in Lake Victoria water levels. The decreased water level is impacting negatively on the fishery, tourism and water transport industries in the Lake Victoria basin and may affect basin-wide activities negatively in future, if the situation is not rectified. Uganda is critical of the 1929 and the 1959 Water Agreements, but believes in cooperation rather than confrontation over Nile basin matters. To this extent she supports the proposed new water agreement for equitable utilization of the Nile resources. Uganda is also actively involved in the planning and implementation of all sub- and basin-wide projects and programmes through the Nile Basin Initiative (NBI). Furthermore, Uganda is involved in drawing up the new treaty – the Nile Commission.

# 8

# Institutional Development and Water Management in the Ethiopian Nile Basin

**Yacob Arsano**

## INTRODUCTION

This chapter examines water planning and development in Ethiopia from a historical perspective. The study gives special attention to the waters of the Ethiopian Nile basin and their significance to the neighboring riparian countries. It introduces the geographic context, socio-economic importance and cultural setting of water use and management in Ethiopia, and explains the links between traditional and modern water-management systems there. The study places water conflict within local, national and regional contexts, and traces the adoption and demise of water management institutions and water sector plans under successive national political regimes.

Given the enormous significance of the Abbay (Blue Nile) to downstream countries, we shall also describe the pertinence of Ethiopia's trans-boundary water development to legal, political and security relations with the co-riparian countries in north-eastern Africa. In this regard a brief analysis of issues involved in the Nile Basin Initiative (NBI), with particular emphasis on the protracted negotiation process of its 'Framework Agreement', will provide the background necessary for some predictions for the future of water use and its management in Ethiopia.

Ethiopia has 123 billion cubic meters (bcm) of surface water on an annual basis. Of this amount a mere 3 per cent remains in the country, while the rest flows out to neighboring countries. By far the largest proportion flows into the Nile system. Ethiopia's western rivers – the Abbay (Blue Nile), Baro-Akobo (Sobat) and Tekeze (Atbara) – together contribute fully 86 per cent of the main Nile's water as measured at Aswan. Master plan studies of the three Ethiopian Nile headwaters have been carried out, and final reports have been published for the Abbay, Tekeze and Baro-Akobo basins, as well as for the Ghibe-Omo and Awash basins. Similar studies are about to begin for the Wabeshibelle and Ganale-Dawa basins in the eastern part of the country. It has been estimated that 2.58 million hectares are available for potential irrigation in all of Ethiopia's basins. Thus far only 4.3 per cent has been developed.[1]

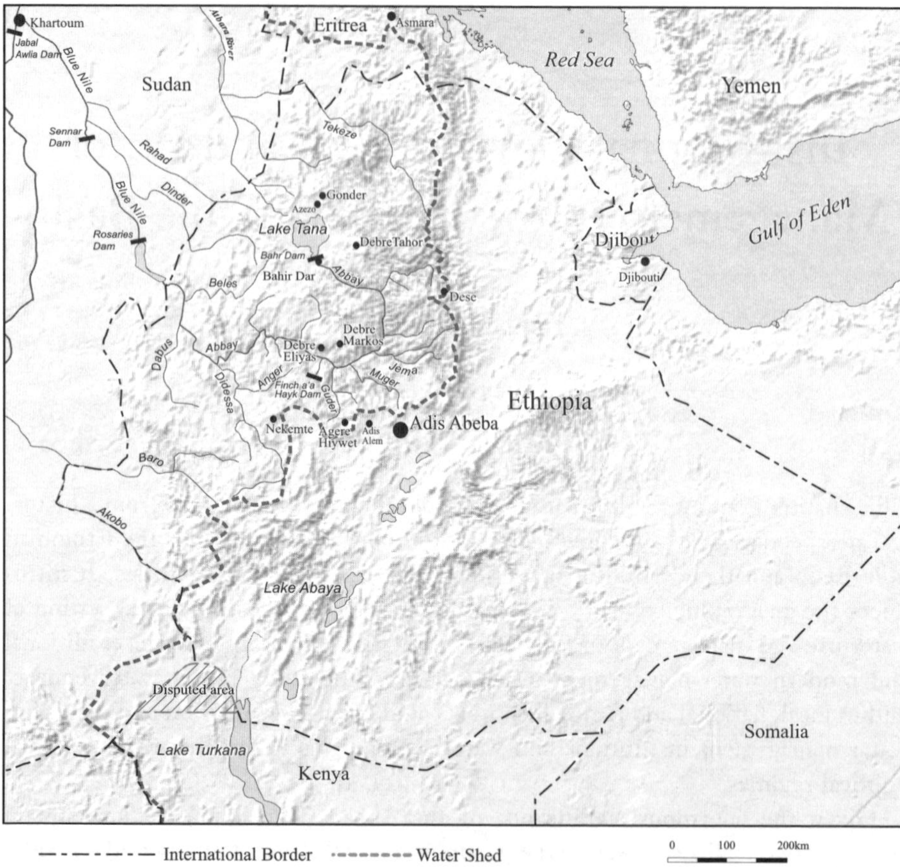


**Map 1:** Ethiopia.

The Ethiopian government enacted a national water resources management policy for the first time in 1999, and issued a water-management proclamation in 2000. The new policy and legislative measures are evidence of a contemporary approach to promote national efforts toward the goals of 'efficient', 'equitable' and 'optimal' utilization of water resources. The Ethiopian national water policy document (1999) specifically acknowledges that previous management failed to solve the problem of spatial and temporal variability of water. It further admits that poor performance in the nation's water development sector was responsible for its slow agricultural development, low crop productivity and lingering poverty. Further evidence of poor performance is that only 72 per cent of the urban population and 23 per cent of rural people in Ethiopia have access to clean water, and only 13 per cent of the country's total population has access to electrical power.[2]

## *The conceptual setting*

Myths and perceptions surrounding the Nile are much older than the realm of water development, management and protection. The Abbay river in Ethiopia, for instance, is the symbol of the nation. Ethiopia's attachment to the Abbay has been deeply spiritual, and the river is held in great reverence. Songs, poems, folk stories and ritual religious performances all extol the Abbay, 'the great father of rivers'.[3] But the Abbay has not been harnessed or put to any meaningful economic use in Ethiopia, and the inhabitants of its 200,000 km$^2$ basin have experienced nature's might through it rather than deriving any economic benefit from it. The river has been seen as a source of identity, of life and of healing power, of unity for societies along its course, of physical destruction, and of potential economic development.

### *Shared identity of the Nile*

Stories and myths idolize the Nile. One may observe a striking similarity in the perceptions of the ancient Ethiopians and Egyptians: both identified with the great river and the power it symbolized. The Egyptians did not know whence the Nile came to their land, nor did the Ethiopians know where it went. In that ancient context there was no upstream–downstream dichotomy.

The ancient Ethiopians knew that the Abbay came from the *amba* (plateaus) of their land; and that the little river (the *gilgal* or child Abbay) and other streams pooled into Lake Tana. They knew that the Abbay emerged from the calm lake, then tumbled over the Tiss-Issat Falls (literally the 'smoke-fire' falls). They knew that the river proceeded westwards through deep canyons. But the ancient Ethiopians were ignorant of the river's destiny; in the words of a popular saying: 'The Abbay has no home, but where does it take all those logs it carries?'

The great Ethiopian poet Tsegaye Gebre Medhin explained that the ancient people of Cush were bound together by the bounties of the Nile: the river is the cradle of the black race and 'the blood of Ethiopia, mother of Cush'. The Nile is thus a spiritual heritage of the riparian communities and countries, and remains a source of hope for emergence from the stark poverty, lingering misery and creeping discord in the basin. Damaka Mattaferia, who headed the Ethiopian engineers during the Abbay Basin Study Program (1958–64), wrote of the unifying element of the river after having witnessed firsthand its twists and turns. He describes the imagination of prophet Isaiah, who seems to have believed that the Nile was used as a highway for 'ambassadors' between the upstream and downstream nations.[4]

## INSTITUTIONAL DEVELOPMENT

### *Water rules under the medieval* Fetha Negast *('Justice of the Kings')*

The origins of the *Fetha Negast* are obscure, but this code was translated from Arabic into *Ge'ez* (the old Ethiopian language) during the reign of Emperor Zar'a Yacob (1444–68) and was applied in adjudicating both civil and religious matters. It remained the law of the land until the Ethiopian Civil Code was enacted in

1960. The *Fetha Negast* embodied the following principles in relation to the use and management of water resources. It recognizes shared rights among the riparians with regard to water use as well as the protection of the environment of the river course. For instance, it goes to the extent of prescribing that, in the case of shared river banks, if soil is eroded from one holding to another, compensation is due. The *Fetha Negast* prescribes principles of management and dispute handling over shared water resources.

### *Water in the Ethiopian constitutions*

Since 1931, Ethiopia has had three constitutions, representing three political eras with distinct political systems. Article 130 of the 1955 Revised Constitution of the Imperial Ethiopian Government provided that the natural resources of the waters, forests, land, air, lakes, rivers and ports of the empire were held in sacred trust for the benefit of present and succeeding generations of the Ethiopian people. The conservation of resources was stated as essential for the preservation of the empire. Article 13 of the 1987 Constitution of the People's Democratic Republic of Ethiopia, however, stipulates that 'Natural resources, especially land, mineral resources, waters and forests are the property of the state'. Splitting the difference, the 1995 Constitution of the Federal Democratic Republic of Ethiopia provides that the right to ownership of all natural resources is exclusively vested in the state and peoples of Ethiopia.

All three constitutions are consistent in upholding the principle of public ownership of water. The constitutional provisions assume that use, management and protection of water resources should be in line with the general public interest.

### *Water rules under the Ethiopian Civil Code*

The 1960 Civil Code of Ethiopia deals with water resource ownership. Unlike the constitutions, the code remained unchanged after regime changes. Chapter 2, Section 3, articles 1228–56 of the code articulate rights and obligations with regard to ownership of and access to water. Issues including priority of use, irrigation, domestic use and compensation are given special attention. Article 1228 recognizes the community's priority rights to all running and still water, and provides for water's control and protection by competent authorities. Article 1236 guarantees a landowner's right to use water that crosses or borders his land, but not to the detriment of those downstream. Article 1237 establishes priority for domestic use over irrigation or other consumptive uses. Article 1239 provides for compensation for a landowner whose ability to exploit his rights to water is compromised.

The provisions of the civil code represent an effort to consolidate rights enjoyed under the *Fetha Negast* and other customary water rules. The scope of the provisions seems to aim at regulating upstream–downstream water use rights and obligations on a small-scale level. The code does not assume large-scale water management problems.

### *Customary water rules*

There are two primary sources for the study of customary water rules in Ethiopia. One is a compilation of old rulings up to 1935, which was collected from 7,296 cases and contained in a multi-volume collection known as *Digest of old Ethiopian judgments*.[5] Customary water rules are also to be found among the unwritten traditional principles and practices of various geographically isolated or socio-economically marginalized communities.

From these rules we learn that those geographically closest to a river or stream, or those who may physically control a watering point, cannot deny access to those who may be situated at a distance but depend on the water. Here again, longstanding traditions underscore that water is public property, unless a particular type and amount of it results from the labor of an individual or group.

Eighty-five per cent of the Ethiopian population is agrarian and lives in rural settings. Customary rules determine much of the social relations in these rural societies. Their significance in determining the traditional management of natural resources, including water, has not been sufficiently recognized and incorporated into state laws. And because these customary rules have been observed from time immemorial, they are still effective in both mitigating and transforming conflict that arises from competition over natural resources, including water. By contrast, the formal approach to conflict management through litigation in courts of law can be lengthy, expensive and inconsistent with local traditions and, therefore, not preferred over customary law.

Formally enacted water laws, together with customary laws and practices, do indeed need to be consolidated into a single national water code, on the basis of which water rights and management can be firmly established and exercised. Traditional laws have the advantage of being based on experience, tested over centuries, and applied in specific environmental and economic situations. A consolidated national water code would have avoided overlapping jurisdictions. There would have also been an acknowledgement for traditional law as appropriate for local situations.

### *Planning and development of the water sector*

> [I]t is of paramount importance to Ethiopia, a problem of first order that the waters of the Nile be made to serve the life and the needs of our beloved people now living and those who will follow us in centuries to come. However, generally, Ethiopia may be prepared to share this tremendous God given wealth of hers with friendly nations neighboring upon her, for

> the life and welfare of their people, it is Ethiopia's sacred duty to develop the great watershed which she possesses in the interests of her own rapidly expanding population and economy. To fulfill this task, we have arranged for the problem to be studied in all its aspects by experts in the field. Ethiopia has time and again set this forth as her position regarding the utilization of the Nile waters.[6]

These words of Emperor Haile Selassie I in the 1950s still express the principle Ethiopia wishes to follow in the use and management of water resources which are shared with neighboring states. That principle has been maintained ever since, notwithstanding change of regimes, political systems, and allegiance in the arena of international politics.

## WATER DEVELOPMENT OF ETHIOPIA IN THE CONTEXT OF THE EASTERN NILE: A HISTORICAL PERSPECTIVE

During the colonial era, Ethiopia was the only independent state in the Nile basin. Britain, the colonial power that dominated the two major downstream countries, the Sudan and Egypt, did everything possible to ensure an unobstructed flow of water to its dependencies. Ethiopia's priority was survival as a sovereign entity; water development was hardly an immediate agenda under the political and economic circumstances of the day. However, the geographical vantage position provided Ethiopia to use national water planning for development and to use this also as a diplomatic leverage. One way or other, the following water development schemes were conceived from inside or perceived from outside as having not only the economic objective but also a political benefit.

### *The Lake Tana project, 1929–34*

The Ethiopian authorities were prompted to embark on the Lake Tana project by the Anglo-Italian agreement of December 1925, which would allow Britain to build a dam at Lake Tana, and Italy to construct a railway through Ethiopia to connect the two colonial territories of Somalia and Eritrea. The Lake Tana project was conceived as a leverage against renewed colonial encroachment on Ethiopia's territorial integrity. The project envisaged a barrage at Lake Tana's outlet and selling water for large-scale irrigation of cotton in the Sudan and Egypt. The Ethiopian authorities calculated that the Americans would be interested in the enterprise, as they would see this as an opportunity to regain their lost share of the world cotton market. Hence, G.J. White Engineering Corporation of New York was commissioned as a partner in the project. A concession was signed, by which the corporation would construct the barrage, sell water to the Sudan, and pay royalties to the Ethiopian government.[7]

The White Engineering company carried out a complete survey of the project area in 1930 and 1934. In terms of related infrastructure, a motor road was to be built from Addis Ababa to Lake Tana. The cost of the project was estimated at $10 million. But the Lake Tana project was never implemented. The British government intervened with the USA, so that White dropped out, and as the Italian invasion of Ethiopia was already looming in the horizon, the Ethiopian government was preoccupied with the impending disaster.

### *The Abbay (Blue Nile) master plan study (1958–64)*

In 1958, Ethiopia embarked on the Abbay (Blue Nile) basin master plan study program. The three aims of the study were to compile a complete inventory of the water and other natural resources of the basin; to establish a model for other basin studies in the country; and to create and develop human resource capacity for development of the water sector. Three major benefits were envisaged from the program: a regulated water supply for a hydroelectric power station; a regulated supply of water for irrigation schemes planned for farther downstream; and a model for water sharing in the event of negotiation with downstream nations.[8]

The Bureau of Reclamation of the US Department of Interior was commissioned to participate in a joint project: the USA-Ethiopia Cooperative Program for the Study of the [Blue Nile] Basin. It was agreed that the two parties would split the cost of the program and pay their respective personnel employed during the study. The Ethiopian government alone spent 42 million Ethiopian birr (the exchange rate at the time was $1 to 2 birr) for the duration of the study.[9]

The intensive study of the Abbay (Blue Nile) basin took five years and returned results in the fields of hydrology, water quality, hypsography, geology, sedimentation, mineral resources, land use, groundwater and the local socio-economic situation. In spite of the relatively brief time it took to complete the study, it outweighed the earlier British one both in volume and in substantive outcome.[10] The study proposed four dams downstream with a total annual holding capacity of 51 bcm. But like the Tana plan before it, the Abbay basin projects were never implemented, with the exception of the Fincha agro-industrial plan that was taken up in the 1980s in an entirely different context. The country could not afford the vast program, and international financial support was not forthcoming.

### *The Gilgal Abbay project (1960s)*

The Gilgal Abbay – meaning 'baby Abbay' – is the source of the Abbay (Blue Nile). It rises in the south-west of Lake Tana and perennially feeds the lake with fresh water. In 1962 a German engineering team carried out an extensive study of the Gilgal Abbay basin to determine its development possibilities. The study identified great potential for producing oil seeds, pulses and fodder on a commercial scale, and in quantities sufficient to support an export industry with substantial foreign exchange earnings.[11] The findings of the study, however, could not be translated into actual investment ventures because the Gilgal Abbay development project was subsumed in the larger Abbay basin study carried out during 1958–64.

### *The Tana–Beles development project (mid-1980s)*

The Beles river originates on the north-western side of Lake Tana. It then flows in a westerly direction and empties into the Abbay just before the latter crosses the Ethiopia–Sudan frontier. The potential of the Beles river was first recognized by C.E. Dupuis as early as 1906, and was confirmed in 1921 by R.P. Black, who discovered its phenomenal potential for generating hydroelectric power if water could be transferred from Tana to the Beles gorge only a dozen kilometers west of the lake. In the Abbay basin master plan study report (1958–64) it was suggested that water be so transferred from Lake Tana to the Beles by a tunnel, then made available for large-scale irrigation in the valley downstream. Five hydroelectric power stations were envisaged in the upper stream, and the study indicated possibilities for irrigated agriculture and resettlement of farmers from the denuded and overcrowded highlands of northern Ethiopia. Settlers from the northern highlands of Wollo, Tigray, Gondar and Gojam would be the largest groups, comprising some 480,000 people.[12]

More recently, a development project in the Beles valley was launched in the context of the 1984–85 droughts and famine. The immediate goal was to resettle people from drought-affected regions of northern Ethiopia and from the overpopulated areas of the south-central provinces. The resettlement program was carried out under the auspices of the Ethiopian Relief and Rehabilitation Commission and with the Italian government's cooperation. An investment of $300 million was projected for irrigated agriculture and resettlement schemes downstream and for hydroelectric power production in the upper course of the Beles. By and large, the new project drew on the previous studies of 1906, 1921 and 1958–64. As part of the infrastructure for the project, an all-weather road from Dangla to Pawe was constructed; an airport was built at the town of Pawe, which grew and attracted a lot of activities and services.[13]

The total area designated for the Tana–Beles project was 220,000 ha. This was organized into six districts and 42 villages. The newcomers would overwhelm the indigenous population of 15–20,000 Begga, whose livelihood was based on hunting, gathering and shifting cultivation. Once the Tana–Beles project was under way, most of the Begga left their homeland.

The settlers comprised Amhara and Tigray from the north, and Kambata, Hadiyya, Wolaiyta and Guragae from south-central Ethiopia. Although documents evaluating the Tana–Beles project have been kept classified and inaccessible to public examination, there are indications that the project had already failed by the late 1980s. Among the major problems the project faced were anti-government forces operating in the area and poor management of the resettlement.

### *Other water development projects in the Ethiopian Nile basin*

Several water development projects are under way in Ethiopia. The following are those within the Nile basin:

- The Fincha hydropower development project started in 2002, financed by the Ethiopian Electric Power Corporation (EEPCo) and by a loan from the Chinese government.
- The Beles river multi-purpose development project was begun in 2006 by EEPCo and is expected to be completed in 2009, at a total cost of 1.2 billion Birr. It is hoped that the finances for the project will come from the Italian government and European Investment Bank.
- The Tekeze hydropower project was started in 2001. EEPCo is the sole backer of the 1.4 billion Birr for the financing of the project. The expected date of completion is 2009.
- The Tana–Beles integrated water resources development project, financed by the Ethiopian government at a cost of 29.82 million Birr, has been under way since 2005–06 and should be completed by 2009.
- The Abbay basin management organizing project, begun in 2003, encompasses the Amara, Oromiya and Benishangul-Gumz regions. The project is expected to be completed by 2008 at a cost estimated at 22.80 million Birr.
- ENTRO's (Eastern Nile Technical Regional Office) small (6.4 million Birr) Ethiopian Nile basin integrated environmental rehabilitation project, which is being carried out in the Amara region, is expected to be completed in 2008.
- In the Lake Tana sub-basin, water resources development projects are under way in the Megach, Reb, Gilgel Abbay, Jamma and Gumera headwater areas.

There are unresolved constraints in the implementation of most of the Ethiopian Nile Basin projects. Four constraints can be identified for this. Ethiopia did not have sufficient financial resources of its own, and neither did it have readily available access to external finances to carry out water-resources development. The country also lacked the institutional capability for studying, planning and implementing projects. Ethiopia's vulnerability to the upstream–downstream hydropolitical confrontations in the context of the Cold War complicated matters, as did the related and recurrent political upheavals. Presently, however, the Ethiopian government has committed itself to developing the country's water resources with locally-generated finances.

## IRRIGABLE LANDS IN THE ETHIOPIAN WATER BASINS

Of the nine major water basins in Ethiopia, the three in the Nile basin have the lion's share of potential irrigable land. The total potential for the nine basins is 2,583,000 ha, while the area for the three Nile headwater basins of Ethiopia is 1,496,000 ha, or about 58 per cent of the total. Thus far only 6,200 ha have been developed for large-scale agriculture in the Abbay basin, and there are no large-scale irrigation undertakings in the Baro-Akobo and Tekeze river basins. Details and comparisons with other basins are given in Table 1.

### *Irrigation targets*

The Ethiopian irrigation plan until 2016 aims at irrigating 122,000 ha in large-scale projects and 122,000 ha in small-scale projects (Table 2).

The average projected growth rates are therefore of 4.5 per cent, 5.5 per cent and 6.5 per cent respectively during the first, second and third phases of the development periods.

## HYDROELECTRIC POWER DEVELOPMENT

Electric power was first introduced in Ethiopia in 1898 – to provide light for the imperial palace in Addis Ababa. In 1931 a small (6.6 MW) hydroelectric power

**Table 1:** Potentially irrigable land in the Ethiopian water basins.

| *Name of basin* | *Potentially irrigable land (ha)* |
|---|---|
| Abbay | 711,000 |
| Awash | 206,400 |
| Baro-Akobo | 483,000 |
| Ganale-Dawa | 326,000 |
| Mereb | 38,000 |
| Omo-ghibe | 348,100 |
| Rift Valley | 46,500 |
| Tekeze | 302,000 |
| Wabeshibelle | 122,000 |
| Total | 2,583,000 |

*Source:* Aberra Mekonnen and Deksios Tarekegn, 'Ethiopia's water resources', in *Water & Development*, quarterly magazine of the Ethiopian Ministry of Water Resources (in Amharic), 2001, 18.

**Table 2:** Irrigation targets planned for 2002–16.

| *No.* | *Description* | *Large- and medium-scale (ha)* | *Small-scale (ha)* | *Total (ha)* |
|---|---|---|---|---|
| 1 | Presently developed | 98,625 | 98,625 | 197,250 |
| 2 | Short-term plan (1st 5 years) | 8,700 | 40,300 | 49,000 |
| 3 | Medium-term plan (2nd 5 years) | 41,057 | 34,532 | 75,589 |
| 4 | Long-term plan (3rd 5 years) | 72,091 | 47,016 | 119,107 |
| 5 | Grand total | 220,473 | 220,473 | 440,946 |
| 6 | Total area to be developed during the program period | 121,848 | 121,848 | 243,696 |

*Source:* WRDP/ETH/98/001, Water Works Design & Supervision, *Executive Summary, Final Draft*, December 2001, Addis Ababa, 20.

plant was installed on the Akaki river, south of the capital. The power supply from there was extended to a few residential areas and to light up some streets of the city. In 1955 the Ethiopian Electric Light Power Authority was established with a mandate to produce, market and distribute electric power in Ethiopia. In 1960 another 43.2 MW hydroelectric power plant was installed at Koka on the Awash river. To date, Ethiopia has been able to install hydropower plants with a total capacity of 453 MW, which is only 2 per cent of the estimated total potential. At the same time, only 13 per cent of the population has access to electricity today. In the framework of the Eastern Nile Subsidiary Action Program, Ethiopia's priority has been to produce hydroelectric power both to increase its domestic supply and to export power to neighboring countries. In March 2001 the Electric Power Corporation signed agreements with Djibouti and the Sudan to export electric power.[14]

### *Hydroelectric power development target, 2002–16*

Although Ethiopia has huge hydroelectric potential, the country's energy development and consumption are astonishingly low: 91 per cent of its total energy is still obtained from traditional sources such as wood, cow dung, crop residue, and human and animal power. The remaining 9 per cent comes from modern sources. It is estimated that 25 per cent of potential hydroelectric power could be harnessed at an economic cost.

Table 3 is based on a forecast by the Ethiopian Electric Power Corporation from June 2000. The rate of population growth and an estimated annual average GDP growth of 6.6 per cent for the period 2002–16 has also been taken into account.

**Table 3:** Electricity demand target (GWh) for 2006, 2011, 2016.

| *Target/ consumer categories* | *Existing situation* | *End of short term (2006)* | *End of medium term (2011)* | *End of long term (2016)* |
|---|---|---|---|---|
| Domestic | 441 | 636 | 857 | 1,152 |
| Services | 311 | 507 | 750 | 1,124 |
| Industries | 562 | 836 | 1,163 | 1,619 |
| Rural | — | 24 | 70 | 145 |
| Total | 1,314 | 2,003 | 2,840 | 4,040 |

*Source:* WRDP/ETH/98/001, Water Works Design & Supervision, *Executive Summary*, 25.

Until recently, Ethiopia has been able to install a generation capacity of less than 1,000 MW – 92 per cent from hydroelectric power. The total potential capacity is estimated at 30,000 MW, and only 2 per cent has been harnessed so far.

Water remains the key to Ethiopia's economic development. Of 123 bcm/year of fresh water runoff, only 3 per cent is retained in the country, while the rest finds its way out of the country via international rivers.

## ETHIOPIA AND DOWNSTREAM RIPARIAN STATES

### *The problem of agreement on the shared waters*

The issue of ownership arises from the basic assumption that riparian states have juridical rights to shared water resources. Establishing and maintaining legitimate and sustainable solutions for shared water resources requires short-term sacrifice for long-term benefit. In the absence of clear rules and mutually agreed procedures, the utilization and management of shared water resources and their proper development can be severely constrained. In the absence of defined and mutually agreed rules and procedures, riparian states may resort to invoking principles they think most advantageous to their respective national interests. These principles include the doctrine of absolute territorial sovereignty and the doctrine of absolute territorial integrity.[15] Actually the hydropolitical discourse between Ethiopia and the downstream nations revolves around the two doctrines.

In an actual dispute riparian states may not hold to one or another water-use doctrine. These are theoretical schema, usable in a body of water law only when supported by bilateral or multilateral accords. In the absence of these, contesting riparian states rarely opt for reference to the International Court of Justice, preferring to adhere to the particular legal theory that best justifies their exclusive demands, then applying it as a bargaining chip. Hence it is not surprising that downstream states reject the doctrine of absolute territorial sovereignty, while upstream states reject the doctrine of absolute territorial integrity. The basic problem remains to be the absence of a binding legal framework to govern the general and specific terms of sharing trans-boundary waters, and the absence of inter-riparian institutions to assure compliance among users of a watercourse.[16]

In the world, supra-national institutions have been evolving, but slowly. The first attempt in this regard was the International Law Association's Helsinki Rules of 1966 on the use of the waters of international rivers. Some of the rules, however, created more controversy as to their own meaning and interpretation. Provisions that embody such notions as 'reasonable' and 'equitable' sharing of water, and 'international drainage basins', have been contested. The result has been less than universal support for the Helsinki Rules. Some states support the concept of 'international watercourse', others the concept of 'international drainage basin', obviously in pursuit of respective national strategies for dealing with co-riparian states.

A subsequent effort at codifying international water law has been the Convention on the Law of the Non-Navigational Use of International Water Courses adopted by the UN General Assembly on 21 May 1997. Its significance lies in its aim to shift international water disputes from contests of power to objective rights and obligations. The responsibility of each state is inherent in the provisions to use water resources efficiently and to avoid damaging the interests of co-riparian states. The International Law Commission, an autonomous body which was entrusted by the UN to promote international water law, had been working on this task since 1970. It is noteworthy that the two main principles of the convention, of 'equitable use' and 'not causing appreciable harm', are similar to the two doctrines of absolute territorial sovereignty and absolute territorial integrity. But the efficacy of international conventions depends on the willingness of riparian states both to accept them and to be bound by them.

It is for this reason that legal and institutional frameworks are needed for guiding and regulating inter-riparian cooperation over the utilization of shared water resources. In the actual case of upstream Ethiopia and downstream Egypt and the Sudan there is no such mutually agreed – or mutually acceptable – framework. Agreements between the colonial powers, between colonial powers and Egypt, and between the two downstream states of Egypt and the Sudan, have only served to exacerbate contemporary disagreement, especially between Ethiopia and Egypt.

Upstream Ethiopia was a party to none of the colonial or post-colonial agreements, with the exception of the 1902 boundary agreement with Britain. The main point of that protocol was to delimit the border between the Anglo-Egyptian Sudan and western Ethiopia. It was also agreed, however, that Ethiopia would not barrage the flow of water from bank to bank on the Abbay, Baro or other tributaries of the Nile. The plain meaning of this provision is that Ethiopia would not stop or divert the flow of the Abbay or other tributaries of the Nile. But that agreement did not imply that Ethiopia would refrain from utilizing the waters of the Abbay or other tributaries of the Nile. And Ethiopia has never accepted any obligation arising from the Nile agreements to which it is not a party. Hence, any future agreement has to be based on negotiations from respective national interest positions. The following is an outline of the negotiating strategies of the three states of the eastern Nile basin.

### *Negotiating strategies of Ethiopia, Egypt and the Sudan*

#### *Ethiopia*

Ethiopia's consistent position has been that it must make use of its water resources in general and the Nile waters in particular to develop an impoverished country. In the circumstances of cyclical droughts and traditional agriculture, it must aim at sustainable economic development for a growing population. Ethiopia has, however, shown willingness to share with downstream states the water resources that arise within the country's boundaries.

In light of this basic position, Ethiopia has argued for replacement of the status quo, as embedded in agreements to which it is not a party, by a comprehensive legal and institutional regime, whereby the riparian countries will have 'equitable' and 'judicious' use of the water resources that traverse their respective territories. The downstream states have a special responsibility not only to understand this point of view, but also to make positive efforts toward creating conditions conducive for collaborative water use. The present mismanagement of the Nile waters – excessive exposure to evaporation, long-distance transfer of the water outside the basin – by downstream states, especially Egypt, will have to give way to cooperative planning.

Pending adoption of amicable water use and management arrangements, Ethiopia reserves its sovereign rights and stands duty-bound to harness the water resources within its jurisdiction. Ethiopia thus clings to the doctrine of 'absolute territorial sovereignty', while enjoying the overwhelming advantage that fully 86 per cent of the main Nile's waters originate in the country. The drawback of that position remains a lack of investment capacity to exploit the very resource Ethiopia claims to need.

#### *Egypt*

Egypt has employed a multi-layer negotiating strategy. It takes active diplomatic initiatives within the Nile basin, presumably to isolate Ethiopia. The first of these was of course the bilateral agreement with the Sudan, on 'the full utilization of the Nile waters', which was signed in 1959. Ethiopia, the only other sovereign state in the basin at the time, was excluded from the negotiation process and indeed was not even consulted. Egypt has since made the utmost effort to initiate ad hoc organizations, such as Hydromet and Undugu, with the intent of maintaining the status quo of imperial, unequal and exclusionist agreements.

The Egyptian side maintains and elaborates an argument that Ethiopia has abundant alternative water sources from seasonal rains and other (non-Nile) water basins, and therefore has no need to focus its development plans on the Nile, to which Egypt, on the other hand, has no alternative. Both upstream and downstream countries should work from positions of comparative advantage in utilizing the Nile waters. Meanwhile, since in Egypt's view there is no surplus Nile water, upstream states should refrain from new consumption. Recently, Egypt's position has seemed to soften: reallocation of Nile waters may be possible after all, but only to the extent that additional water resources are made available to balance such reallocation.

Egypt's best alternative to negotiated agreement is to cling to the 'absolute riparian doctrine' supported by an implicit threat of force reinforced by diplomatic means. The aim is to deter upstream countries from using the Nile waters. Egypt has always had a greater capacity than they, better access to external sources of investment, and the diplomatic support of the Arab world. However, the country's absolute limitation remains as a major and permanent factor: Egypt's entire water resources come from outside its borders, and mainly from Ethiopia.

### *The Sudan*

The Sudan is between Egypt and Ethiopia geographically. Its historical link with Egypt is understandably strong; its hydrologic link with Ethiopia is indestructible.

The Sudan has taken the position that each country in the eastern Nile basin should specialize in what it does best – for instance, the Sudan in irrigated agriculture, Ethiopia in hydropower generation, and Egypt in industry. The Sudan has also advanced the idea of increasing the supply of water available and sharing it, thus mitigating the side effects of water utilization and management. The Sudan has favored cooperation in watershed development and in flood and silt control, a position that has made close cooperation with Ethiopia, whence destructive floods have arisen, highly desirable.

The Sudan's dilemma has thus been characteristic of the wider hydropolitical issue of north-eastern Africa. It is likely that the Sudan will remain allied to Egypt, its most powerful neighbor, at least in the short term, and maintain its role in the broader Middle East.

## ETHIOPIA IN THE CONTEXT OF THE NILE BASIN INITIATIVE

The riparian states of the eastern Nile basin have pursued unilateral and conflicting approaches to the utilization of shared water resources, avowedly driven by their respective national interests. The political expressions of these national interests have resulted in controversies and tense relations between Ethiopia and her two downstream neighbors. Throughout the twentieth century the two downstream states were the sole beneficiaries from the Nile waters, basing their respective claims on 'historical' and 'natural' rights. Upstream Ethiopia, on the other hand, not only repudiates the status quo established by colonial and post-colonial accords to which it was not a party, but also, and in reaction to this, holds to the doctrine of 'absolute territorial sovereignty'. Ethiopia's position is shared by the Equatorial Lakes states, particularly Tanzania, Uganda and Kenya. Hence, the hydropolitics of the eastern Nile basin involves a dispute over the status quo versus an alternative ground for mutual interests and cooperation.

At present there are no legal or institutional arrangements to harmonize upstream–downstream water utilization interests at the sub-basin or basin level. Nor are there any mutually acceptable customary modalities that might be acceptable for regulating inter-riparian water utilization and management. The lack of active engagement to mitigate numerous water-related problems has given rise to environmental and security concerns in the riparian states. The Nile Waters Agreement of 1959 was only bilateral, solely involving the two most downstream states. That agreement effectively excluded upstream states even from the negotiation process, while the two most downstream countries agreed between themselves on 'full utilization' of the Nile waters.

Owing to the lack of adequate upstream–downstream water utilization and management, all three riparian countries of the eastern Nile have been challenged by various complex problems: excessive erosion and land-cover loss in upstream Ethiopia; floods and silt accumulation in the midstream Sudan; and excessive water loss through evaporation in downstream Egypt. Watershed management and environmental protection can best be handled at a regional level. This includes flood and drought management; mitigation of erosion and sedimentation; irrigation and drainage development; and hydroelectric power development and power pooling.

The relevance of the continuing Nile Basin Initiative (NBI) must be viewed in this context. The rather ambitious initiative aims to establish regional cooperation and build mutually beneficial relations among the ten riparian states of Burundi, the Democratic Republic of Congo, Egypt, Eritrea, Ethiopia, Kenya, Rwanda, the Sudan, Tanzania and Uganda. The Nile basin countries, as well as the international community, seem to have realized the intimate linkage between environmental factors and regional security in the Nile basin. Hence, the states of the Nile basin and international agencies have recognized the importance of developing regional relations through the NBI.

With the objective of making the NBI strategically more meaningful, the basin is divided into two sub-basins: namely, the Eastern Nile basin (comprising Ethiopia, Eritrea, the Sudan and Egypt) and the Nile Equatorial Lakes basin (comprising Burundi, Egypt, the DRC, Kenya, Rwanda, the Sudan, Tanzania and Uganda). The two sub-regions are meant to subscribe to respective strategic action programs known as ENSAP (Eastern Nile Strategic Action Program) and NELSAP (Nile Equatorial Lakes Strategic Action Program).

The NBI has set out two ambitious goals. On the one hand it aims to enhance the socio-economic development of riparian countries through sub-basin cooperation in a variety of fields. On the other hand, it works toward establishing a legal and institutional framework that will have the effect of regulating inter-state utilization and management of shared water resources. While the first goal has achieved slower but smoother steps in its progress toward mutually beneficial arrangements, the second goal has been locked up in slow-moving and conflicting national positions.

The NBI was formally set up in February 1999, with all water ministers of the riparian countries agreeing to come up with a Subsidiary Action Program and Institutional and Legal Framework referred to as the 'D–3 Project'. A provisional structure, comprising the Council of Water Ministers (Nile COM), a Technical Advisory Commission (Nile TAC) and the Nile Secretariat (Nile Sec) was set up.

Significant decisions have been taken on both basin-wide and subsidiary levels. Shared vision projects have been prepared and presented to the International Consortium for Cooperation on the Nile (ICCON), meeting at Geneva in June 2001. (ICCON is an international forum of bilateral, multilateral and private funding agencies from which the Nile basin countries seek funding.) Initially Ethiopia, the Sudan and Egypt proposed 46, six and five projects respectively. A short list of projects was jointly prioritized by the three countries. These are in the areas of water conservation, early flood warning, power pooling and interconnection, basin development simulation, hydropower generation, and regional watershed management and irrigation projects.

Consensus over the projects of shared vision and the Subsidiary Action Program was not achieved without difficulty. But a mutually acceptable legal and institutional framework has been still more difficult to reach. Negotiations have been going on since 1999, with all Nile basin states participating. Upstream countries have insisted that a new framework must disregard all previous agreements to which they were not parties. Conversely, the downstream countries seek a new framework to incorporate such earlier agreements. Pending consensus the riparian countries have proceeded with 'Shared Vision' and SAP projects both in the eastern Nile and equatorial Nile basins, separately but simultaneously.

Ethiopia's joining of the Nile Basin Initiative was predicated on the expectation that a cooperative framework acceptable to all riparian states would be created, and that equitable and reasonable water use rights would be established for all basin countries. Ethiopia's willingness to take part in the process of preparing the Nile Basin Cooperative Framework anticipated a change in the status quo in upstream–downstream relations, particularly in the eastern Nile basin.

Ethiopia argues that establishing a legal and institutional regime for Nile waters utilization will create a sustainable basis for equitable and reasonable entitlement to water resources, and that with rights and obligations firmly established the longstanding obstacles to collaborative investment and development will be done away with. Ethiopia agrees that current activities within the Nile Basin Initiative can continue on an interim basis, especially since these benefit the riparian states, but the ultimate goal is establishment of a firm legal and institutional framework.

In 1997, with funding and technical support from the UNDP, a panel of experts comprising three senior experts from each of the basin countries prepared a draft 'Nile Basin Cooperative Framework'. When the document was presented to Nile COM in August 2000, it included 19 points on which the experts had reached consensus, and 17 on which they had not. Nile COM thereupon set up a 'transitional' committee with a mandate to prepare further recommendations on the points of disagreement, for subsequent presentation to the Negotiation Committee.

The most intractable provision in the draft Cooperative Framework proved to be article 15 of the general principles. The representatives of six upstream states, Ethiopia included, upheld the draft article, which stipulated that existing agreements – meaning colonial-era instruments and the 1959 Nile Waters Agreement – must conform to the framework. The Egyptian and Sudanese delegations proposed an amendment that 'the principles and framework are without prejudice to existing agreements'.[17] This and other disputed articles have not been resolved and progress in resolving them has been slow. Up to now negotiations have been handled at multiple levels: the panel of experts; transition committee; negotiating committee; the ministerial forum; and the formal ministerial committee. Although it remains committed to the NBI efforts, Ethiopia has serious concerns regarding the lack of progress in the pivotally important matter of achieving cooperation at the sub-basin as well as at the Nile basin levels.

# 9

# The History of the Nile Waters in the Sudan

**Fadwa Taha**

## THE GEOGRAPHICAL AND HYDROLOGICAL POSITION

The Sudan occupies a middle position of the Nile basin between the upstream countries and Egypt. All major tributaries of the main Nile – the White, the Blue and the Atbara – flow through the country. For millennia, agriculturalists have taken advantage of the Nile's annual flood and traditional irrigation, and in modern times Nile-based irrigation schemes have dominated the country's agricultural development. Some writers have gone so far as to view the opening of the Sennar Dam in 1925 as the Sudan's ecumenical baptism as a modern state.[1]

Irrigated projects support the Sudan's domestic agricultural economy, trade and foreign exchange earnings. The country has the second greatest irrigated area in Africa, after Egypt, and irrigated agriculture has become even more important in recent decades as a result of drought and the variability and uncertainty of rainfall. Even though the Sudan has lately become a major oil producer, irrigation remains central to the economy in general for hopes of increasing the living standards of the majority of the population.

Using Sudanese, Egyptian, British and other sources, this chapter traces the history of the Nile waters in the Sudan through successive political regimes since independence in 1956. Special attention will be paid to relations between the Sudan and the other riparian states; the Sudan's position upstream from Egypt has been a central factor in its history, never more so than during the colonial era, when both countries were subject to British control. Owing in part to the centrality of the Nile to Britain's imperial designs, there is an enormous literature, both official and general, in Arabic, English and other languages, and in a variety of disciplines, dealing with the Nile waters in the Sudan and related issues.

The Sudan, with a territory of more than 2,500,000 km$^2$, lies between latitude 21°55'N and 3°53'N, and longitude 21°54'E and 38°30'E. It is bounded on the north by Egypt and Libya, on the north-east by the Red Sea and Eritrea, on the east by Ethiopia, on the south by Kenya, Uganda and the Democratic Republic of the Congo, and on the west by the Central African Republic and Chad. All of the

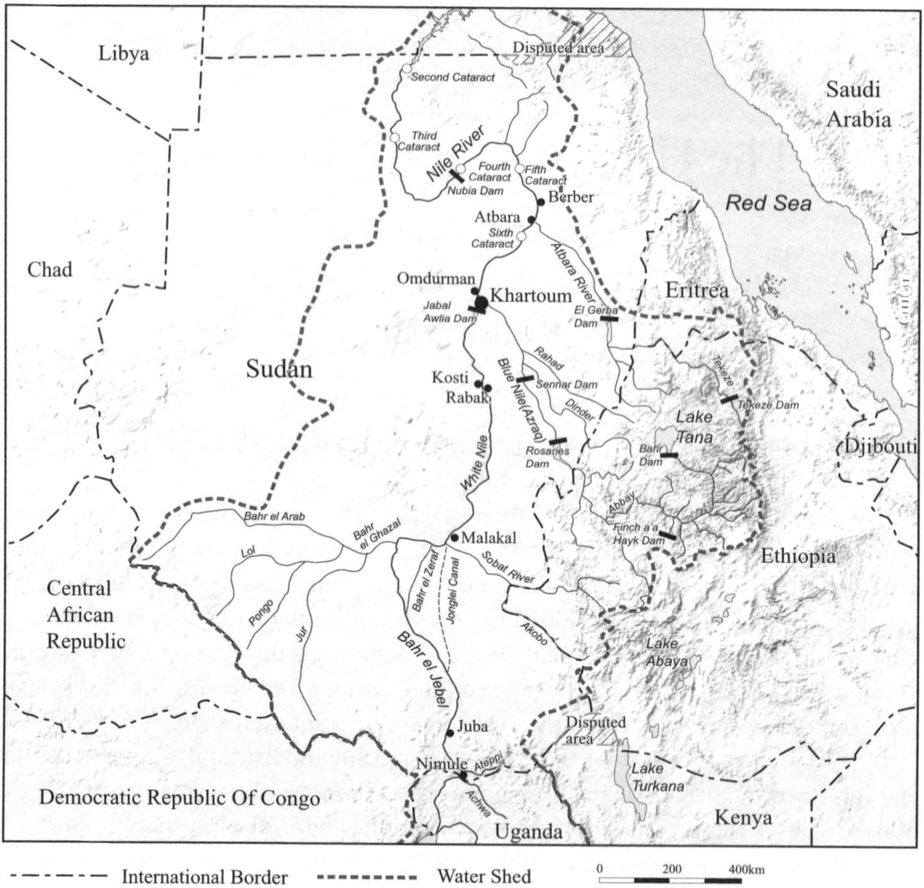


**Map 1:** The Sudan and its neighbors.

Nile countries except Tanzania, Rwanda and Burundi are neighbors of the Sudan, 63.6 per cent of the area of which lies within the basin.

The Sudan today comprises 26 states and has three levels of government: federal, state and local. The estimated population is over 40 million, compared with 23 million in 1993, with an average growth rate of 2.6 per cent per year.[2] More than two-thirds of the population work in agriculture and animal husbandry. In spite of the large area of cultivable land in the Sudan, the continuing rapid growth of the population has resulted in pressure on land resources in certain favored areas of the country. The Nile and its tributaries, many of which are navigable, traverse the central plain, and constitute the country's only source of water for irrigated crop production, hydropower potential and fishing.[3]

The land areas along the Nile in the Sudan can be broadly divided into three agro-climatic zones, separated by the 200 mm and 800 mm isohyets:

1. In the north is a desert zone, sparsely populated and with little development potential. Irrigation is confined to ribbon development along the Nile where cereal, vegetable and fruit crops are grown, entirely dependent on the flow of the Nile for their source of water.
2. A savanna zone of the central clay plains lies between 200 and 800 mm isohyets, where the predominant soils are deeply cracking clays. The area supports both rainfed and irrigated agriculture, the main rainfed crop being sorghum and the principal irrigated crops cotton and wheat. To the west lie the *qoz* sands, where natural vegetation is open woodland with grass. Here the economy is largely nomadic pastoralism.
3. The southern zone, which embraces all the tributaries of the White Nile, has two main sub-regions: the southern clay plains and the ironstone plateau. The former are characterized by swamps that expand seasonally as the Nile overflows. The ironstone plateau supports subsistence rainfed agriculture based on a mixed cropping of cereals, vegetables, cassava and some livestock.[4]

The Sudan's hydrology and surface water consist of the White Nile and its tributaries, including the Bahr el Jebel, Bahr el Ghazal and Sobat; the Blue Nile basin, comprising the Blue Nile itself and its two principal tributaries, the Dinder and Rahad, the flows of all of which are derived almost entirely from rainfall in the Ethiopian highlands and are therefore highly seasonal; the main Nile, formed by the confluence of the White Nile and the Blue Nile at Khartoum; and the Atbara, which rises in the northern part of the Ethiopian plateau, joins the Nile 300 km downstream from Khartoum, and has a seasonal flow pattern similar to that of the Blue Nile.[5] The Nile's total annual flow at the Egyptian border has been estimated historically at 84 bcm. In the average year, the contribution to the total flow of the Nile of its major tributaries is 59 per cent from the Blue Nile, 14 per cent from the Sobat, 14 per cent from the Bahr El Jebel and 13 per cent from the Atbara.

## IRRIGABLE AREAS

A major survey published in 1954 divided the country's land irrigable by gravity or pumped water projects as follows: areas in the catchments of the Rahad and Dinder rivers, from the Ethiopian border to their confluence with the Nile; areas between Sennar and the Abyssinian border irrigable from the Blue Nile; and areas irrigable from the White Nile between Kosti and the mouth of the Sobat.[6] All of these irrigable areas would thus be watered from either the Blue or White Nile, to a total extent estimated at 1,771,000 *feddans*.[7] The regions covered by the report extend for about 400 km from north to south and 30 km from east to west. The full development of such a huge area would plainly be a long-term operation.[8]

The main engineering limitation to pumped irrigation in the Blue Nile region has been topographical. Ideal conditions exist where the ground immediately behind the river bank is as high as the highest parts of the area to be irrigated, but such conditions rarely exist and rising mains, high embankments and other structures are therefore required. Gravity irrigation is ideal from the Blue Nile on land east and west of the river and irrigable from a reservoir at Roseires, and in the lower Rahad and Dinder basins irrigable from the Sennar Reservoir.[9] According to the 1954 report the White Nile, with the very slight slope of about 1.5 cm/km, while unsuitable for gravity irrigation, should pose no difficulty in the accurate interpolation of water levels between existing gauging stations for the design of pumping installations.[10]

## RAINFALL

The arid region of the Nile Basin in the Sudan can be divided into three rainfall zones: the extreme south, where rainfall ranges from 1,200 to 1,500 mm per year; the fertile clay-plains, where 400 to 800 mm of rain falls annually; and the desert northern third of the country, where rainfall averages only 20 mm per year. Rainfall for purposes of agriculture is adequate only in the southern parts of the central clay plains and the ironstone plateau of the southern region.[11]

The value of rainfall depends not only on its quantity but also on its distribution.[12] In northern Sudan, rainfall is insufficient for agriculture without the help of irrigation. Over most of central Sudan the rainfall, though higher than in the north, is extremely variable and uncertain: much comes during violent storms and cannot be efficiently used by farmers, and in bad years total rainfall will not support crops without irrigation. A large proportion of the population nevertheless depends on rainfed agriculture.

## THE ANGLO-EGYPTIAN COLONIAL ERA, 1899–1956

### *The Sennar Dam and the Gezira Scheme*

Most of the projects the British recommended for increasing Egypt's water supply during the colonial era would be built in the Sudan, Uganda and Ethiopia. The first of these upstream projects was the Sennar Dam on the Blue Nile, construction of which was held up during the war years of 1914–18. By 1918 the whole project had been reconsidered in the light of additional data, and it was decided to build the dam to its full height from the start, in order to meet the Sudan's whole annual requirements after 18 January. Construction of the 3,025-m long dam was completed in 1925. Almost 80 per cent of the dam's water would be supplied to Egypt.[13] The dam had two functions: as a regulator it would raise the water level to command the Gezira irrigation canal during the time of the high Nile; as a reservoir it would store water during the period of falling flood for use in the low-water season.[14] By storing 800 million $m^3$, the Sennar Dam enabled the cultivation of cotton on a commercial basis in the Gezira plain. Maintenance and operation of reservoirs in the Sudan were the responsibility of the Irrigation Department, with headquarters at Wad Medani. After independence in 1956, stages one and two of the Sennar Hydro Electric Power project were completed and linked to the thermal power plants in Khartoum and Wad Medani.[15]

British agricultural policy was characterized by the promotion of cotton production in the Nile basin. In this the Lancashire textile industry wielded great influence.[16] In 1912 the British Cotton Growing Association described the Gezira area as 'about 5 million *feddans*, of which probably 3 to 4 million are irrigable, a district nearly as large as the Egyptian Delta'.[17] Lancashire's needs complemented the Sudan government's views. The main objective of the colonial state was to encourage and promote particular agricultural items for export. In the agricultural sector this policy meant increasing the cultivation of crops for export and expanding irrigation systems and communication networks.[18]

The Gezira Scheme was the first and largest irrigation project in the Sudan. The Gezira area is generally flat, with a gentle slope to the north and west, permitting natural gravity irrigation, for which its fertile cracking clay soils are well suited. The region's potential had long been recognized, but full investigation and preparation of definite schemes date from 1904, when William Garstin, of the Egyptian Ministry of Public Works, published his comprehensive report on the basin of the Upper Nile.[19] A complete project for an original gross area of 500,000 *feddans* was prepared; cotton was to be the principal crop, though millet, wheat, sesame and legumes were also to be grown. The building of the dam and the canalization were carried out in phases. By 1931 the gross area was 667,000 *feddans*; by 1938, some 852,000 *feddans*; and, in 1944, 876,000 *feddans*. Before the scheme began, the Sudan government determined to nationalize the use of land intended for irrigation; benefits would go to peasant farmers, not landowners. They rented the land from the landowner for 40 years at the highest rate paid before irrigation.[20]

From the start it was decided that to ensure satisfactory working and economy in water the system of water distribution in the Gezira should be quantitative, and this involved an exact knowledge and control of the discharges passing each regulator at all times.[21] The system, though simple, has one serious defect. A severe check to the velocity of flow of water is caused by considerable deposits of silt, especially in the head reaches of minor canals, which have to be cleared every few years if the flow capacity of channels is to be maintained.[22] As will be shown later, silting remains a chronic problem.

The Sudan government's irrigation department, with headquarters at Wad Medani, was in charge of canalization. The whole scheme was divided into four administrative areas: the Sennar Dam, Wad Medani, Abu Ushar and the Western

Gezira. The first was responsible for operation of the Sennar Dam, in accordance with the 1929 Nile Waters Agreement, and for the upper reach of the main canal. The others were canalization divisions, under which the scheme was further subdivided. Each of these units was responsible for water control and for maintenance in its own area.[23]

The Gezira Scheme was run on a unique system of triple partnership, based on a share-farming system. Farmers operated the scheme in partnership with the government and the Sudan Plantations Syndicate, which provided administration, credit and marketing services. The role of the syndicate was taken over by the Sudan Gezira Board, when the scheme's lease expired, and it was nationalized in 1950. The scheme has played an important role in the economic development of the Sudan, serving as a major source of foreign exchange and government revenue. It has also contributed to national food security and in generating a livelihood for the 2.7 million people who now live in the command area of the scheme.

In spite of this revolution in irrigation, the traditional *shaduf* (a device to raise water) and *saqiya*[24] (or water-wheel) continued in use long after gravity systems and pump irrigation were introduced. They were not subject to any restrictions as far as water abstraction was concerned.[25] The ancient method of basin cultivation practiced in depressions watered at the peak of the Nile flood also continued.

### *The Jebel Auliya Dam: 'The Egyptian reservoir'*

The Jebel Auliya Dam on the White Nile was the second dam built in the Sudan for the benefit of Egypt. Its construction was mooted in connection with the Gezira Scheme; together with the Sennar Dam it would assure Egypt's supply for a long time to come, while allowing also a possible increase in the scope of the Gezira Scheme.[26] The Nile Waters Commission confirmed this conclusion in 1921 and recommended that the Jebel Auliya, Gezira and Sennar works should be carried out at once. Political problems in Egypt in 1921 and financial difficulties postponed work on the dam, which was completed only in 1937. Egypt paid the Sudan 750,000 pounds as compensation for local repercussions of the dam, including the cost of flood protection for the town of Kosti and the railway there, and for alternative means of livelihood for the population dispossessed where the dam was built.[27]

The dam was designed to hold back the steady flow of the White Nile for summer cultivation in Egypt. It would store water to the level of 377.20 m, to a total of approximately 3,500 million cubic meters (mcm). The filling of the dam thus affected the livelihood of a large number of people living along both sides of the White Nile. Their economy was based on a combination of cattle, rain cultivation and crops grown on the lands uncovered by the falling flood: on the breaking of the rains the people moved inland and occupied semi-permanent villages near their areas of rain cultivation, raised their sorghum crops and grazed their cattle, and then moved back to the river in about October, after the rain crops had been harvested. The changes resulting from the filling of the reservoir cut right across this mode of life.[28]

Storage of water began in 1937. In order to give time for the readjustment of the population to new ways of life, and for the development of various schemes, it was decided that the dam should be brought into full operation gradually over a period of six years, during each of which the storage level would be slightly higher than that of the previous year.[29] After the building of the Aswan High Dam the storage capacity of Jebel Auliya became less important than the potential problem of evaporation losses there. However, irrigated areas along the reservoir and also along the main Nile had by then adjusted themselves to the new options created by the reservoir and the improved regulation of the main Nile. The pumping schemes now dependent on the reservoir have an area of 200,000 ha, and the Sudan cannot afford to forego the storage at Jebel Auliya.[30]

### *Pump irrigation*

Irrigation with pumps dates to the beginning of the twentieth century, and to a great extent has replaced traditional flood irrigation and water-wheel techniques. During the colonial period, in what was then called the Northern Province, there were both government-run and private pump schemes. Schemes were equipped with modern pumping installations driven by diesel engines. Religious and tribal leaders were encouraged to set up private pumps, in addition to those of local merchants and foreign enterprise.[31] During and immediately after the First World War the government opened up seven irrigation schemes in the province, intended primarily to increase the production of local crops and to meet the demands for fodder made by the British regiments then stationed in Egypt.[32] Soon after the outbreak of war in 1939, cotton growing was abandoned in all government schemes, which then concentrated on producing a maximum output of food crops. Improved varieties of crops were cultivated.

Soon after the end of the First World War a few foreigners acquired estates in the province and opened up pumping schemes. The Zeidab estate, originally a settlement for freed slaves, was eventually acquired by the Sudan Plantations Syndicate as a cotton-growing project. Most of the other foreign-owned estates were purchased by local Sudanese. In the 1920s private irrigation projects using diesel pumps also began to appear in Khartoum Province, mainly along the White Nile, to provide vegetables, fruit and other food to the capital area.

Pump licenses were limited so as not to interfere with the flow of water to Egypt. From the middle of July until the end of December the volume of water that passed Khartoum was, in a normal year, sufficient for the needs of Egypt when it reached that country. However, from the beginning of January until the middle of July the volume passing Khartoum was normally too little to satisfy the irrigation requirements of Egypt. During the period of shortage, Egypt supplemented the natural flow of the river by drawing the water stored earlier in the season behind the dams at Jebel Auliya and Aswan, and the Sudan was limited by the Nile Waters Agreement of 1929 to the volume of water stored behind the Sennar Dam.[33]

### *Flood and flush irrigation in the Tokar delta and Gash river*

The seasonal rivers of the Gash and Baraka in eastern Sudan are set apart from the Nile system. Both arise in the northern part of the Ethiopian plateau and have characteristics very similar to the Nile's.[34] Among the first efforts ever to employ irrigation for modern commercial cropping was to use the floodwaters of the Gash and Baraka to grow cotton on their deltas. This project started under the Turco-Egyptian regime in the late 1860s, but was interrupted by the Egyptian withdrawal from the Sudan following the Mahdist revolution. Cultivation was resumed in 1899 in the Baraka delta in the Tokar area, but in the Gash delta only after the First World War. Between 1924 and 1926 canals were built in the latter delta to control the flood; sandstorms made canals unfeasible for the Baraka.

The Gash delta consists mainly of alluvial soil deposited by the Gash's annual flood, which reaches Kassala in a normal year between 20 June and 10 July. The average flow continues over a period of about three months.[35] The growing of commercial types of cotton in the Gash delta probably began in about 1860 and the construction of a ginnery was planned as early as 1874. No serious attempt to expand this crop was made until 1918, and the first large canal, with head regulator and controlled off-takes, was not built until 1924. In that year an agreement was signed between the United Kingdom and Italy dividing the waters of the Gash between Eritrea and the Sudan.[36] According to this agreement the Sudan government paid annually to the administration in Eritrea 20 per cent of the profits produced by exploiting the Gash waters in the Sudan.

### *The 1929 Nile Waters Agreement*

In order to overcome European, primarily French, objections to its domination of the Nile, Britain adopted the so-called 'dual flag' policy, and the Anglo-Egyptian Condominium came into being.[37] The Sudan was ruled as a condominium by Britain and Egypt from 1899 to 1956. The 1899 Condominium Agreement, which became the basis for the administration of the Sudan, was silent about the Nile waters. Although Egypt's role in the administration of the Sudan was nominal, her role in determining Nile policy with regard to the Sudan was essential. While the British government was adamant in not recognizing any claims of Egyptian sovereignty over the Sudan, they showed readiness to satisfy Egyptian interests with regard to the Nile waters. This was a systematic policy. When the Milner Mission visited Egypt after the 1919 revolution, for example, Milner stated that the paramount interest of Egypt with regard to the Sudan was the Nile; Milner even suggested a permanent commission with representatives from Egypt, the Sudan, Uganda and other countries to deal with all issues connected to the Nile and ensure a fair division of its waters.[38] An exception to the British rule occurred only with Lord Allenby's ultimatum of November 1924 following the murder of the governor-general of the Sudan in Cairo. In this ultimatum Allenby, then British high commissioner in Egypt, told the Egyptian prime minister that the area cultivated in the Gezira Scheme would be extended from 300,000 *feddans* to an unlimited area as needed.[39] This threat was never carried out.

Following this crisis a commission was formed in 1925 to make recommendations with regard to the Nile waters.[40] Its report, which preserved Egypt's historic rights in the Nile waters, was annexed to the 1929 Nile Waters Agreement. That agreement made any irrigation or power works or measures to be constructed or taken on the Nile and its branches, or on the lakes from which it flowed, so far as all these were in the Sudan or in countries under British administration, conditional on the prior agreement of the Egyptian government.[41] The 1929 agreement remains a source of conflict between Egypt and the former British dependencies of Uganda, Kenya and Tanzania; Egypt would later refer to her rights according to the 1929 agreement when she refused to permit the building of the Roseries Dam on the Blue Nile unless the Sudan agreed to the construction of the High Dam at Aswan and the consequent flooding of Sudanese Nubia.

According to Tvedt, the 1929 agreement did not define water rights in quantitative terms, but the 1920 report of the Nile Projects Commission suggested that Egypt should be guaranteed sufficient water to irrigate the maximum acreage cultivated up to that time, 5 million *feddans*. On that basis quantitative estimates were derived that gave Egypt acquired rights to 48 bcm.[42] The Sudan's share was recognized as 4 bcm. The 1929 Nile Waters Agreement stated that the Sudan could take water from the river and its tributaries as follows: from 15 July to 31 December, without limitation; from 1 January to 28 February, enough to irrigate 38,500 *feddans*; and from 1 March to 15 July, enough for 22,500 *feddans*.[43] Until a new Nile agreement was signed in 1959, the Sudan's needs were therefore limited by Egypt's.

## THE POST-INDEPENDENCE PERIOD

### *The 1959 Nile Waters Agreement*[44]

Development in the Sudan was in its infancy when the Nile Waters Agreement was signed in 1929. Wholly different factors had to be taken into consideration in the early 1950s. The Sudan's need for water at that time was fully as great as Egypt's had been in 1929. Moreover, the population ratios of the two states had altered. Egypt's population of 14 million in 1929 has increased to some 19 million in the 1950s, but the Sudan's population had almost doubled itself, from

5.5 million to 10 million. The 1929 Nile Waters Agreement had to be adapted to these changes in circumstances.[45]

Sudanese opposition to the 1929 Nile Waters Agreement had been expressed in the Sudan Legislative Assembly and Executive Council in December 1948. The council discussed the Sudan's need for additional water and to negotiate with Egypt on this matter.[46] Raising the height of the Sennar Dam was suggested so that the Sudan would be able to store an additional quantity of water estimated at 20 mcm. The Sudanese under-secretary for irrigation told the assembly that the 1929 agreement was based on the conditions prevailing at the time when it was signed, without consideration for the remote future. He made it clear that the Sudan was now in need of more water than that provided by the agreement.[47]

Once a Sudanese government was formed in January 1954 (but before independence), demands for negotiating a new Nile Waters Agreement were presented to Egypt.[48] Revision was necessary if the Sudan was to construct the Roseires Dam and the Managil extension of the Gezira Scheme, and to extend private pump irrigation. The Sudanese were provoked in February 1954 by statements in the Egyptian press by Salah Salem, the Egyptian Minister for Sudanese Affairs, to the effect that the Sudan's quota of Nile waters had not been exhausted. Abdel Rahman Abdun, former under-secretary for irrigation, publicly denounced Salem for trying to mislead the Sudanese.[49]

Negotiations began in September 1954, but with Sudanese misgivings. Although the general idea of the High Dam at Aswan had first been put forward at the end of 1952, Egypt had not supplied technical particulars to the Sudan government until September 1954, despite repeated requests.[50] Now the Sudan government offered to agree to the construction of the dam on the following conditions: the Sudan's ultimate share in the natural flow of the Nile as measured at Aswan must be determined before work was started; the Sudan would then have the right to build on the Nile or on any of its tributaries such dams or other control works necessary for the effective utilization of her share in the natural flow of the Nile; and the population of the town and district of Wadi Halfa must be provided with an adequate alternative livelihood in some other part of the Sudan before the water level at Halfa was raised above its present maximum. Egypt must bear the entire cost of providing this alternative livelihood and of transferring the population.[51] The negotiators agreed that the average total annual flow of the natural river at Aswan could safely be taken as 84 bcm; and that the present established right of Egypt was 48 bcm and of the Sudan 4 bcm, as measured at Aswan.[52]

Although the Sudan's delegates upheld the priority of 'established rights' in negotiating future arrangements, they demanded that those be based on the Cory Award of 1920,[53] when Egypt and the Sudan's 'established rights' were to 40 and 2 bcm respectively. The Cory Award would therefore now entitle the Sudan to 23 bcm. The limiting factor in the Sudan's development was water, not land: an area of 5.5 million feddans had been found suitable for immediate development, with a water requirement estimated at 35 bcm. In any case, the Sudan government insisted that the full 84 bcm of available water must be equitably divided before work started on the High Dam: 35 bcm if the area principle obtained, 28 bcm if population ratio was taken into account, or 23 bcm according to the Cory Award. The Egyptian delegates held that the High Dam alone should be the basis for discussion and that evaporation losses from storage there should be discounted before the remaining water was allocated. On this basis, according to Mohammed Amin, the Sudan would get 5.9 bcm only.

Neither side would budge, and no progress was made. When talks resumed on 6 April 1955, Egypt presented new figures at odds with those on which it had based its previous stance, and put the average annual flow of available water at 80 instead of 84 bcm. They offered the Sudanese only 4 bcm in addition to the 4 bcm previously conceded. The Sudanese delegation considered this tantamount to closing discussions, and no progress was made. In July 1955, Egypt suggested division of the surplus on a 50:50 basis, but in all other respects maintained its previous position.[54]

In its reply of 20 November 1955 the Sudanese government drew attention to the fact that the Sudan was making arrangements for the design and construction of the Roseires Dam, to which the Egyptian government had already agreed in principle.[55] The Egyptian response was aggressive. On 15 December the minister of public works notified Khartoum that Egypt reserved its right under the 1929 Nile Waters Agreement to prevent construction of the Roseires Dam.[56] (This was a reference to the second clause in that agreement that restricted the right of the Sudan to establish reservoirs without the approval of the Egyptian government.)

After the Sudan's independence on 1 January 1956, the Nile waters issue remained unsolved. Relations deteriorated further during 1958[57] and the atmosphere was not suitable to resuming negotiations. In July 1958, to obtain water for the Managil extension of the Gezira Scheme, the Sudanese ministry of irrigation limited the flow from the Sennar Dam at an earlier date than usual in order to raise the level in the reservoir. Although this action did not affect the flow of water to Egypt, Cairo accused the Sudan of breaching the 1929 Nile Waters Agreement. The Sudanese declared that they had not been a party to that agreement and were not bound by it.[58]

### *The military coup of 17 November 1958 and the making of the Nile Waters Agreement*

On 17 November 1958 the Sudanese government was overthrown in a bloodless coup. In a radio address, General Ibrahim Abboud, the new leader, dismissed issues between Egypt and the Sudan as artificial. In March 1959 he formally requested the reopening of negotiations on the Nile waters. The Egyptians welcomed this step: they needed an agreement with the Sudan to construct the Aswan Dam, which would flood much of Sudanese Nubia. Notes were exchanged,[59] and on 7 October a Sudanese delegation arrived in Cairo. Egypt had already made goodwill gestures by withdrawing troops from the border and allowing the abstraction of water from the Sennar Dam for Managil without complaint.[60]

Egypt thus seized the opportunity of the military coup to achieve a new Nile Waters Agreement. On 8 November 1959, after less than a month of negotiations, an agreement was signed for full utilization of the Nile waters. Unfortunately the military men had not found time to read the old papers from previous negotiations, or to take advantage of the expertise of engineers in the ministry of irrigation who had been present during those talks. In a note suggesting an answer to the Egyptian government's letter of May 1955, they stated: 'Speaking purely as technicians, we cannot regard a share of only 20 bcm as satisfactory for the Sudan. In practice, however, the solution to the vital problem of dividing the Nile fairly between Egypt and the Sudan cannot be based solely on technical considerations. It may be advisable for political reasons to accept a smaller share than would seem reasonable to us on purely technical grounds, but the wisdom of so doing is a matter on which we are not competent to judge.'[61]

According to the terms of the agreement, Egypt won her earlier demand that established rights must be subtracted from the net unusable amount and that divisions should apply only to the remainder.[62] This ignored the great differences in the development and use of the Nile waters between Egypt and Sudan since 1929. The Sudan was also to bear the losses of evaporation in the High Dam's reservoir, a point that previous Sudan negotiators had adamantly refused to concede.

For its part, Egypt abandoned the demand that her figure for established rights should be increased by 3 bcm to 51 bcm, to include losses resulting from the storage of water in the Jebel Auliya reservoir. Egypt thus increased her share to 55.5 bcm and the Sudan received 18.5 bcm. The established rights of both were fixed at 48 and 4 bcm. The remaining 22 bcm was divided, with 14.5 for the Sudan and 7.5 for Egypt. The Egyptians considered this division of the net a concession from their previous stand. The agreement stipulated that any additional water in the future would be shared, as would the cost of any works it required.

Compensation to the Sudan for the flooding of its territory behind the High Dam was fixed at 15 million Egyptian pounds, or only half of the initial request by previous negotiators. The actual costs of resettlement (and inflation) were not taken into consideration. The Sudan had to depend on its own funds to complete the resettlement of the people of Wadi Halfa.

The agreement also included a provision for Egypt to borrow 1.5 bcm from the Sudan's share of waters. The Sudanese criticized this on the grounds that once Egypt was using this water it would prove impossible to deprive her cultivators of it. This has never since become an issue between the two downstream countries because the Sudan has failed so far to exploit her whole share.

### *Post-independence water management institutions*

Garstin had suggested as early as 1904 a separate Sudan irrigation department to oversee the collection of hydrological information in Khartoum, but he found little support from the Anglo-Egyptian administration in the Sudan and none in Cairo.[63] A Sudan branch of the Egyptian irrigation service was established, reporting directly to Cairo. With the creation of the Sudan branch, scientific study of the Nile basin became the sole responsibility of the Egyptian irrigation service.[64] Following the assassination of Sir Lee Stack in 1924 the Sudan government set up its own irrigation department in Khartoum, and made it responsible for the maintenance and operation of the irrigation systems in the Gezira and the Gash Scheme.[65]

The Sudanese prime minister in his inaugural address to parliament in March 1954 said that, since irrigation and agriculture were tightly connected, his government had decided to join them in one ministry,[66] but this merger was soon repealed. The name of the Ministry of Irrigation and Hydro-Electric Power was changed in the early 1970s by substituting the word 'Energy' for 'Power'. In 1973, in order to streamline its structure, the government combined the ministries of Irrigation and Hydro-Electric Power, Agriculture and Animal Wealth,[67] only soon again to reinstate separate ministries for Agriculture and Irrigation. In the early 1990s the latter's name changed to Ministry of Irrigation and Water Resources.

The Ministry of Irrigation and Water Resources is the federal body in the Sudan legally responsible for all water affairs. It offers technical advice and assistance to water projects within the state and private sectors. It is in charge of groundwater and the non-Nilotic streams and valleys under a Groundwater and Wadis Directorate. It works in coordination with relevant departments and technical offices. The Ministry of Agriculture and Natural Resources (MANR) supervises the agricultural corporations that manage the large irrigation schemes.

The duties of the Ministry of Irrigation are to develop, improve, maintain and operate irrigation schemes; implement projects for the increase of the yield of the water resources available in the country, with the object of meeting the water requirements for agricultural expansion as well as improving irrigation techniques in the country; discern the beneficial use of the share of the Sudan in Nile waters and pursue studies for the increase of Nile yield, the development of hydropower potential for industrial and domestic use, and satisfaction of the water requirements of the various users; water resources planning, management and development; construction of new irrigation works; international and regional cooperation concerning shared water sources; operation and maintenance of all large-scale irrigation structures and drinking water facilities; and provision of the means for hydropower generation and protection of the water-related environment.[68]

To execute these many duties the ministry needs strong infrastructure, capacity building, extensive funding, political and social stability, balanced international relations, and sound urban and rural development. Some or all of these have been lacking for most of the period since independence, as has adequate finance. The ministry's inability to use the Sudan's share of the Nile waters, to undertake large-scale development of water resources and to pursue a sound irrigational policy could be summarized as owing to a lack of economic and organizational capacity, inadequate planning, civil war, and the priorities of successive governments.

## MAJOR IRRIGATION AND WATER WORKS

### *Water conditions before the construction of the Roseires Dam*

Following independence the Sudan continued to look to agriculture, and especially irrigated agriculture, for its future development, and to this end the first of the massive extensions to the irrigated area began. The government's main objective was increasing production of cotton. Most of these projects had been conceived during the colonial period and involved no change in plans. Long-term national development planning was a characteristic of the 1960s–90s. Its declared objective was to improve the socio-economic conditions of the country. In development planning, irrigation and irrigated agriculture had a central position.

The signing of the 1959 Nile Waters Agreement paved the way for the Sudanese government to execute its irrigation plans. The first priorities were construction of the Roseires and Khashm el Girba Dams and the laying out of the Khashm el Girba Irrigation Scheme. The most profitable parts of Sudanese agriculture and the electrification of the country are based on the utilization of the Blue Nile,[69] and until the Roseires Dam was built the country faced the prospect of a water shortage. The Sudan therefore asked Egypt to increase its water loan. This would mean raising the storage level of the Jebel Auliya dam. The Egyptian government agreed in principle, leaving discussion of the technical details to the joint commission established under the 1959 agreement, which in turn approved a plan to enable the Sudan to make use of an additional amount of 450 mcm for agricultural expansion during the transitional period before completion of the High Dam. The Sudanese committed to take precautions against any danger resulting from raising the storage level, and the Egyptian government would bear no responsibility for injury to people or property in this connection.[70] The raising of the water level behind the Jebel Auliya Dam would be completed in 1960, but because of complaints raised by the local inhabitants about the effect on their economy, the level was increased more slowly, and was halted on 1 October 1961, before the maximum had been reached.

The raising of the water level in the Jebel Auliya reservoir would compensate Egypt for additional water used in the Sudan that had not previously been stored in the Sennar reservoir. This increased the volume of stored water in the latter reservoir to 6,300 mcm, and allowed the Sudan to draw more water from the Blue Nile in low flood while maintaining the flow of water to Egypt by release from the reservoir. The quantity of 450 mcm of water was a loan to the Sudan from Egypt until the first phase of the Roseires Dam was completed.[71] Through this temporary solution the irrigation season of the last three phases of the original Managil extension was not curtailed at the end of December every year, but could continue until the beginning of March. Thus production losses estimated at 10 per cent were avoided.[72]

### *The Managil Extension*

The Managil Extension to the original Gezira Scheme was part of a wider program for developing gravity irrigation in the Sudan. Lack of financing and a shortage of water resulted in the extension's development in four stages, two of which were divided into two parts. Each stage involved an area of 200,000 *feddans*. Stage 1 was put under cultivation in 1958–59. The last section was ready for cultivation in July 1962. As in the Gezira's main area, the general westward and northward slope of the land made the extension easily adaptable to gravity irrigation.[73]

All the physical and human factors facilitating the Gezira Scheme were again operative, with the added advantage that in 1958, when the first Managil phase was completed, the local population was more amenable to economic change, since many had already gained experience of picking and pulling stalks.[74] In 1959 a new main canal was constructed parallel to, and slightly larger than, the former Main Canal of the Gezira Scheme, leading from the Sennar Dam, where a new head regulator was built. From this canal, water is diverted into the Managil Branch Canal. Little experimentation was required, and unlike the Gezira Scheme, which took about 30 years to complete, the last of the four Managil phases came into operation in the 1961–62 season. A fifth phase, designed to join the Managil and the Gezira properly together, came into production in 1963–64.

Distribution of the tenancies in the extension was governed by the Gezira Land Ordinance of 1927. Land was compulsorily rented from registered landowners for 40 years, as in the Gezira Scheme.[75] The Gezira Board administers the agricultural aspects of the Manaqil Extension. The Board, established in June 1950, is an 'independent body responsible to the Government through the Minister of Finance for the efficient management of the Gezira scheme'.[76] Cotton became the main crop, and in fact all the canals were designed on the assumption that about one-third of the scheme would be given to that crop in any one season. The significance of the extension to the economy of the Sudan may be gauged from the fact that, owing to the introduction of tighter crop rotation, it more than doubles the area devoted to cotton in the Gezira.[77]

### *Khashm El Girba: the dam of the displaced*

Interest in the Atbara had been aroused in 1953. The Gash, flowing some 60 km east of the Atbara, had over time formed an inland delta of rich soil, part of which it watered every year. Sometimes it watered very little land effectively and at other times it washed away the crops that had been recently sown. It always failed to reach much of its own delta. Thus the idea was born of diverting water from the Atbara to irrigate those parts of the Gash delta that could not be reached by the Gash flood.[78] It was with this thought uppermost that a reconnaissance began in 1954. But other ways of using the Atbara's water had not been excluded, and attention was directed to a large area of clay plain on the left bank of the Atbara, northwest of Khashm El Girba.

The Khashm El Girba gorge appeared to be the best site for a dam on the Atbara, because of its steep banks, the nature of the rock there, and its geographical position at the northern extremity of the rain belt. Irrigation of the right bank is made very difficult by large tributaries which any canal would have to cross. Irrigation on the left bank is comparatively easy, with a short canal and no cross-drainage. The soil is clay of first-class quality, except in the north. It was estimated that 500,000 *feddans* could be irrigated along the general lines of the Gezira Scheme. There was also hydroelectric potential. If large-scale development of the Atbara was contemplated, the area on the left bank north of Khashm El Girba served by a dam there would be the best location.[79]

Work on the dam was completed in 1964 with a gross reservoir capacity of 1.3 bcm, and the New Halfa irrigation project of 500,000 *feddans* was established. An area of 45,000 *feddans* was developed for growing sugar cane. The main commercial crops initially introduced included cotton, peanuts and wheat. Water is conveyed to the project through a 16-km long main canal and distributed through a network of smaller canals. The irrigated area is a strip of land 30 km wide and 80 km long, to the west of the Atbara and parallel to it.[80]

The dam was the only large irrigation project in the country that did not use the waters of the Blue Nile or White Nile. Resettlement of the people displaced

**Table 1:** Sudan dams.

| *Name* | *Year of completion* | *Utilization* | *River* | *Original storage capacity ($km^3$)* | *Present storage capacity ($km^3$)* |
|---|---|---|---|---|---|
| Sennar | 1925 | Irrigation and hydroelectric power | Blue Nile | 0.93 | 0.37 |
| Jebel Auliya | 1937 | Regulate flow | White Nile | 3.50 | 2.54 |
| Khashm El Girba | 1964 | Irrigation | Atbara | 1.30 | 0.60 |
| Roseires | 1966 | Irrigation and hydroelectric power | Blue Nile | 3.00 | 2.20 |
| Merowi | 2009 | Hydroelectric power | Main Nile | — | 1.25 |

from Wadi Halfa by the High Dam at Aswan was effected mainly after completion of the dam in 1964. Part of the irrigated area at Khashm El Girba was also reserved for settling local nomads, thus bringing together two different and conflicting societies: the Halfa people were settled cultivators with a long history of traditional agriculture; while the nomads had never practiced irrigated agriculture, and their mode of life was dictated by the need for securing pasture and water for animals.[81]

Technical and other problems soon arose. Heavy silting as well as serious problems of drainage and salinity occurred. As a result, by the late 1970s the reservoir had lost more than 40 per cent of its original storage capacity and was unable to meet projected water requirements. Decline in productivity of the agriculture scheme resulted, and in 1980 a plan to rehabilitate the New Halfa project was established by the Sudanese government in cooperation with the World Bank and other agencies. Over a period of five years, funds would be provided for the purchase of machinery and spare parts; to repair the irrigation network; and for fuel storage facilities and other community services.[82] More than two-thirds of the plan's cost was provided through loans from the World Bank/ International Development Association, International Fund for Agricultural Development, and the African Development Fund. Early in 1980, the New Halfa project was made a public corporation independent from the Public Agricultural Production Corporation.

At present the Khashm El Girba Dam has lost 60 per cent of its storage capacity because of the annual accumulation of silt in its lake: some 6 million tonnes of the annual total of 190 million tonnes of silt carried by the Nile. The dam's administration adopted the policy of flushing the lake to lessen the problem, but this led in turn to the annual loss of 300,000 tonnes of fish. No other solution has presented itself except to build a dam upstream to lessen siltation.[83] Officials argue that if the planned Seteit Dam were built it too would lose its storage capacity after about 40 years of siltation. The only way out is to build dams in Ethiopia to control the Nile waters. Contacts are continuing between the Sudan and Ethiopia to execute the Kardobi Dam, with a capacity of 40 bcm and with joint financing. This dam would save the Sudan from the danger of floods, organize the use of water and lessen evaporation.[84]

### *Roseires: the second dam on the Blue Nile*

The first suggestion for a dam at Roseires was made in 1904,[85] but, as we have seen, construction was at the heart of negotiations for the 1959 Nile Waters Agreement and, in the event, the dam was finally built only in the 1960s.

According to a 1961 agreement between the Sudan and the World Bank, the Roseires project consisted of a dam on the Blue Nile, and irrigation works and farm development to enable the water thus stored to be effectively utilized. The dam would be constructed at a site on the Damazin rapids above the town of

Roseires, and would create a reservoir with a useful capacity of about 3,000 $m^3$.[86] The dam would be designed and constructed so as to permit future heightening by 10 m without interference with normal reservoir operations, and the central section would be provided with intakes for the future addition of an irrigation canal and a powerhouse. The canal and powerhouse do not, however, form part of the project. It was estimated that the construction of the dam would be completed by 1 July 1967.[87]

A further agreement incorporated in the dam project the head works for a future gravity irrigation canal on the west bank of the river, with a maximum capacity of 360 $m^3/s$ for reservoir levels at or above 467.2 m. Provision was also made for intakes for a future hydroelectric power station with an ultimate capacity of seven sets of a nominal power output of 25 MW each.[88] During the course of the construction of the dam it was decided to incorporate head works for a future gravity canal on the east bank, of similar capacity to that on the west bank.[89]

A memorandum on utilization of water prepared by Alexander Gibb stated that stage one of the Roseires reservoir, in conjunction with the Sennar reservoir, could support a total net irrigable area of 3,512,000 *feddans* watered from the Blue Nile.[90] Net storage from the first phase of the Roseires dam was estimated at 2 bcm; the capacity of the dam when completed would be 7.6 bcm. (The second stage has not yet been accomplished.) It was anticipated that the water thus stored would be used to irrigate by gravity direct from the dam a tract of some 1,200,000 *feddans* at Kenana.[91] The dam is the site of the largest hydroelectric plant in the Sudan, with a capacity of 250 MW.

The 1959 Nile Waters Agreement states: 'In order to enable the Sudan to utilize its share of the water, the two Republics agree that the Republic of the Sudan shall construct the Roseires Dam on the Blue Nile and any other works which the Republic of the Sudan considers essential for the utilization of its share.'[92] Egyptian agreement is therefore not needed for raising the dam's height. Furthermore, again in consonance with the 1959 agreement, there is to be no cost-sharing in this project because it does not involve utilization of waters lost in the swamps of southern Sudan. Besides, the stored water resulting from heightening would not go beyond the Sudan's share in the Nile. The heightening of the Roseires Dam is a matter of urgency, and fears have been raised about construction of the Merowi Dam before that heightening takes place.[93] In any case, by 1990 the total capacity of the three Blue Nile dams – at Sennar, Roseires and Khashm El Girba – was at most 2.5 billion cubic meters (bcm). Heightening of the Roseires dam would increase the volume stored from 2.4 bcm to about 7 bcm,[94] giving it alone 45 per cent of the Sudan's current share.

### *The Merowi Dam*

The Merowi Dam, also known as the Hamdab Dam, is a large project about 350 km north of Khartoum. It is situated on the main Nile, close to the fourth cataract where the river divides into multiple smaller branches with large islands in between. Merowi is a city about 40 km downstream from the construction site at Hamdab. The main purpose of the dam is the generation of electricity. Its dimensions make it the largest contemporary hydropower project in Africa.

The idea of a Nile dam at the fourth cataract is quite old. The authorities of the Anglo-Egyptian Sudan proposed it several times during the first half of the twentieth century as a way of equalizing the large annual Nile flow fluctuations, creating the possibility of growing cotton and providing flood protection for the lower Nile valley. The new Egyptian government of July 1952 finally decided instead to control the Nile floods with a dam and reservoir on its own territory: the Aswan High Dam and Lake Nasser. As early as 1955 the Sudanese government considered electricity generation from a dam at Merowi. In an official statement it conceded 'no good reason why the Fourth Cataract dam (Merowi) reservoir should be used only for flood protection. It is an excellent power site, and could also provide some valuable over-year storage.'[95]

According to Morrice and Allan, former irrigation consultants to the Sudan government, it was an excellent power site.[96] Fulfillment of the Sudan's hydroelectric potential has thus been on the national agenda since independence, but the government has lacked the funds to translate ambition into concrete action.[97] The Merowi plan was revived in 1979. A total of four feasibility studies were produced between 1979 and 1989, but insufficient funding and lack of investor interest effectively stalled the project at the planning stage. The potential for hydropower generation remained the same, however, as did the Sudan's hydrological resources.[98]

The Sudan started exporting oil in commercial quantities in 1999. This brought an influx of foreign investment, and contracts for the construction of the Merowi Dam project were signed in 2002 and 2003. The main contractors are the China International Water & Electric Corporation and the China National Water Resources and Hydropower Engineering Corp., for construction of the dam and hydro-mechanical works; Lahmeyer International of Germany, for planning, project management and civil engineering; the French company Alstom, for generators and turbines; and the Harbin Power Engineering Company and Jilin Province Transmission and Substation Project Company, both Chinese firms, for transmission system extension. The project's total cost, which is reported to be $1.8 billion, is covered by loans from China, various Arab funds and the Sudanese government. The building of this dam on the main Nile should not affect Egypt's allotment from the Nile waters, not least because its purpose is to produce electricity and the Sudan has the right to build works that would enable her to use her own full allotment.[99] Thus the issues of cost and international objection that some commentators anticipated have not arisen.[100] An annual electricity yield of 5.5 TWh is expected at Merowi, corresponding to an average load of 625 MW, or 50 per cent of the rated load. To utilize the extra generation capacity, the Sudanese power grid was upgraded and extended as part of the project. About 500 km of new 500 kV aerial transmission lines across the Bayuda desert to Atbara, continuing to Omdurman and Khartoum were built, as well as about 1,000 km of 220 kV lines eastwards to Port Sudan and westwards along the Nile, connecting to Merowi, Dabba and Dongola.

River diversion and work on the dam began in early 2004. The first generating unit was expected to go online in mid-2007, but work was delayed and only accomplished in May 2009.[101] The Merowi Dam, with its peak output of 1,250 MW would almost double the electric production in the country.

As in the cases of Jebel Auliya and the High Dam, the construction of the Merowi Dam has affected the inhabitants of the area, and rehabilitation had hardly begun when many problems arose between them and the dam's Implementation Unit. Before dam construction began, an estimated 55,000–70,000 people lived in the area that would be covered by the reservoir, mainly members of the Manasir, Hamadab and Amri tribes in small farming villages along the banks of the Nile and on the islands in the cataract. The whole region is relatively isolated, without paved roads or other infrastructure, and communities are largely self-sufficient. In addition to beans and millet the farmers grow vegetables, both for their own consumption and for trading at weekly regional markets. Their most valuable possession are the groves of date palms growing in the fertile silt on the river banks.

While the majority of the farmers would prefer to stay as close to their old homes as possible and build themselves a new existence along the shores of the new lake, the government has decided otherwise and appointed three resettlement sites: al-Multaqah, al-Makabrab and al-Muqadam. At these locations, farmers received plots of land relative in size to their former possessions, in addition to financial compensation for lost assets – their houses and date palms. About 6,000 of the people whose villages were closest to the construction site near Hamdab were resettled at al-Multaqah in the Nubian Desert in 2003–04. A significant portion of the Manasir tribe inhabiting the desert regions close to the Nile valley were transferred to al-Makabrab, although they had the government's agreement for resettlement around the area of the new lake.

### *The Rahad scheme*

The Rahad irrigation project is located east of the Rahad river, a tributary of the Blue Nile. This is a semi-arid region, with annual rainfall ranging from 350 mm in the north to 650 mm in the south. The topography and soil are similar to those of the New Halfa and Gezira schemes.[102] Prior to the project, settlements in the area were confined to the river banks owing to the lack of drinking water. The project area was populated by some 80,000 nomads, who raised camel and sheep and grew sorghum during the short rainy season from July to September; a secondary objective of the project was to settle the nomads.

The Rahad enters the Blue Nile downstream from the Sennar Dam, and during the dry season had an insufficient flow for irrigation purposes. The idea for the Rahad project emerged from two major developments in the late 1950s and 1970s. The 1959 Nile Waters Agreement enabled construction of the Roseires Dam to intensify crop production in the Gezira and to irrigate west and east of the Blue Nile. Technical and financial problems delayed development. By 1970 modifications had been made and priority was given to the Rahad project. But the Five Year Socio-Economic Plan for 1970–75, by proposing a doubling of the previous plan's investment in agriculture, would tilt heavily toward the irrigated sector: 85 per cent of the investment would be in irrigated agriculture; 13 per cent in mechanized rainfed agriculture; and 2 per cent in livestock.[103]

The first phase would involve 300,000 *feddans*. (The initial suggested area for Stage 1 involved 410,000 *feddans* to be irrigated by a gravity canal of 190 km from the Roseires Dam. The World Bank considered this too long and thus unsuitable for financing.[104] The Bank accepted an alternate design in 1973. The area was reduced to 300,000 *feddans* and required a shorter supply canal for water pumped from the Blue Nile at Meina, some 200 km downstream from the Roseires Dam.[105]) Other irrigation works consisted of a supply canal conveying water to an outflow on the Rahad river, a barrage and a head regulator on the Rahad, and a network of canals to carry water from the barrage to the fields. The first half of the project was completed in 1977 and the second in 1981.

**Table 2:** Estimated irrigable area in 1966.

| *Location* | *Feddans* |
|---|---|
| Main Nile | 60,000 |
| White Nile | 207,000 |
| Blue Nile – Sennar to Khartoum | 20,000 |
| Blue Nile – Roseires to Sennar | 189,000 |
| Blue Nile upstream of Roseires Dam | |
| – Roseires Project, Blue Nile Right Bank | 500,000 |
| – Kenana Project, Blue Nile Left Bank | 1,500,000 |
| – Dinder Project | 350,000 |
| Rahad System | |
| – Rahad Project | 430,000 |
| – Guneid Project | 120,000 |
| – Hawata Project | 300,000 |
| Total | 3,676,000 |

*Source:* Memorandum on Utilization of Water from Roseires Reservoir, Sir Alexander Gibb and Partners, published by the Ministry of Finance and Economics, The Republic of the Sudan, January 1966.

Finance was obtained through soft loans from the International Development Association, Kuwait Fund for Arabic Economic Development, Saudi Development Fund, Arab Fund for Economic and Social Development, and USAID. An independent public company, the Rahad Agricultural Corporation, was created to manage the project.[106]

The Rahad project has faced a number of problems. In the second season (1978–79) a shortage of fuel in June reduced the peanut yield. Heavy rains and flooding in August and September 1978 delayed the planting of cotton and drastically reduced cotton cultivation. In October, the month of highest water demand, the pumps at Meina were not operated owing to a lack of electric power; water shortages further reduced the cotton yield. Mechanical problems and fuel shortages continued to reduce yields thereafter.

### *The Jonglei Canal*

The evolution of the Jonglei Canal has involved so many permutations over 70 years of research and study as to earn the project the accolade of most-investigated major development project in the Third World.[107] As early as 1904, Garstin's plan for the Nile mooted a number of ways to reduce the waste of water in southern Sudan. One was excavation of a new channel for the summer water of the Upper Nile, in a line running from Bor to the Sobat junction with the White Nile.[108] He summed up the benefits that would accrue to Egypt and to northern Sudan if 600–700 $m^3/s$ could be added to the volume of the river during a period when water was most required.[109] Although Lord Cromer, the British proconsul in Egypt, supported Garstin's plans, the matter became subject to the calculations of the Sudan government and overall British policy in the Nile Valley. The Foreign Office and the Sudan government considered the Jonglei project as part of a strategy to exert pressure upon Egypt, and to have Egypt pay for southern Sudan's development.

Just before the Second World War, Egypt asked the Sudan government for comments on the Jonglei proposal. It was not until after the war that Khartoum set up a team to study local consequences of the project and potential compensation for Egypt.[110] The final report of the Jonglei Investigation Team (in four volumes) was completed in 1954. This stated that because a vast quantity of water was lost in the *sudd* region, any plan for storage and control further upstream would be useless unless these losses could be eliminated. In order to do so a canal should be dug to bypass the *sudd*. This would leave the Bahr el Jebel at Jonglei, a small Dinka village in the middle of a swampy reach of the river, and rejoin the White Nile near the mouth of the Sobat about 280 km downstream.[111] Submission of the report coincided with the beginning of the period of Sudanese self-government following the signing of the Anglo-Egyptian Agreement of February 1953.

After the July 1952 revolution, Egypt started to consider alternative over-year storage sites, and recommended only one, in Egypt itself.[112] This was nothing new; Egyptian nationalists had opposed the Jebel Auliya Dam because it would be built in the Sudan, beyond the sovereignty of Egypt.[113] And now, with the Sudan likely nearing independence, the Egyptians were focusing on their own Greater Aswan project.[114]

The Sudanese civil war, fought mainly in the south, prevented implementation of the Jonglei project. In 1969 the PJTC formed a sub-committee to prepare a memorandum embodying the broad lines of the project for the reduction of losses in the swamps of the Bahr El Jebel and Bahr El Zeraf. The sub-committee submitted its report in September 1969. The PJTC thereupon presented the two governments with its own draft report containing the proposed first phase of the Jonglei project for increasing the Nile yield in the swampy regions of the Bahr El Jebel and Bahr El Zeraf.[115] No practical steps were taken, however, as the civil war continued in the south and relations between Egypt and the Sudan were tense in the early 1970s.

On 12 February 1974, a day after the signing of the Charter of Political and Economic Integration between Egypt and the Sudan, the presidents of the two countries issued a statement indicating that studies of the total cost of the Jonglei and Upper Nile projects would be finalized so that work could begin.[116] The Jonglei Canal would thus be the first major development scheme jointly operated by Egypt and the Sudan.[117] But work did not start until 1978, owing to delay in the installation of the engraver machine.[118] The project would be divided into two phases: the canal itself and storage projects in the Equatorial Lakes,[119] for which consent of the concerned riparian states would be crucial. After two-thirds of the Jonglei Canal had been dug, a series of attacks by the Sudan People's Liberation Army forced suspension of the work in 1984.

The major advantages the Sudan stands to gain from a completed canal project are as follows: a share of the 2 bcm of saved water for irrigation use in the central Sudan; potential for irrigating millions of *feddans* in the canal area; improved transport and communications through the canal and a (proposed) parallel all-weather road; and benefits to the inhabitants of the area in non-agricultural respects, such as tropical disease control through drainage of marshy areas. However, the adverse effects on flood plain pastures may be considerable.[120] According to John Garang, the relevance of the Jonglei Canal to the local Dinka, Nuer, Shilluk, Anyuak and Murle peoples is not that it interferes with their traditional way of life, but rather that it opens up hitherto dormant opportunities for socio-economic development and real opportunities for regional and national integration of the area.[121]

The possible emergence of the south as an independent state dramatically alters all previous speculation and calculation about the Jonglei Canal. If the south voted for secession, an option set out in the Comprehensive Peace Agreement of January 2005, and complete independence was declared, the south would have several options: completion of work in cooperation with Egypt, in return for Egyptian assistance in development projects for which the southern

Sudan is in need, and cooperation with the northern Sudan by exchanging water for oil; cooperation among the three states to construct the canal and share the benefits; and cooperation between the South and the riparian countries of East and Central Africa. Independence would also increase regional competition for the precious waters of the Nile; how would such a scenario affect the 1959 Nile Waters Agreement between Egypt and the Sudan?[122] In any case, it is most unlikely that the canal would be taken up during the present transitional period. The southern Sudanese government is concentrating on infrastructure and other pressing problems.

Egypt is particularly worried about the possibility of southern Sudanese secession, for obvious reasons, and has maintained contacts with the current regional government in the south. On 9 August 2006, Egypt and southern Sudan signed a memorandum of understanding regarding technical support, assessment of water resources, forecasting of flood and drought, and reviews of studies of the Jonglei Canal.[123] The Egyptian minister of irrigation and water resources visited southern Sudan in April 2007 to confirm his government's commitment to this agreement. Priorities were set: capacity building and the construction of small-scale dams are to come first. The southern minister of irrigation raised some concerns, such as vegetation in southern Sudan's rivers and the need for gauging systems in all of the country's major towns.[124] A joint delegation from the federal ministry of irrigation and the southern Sudanese government has also visited Egypt. According to press accounts, the visit resulted in a memorandum of understanding by which the government of southern Sudan agreed to resumption of work on the Jonglei Canal.[125]

### *Gash and Tokar since independence*

Between the 1940s and 1970s various projects were developed to irrigate more land in the Gash and Tokar areas, but as late as 1982 both deltas still yielded only one crop a year, watered by the flood. Adequate groundwater offered the eventual possibility of using pump irrigation from local wells for additional cropping or during flood shortages. Drought during the 1980s and 1990s had disastrous results: at Tokar in 1990–91, for instance, water was so scarce that for the first time in 100 years the crops failed. The Gash, on the other hand, has been subject to increasingly catastrophic flooding. Recent major floods took place in 1975, 1983, 1988, 1993, 2003 and 2007. Damage was severe at Kassala, the principal town in the area, partly because of lack of cooperation between the Sudan and neighboring Eritrea.

### *Pump schemes after independence*

The period following independence witnessed a dramatic increase in the number of pump schemes and the amounts of land devoted to them. As of 1 January 1956 there were in the Sudan more than 1,500 privately owned pump schemes on the banks of the main Nile, White Nile and Blue Nile. These commanded over half a million *feddans* and made a large contribution to the prosperity of the country. The same was true of the small number of relatively large schemes operated by the Sudanese government, covering over 100,000 *feddans*.[126] By 1958 almost half the country's irrigated cotton was grown under pump irrigation. During the 1960s, however, downward fluctuations in world cotton prices and disputes between entrepreneurs and tenants led to numerous failures of pump projects, and in 1969 the government began to assume ownership and operation of them. The government established the Agricultural Reform Corporation for this purpose, and the takeover began with the larger estates. Subsequently, as leases expired, the corporation acquired smaller projects, until in May 1970 all outstanding leases were revoked. A considerable number of small pump operations that developed on privately owned land, chiefly along the main Nile but also on the Blue Nile, nevertheless continued to operate.

Figures published by the ministry of irrigation in 1956 put the area of land suitable for immediate development at 1.6 million hectares: 1.2 million for gravity irrigation and 400,000 for pump-scheme irrigation. Since the establishment of the Gezira Scheme in 1925, most of the expansion in irrigated lands had in fact been in the field of pump-scheme irrigation. The gross area under pump schemes in the whole country increased from 15,4000 ha in 1925 to 73,000 ha in 1944, and by 1956 the number of pump schemes had exceeded 1,500 irrigating over 253,000 ha.[127] In June 1961 the gross area of pump schemes under restricted licenses was 395,500 *feddans*, and by 1964 some 407,700 *feddans*. After the completion of the Roseires Dam all restrictions to seasonal irrigation were removed. In the late 1970s the area under pump irrigation was estimated at 1.5 million *feddans*, representing about 34 per cent of the total irrigated land and about 8 per cent of the total cultivated land in the country.[128]

There was a particularly rapid rate of growth of private pump schemes in the Blue Nile Province. Water has never been a serious obstacle and the Nile Pumps Control Board had always been ready to grant a license to any applicant with a moderate degree of eligibility.[129] As of 2006, on the other hand, some 23,450 *feddans* were being watered by White Nile pump schemes, out of 98,000 *feddans* available; only 65 of 137 pump schemes were actually working. Planting was delayed because of unavailability of water.[130]

### *The Guneid pump scheme*

The Guneid project, with an irrigated area of about 36,000 ha, began operation in 1955 to provide alternative livelihood for nomads. It produced cotton until 1960, when about 8,400 ha were converted to sugar cane. A sugar factory built to process the crop (with a potential capacity of 60,000 tonnes of sugar a year) opened in 1962. The canalization was designed and executed by the Sudan Irrigation Department, but the pumping station proper was built under the supervision of

Sir Alexander Gibb & Partners.[131] Construction work started in 1953 and by June 1954 the canalization system and most of the water control structures were complete. By the end of August 1954 the pump house was practically finished, and in March 1955 the rising main was completed and tested under pressure. By the end of May 1955 four engines and two pumps had been erected in the powerhouse. Each pump was designed to lift 3 tonnes of water per second to the height of 20 m. The rising main is 1,100 m long and has a diameter of 2.4 m.[132] The initial gross area of the Guneid Scheme was about 32,500 *feddans*.

Another pump scheme, the Suki Scheme, which started operation in 1972, is located in the Blue Nile Province, south of the confluence of the Dinder river and the Blue Nile. The project was totally financed from the national budget. The scheme settled 8,000 nomad families. Today the Suki Scheme faces problems similar to those of the Rahad Scheme, for similar reasons.

### *Sugar farms*

Research into sugar production started in 1952 in southern Sudan and in 1957 in the Gezira.[133] In 1962 sugar-cane production was added to Guneid, and in 1965 to Khashm El Girba. In Khashm El Girba a sugar factory having a design capacity of 60,000 tonnes was built to process it. The project enabled 200,000 ha of land to be irrigated for the first time.

In the 1970s, when the consumption and import of sugar grew rapidly, domestic production became a priority, and two major pump-irrigated sugar plantations were established on the White Nile in the Kosti area. The Hajar Asalaya Sugar Project, begun in 1975, had an irrigated area of about 7,600 ha. Its factory, completed in 1977, had a potential annual capacity of 110,000 tonnes. Production at the factory did not get under way until the 1979–80 season because of cane- and sugar-processing difficulties. Problems have also affected the other three state sugar factories, but, as a result of proposed World Bank management, the output total of these four government operations for the 1984–85 season improved to nearly 200,000 tonnes. Output declined to 159,000 tonnes in 1985–86 because of the drought. In 1989 sugar-cane production reached 400,000 tonnes.

The Kinanah Sugar Project, which had almost 16,200 ha under irrigation in 1981 and had a future potential of over 33,000 ha, was one of the world's largest sugar milling and refining operations. The project, unlike the country's four other government-owned sugar projects, was a joint venture of the governments of the Sudan, Kuwait and Saudi Arabia, and the Arab Investment Company, the Sudan Development Corporation, and local Sudanese banks. An initial trial run in the 1979–80 season produced 20,000 tonnes of sugar. Yield increased to an estimated 135,000 to 150,000 tonnes the following season.[134] In 1985–86 production reached more than 330,000 tonnes. But the project was beset with funding problems and overruns.

## OTHER WATER RESOURCES

### *Groundwater*

In areas with no access to the Nile and its tributaries, 75 per cent of the population depends on groundwater and rainwater for domestic uses. In the Sudan, attention has recently been paid to underground water reserves owing to low rainfalls in arid and semi-arid areas, coupled with difficult access to the Nile in areas with high populations.

The hydrological system of the northern Sudan consists of two aquifers.[135] The major groundwater formations are the Nubian Sandstone Basin and the Umm Ruwaba Basin. Groundwater investigations in the Sudan are still in their early stages. Available information is limited to small areas. Groundwater resources are vitally important to meet both human and animal dry season demands. Extensive groundwater investigation would be costly and require well-developed technical abilities. With the exception of a few locations, the utilization of groundwater resources for crop production is generally expensive and difficult to establish.[136] However, important investigations started during the 1980s, with international financial and technical assistance. These include studies of the savanna belt, the Sahara Nubian aquifer and the Umm Ruwaba formation in conjunction with petroleum oil investigations.[137]

Groundwater is especially important for regions such as Kordofan and Darfur, where the dominant feature is the absence of perennial streams. Even in Khartoum, which lies at the confluence of the Blue and White Niles, the demand for alternative water sources has grown rapidly, simultaneous with the extension of the city far away from the Nile. This demand is mainly owed to the high cost of infrastructure to purify and transfer Nile water.[138]

### *Rainfall versus the Nile*

The colonial-era Sudan government concentrated development on irrigated agriculture, with a clear bias in favor of areas immediately adjacent to the Nile. Long after independence the ministry of irrigation predicted that the agricultural development of the Sudan depended absolutely on the provision of ample supplies of irrigation water, and almost the only source for this water is the Nile, including its tributaries.[139] Irrigated agriculture has so far contributed the bulk of the country's wealth and this role will continue to be dominant for a long time to come.[140] But rain fed, traditional farming is the main subsistence activity almost everywhere else in the Sudan, accounting for an area of 9.2 million ha, mostly in the states of Kordofan, Darfur, White Nile and Blue Nile.

A 1954 report called attention to the effects of rainfall on irrigation water requirements. During the months of July and August in the 400 mm rainfall belt, and from June to September in the 800 mm belt, the mean monthly rainfall represents large and sometimes total proportions of the monthly basic requirements of a crop. However, the high degree of variability of rainfall, the long drought periods and the intensity of storms are all factors that frequently reduce the effective contribution of rainfall to the production of crops.[141]

The vagaries of the annual rainfall in the summer rainfall zone of the central Sudan and the consequent fluctuations in yields from one year to the next prompted the post-independence Sudanese government to plan the revolution of the whole agricultural system through the widespread expansion of irrigation.[142]

The mechanized rainfed agriculture sector has developed since 1944–45, when a government project to cultivate the cracking clays of central Sudan started in the Qadarif area, largely to meet the food needs of British army units stationed in eastern Africa. An average of about 6,000 ha/year was cultivated between 1945 and 1953, producing chiefly sorghum, under a sharecropping arrangement between the government and farmers who had been allocated land in the project. These estates proved costly, however, and in 1954 the government began encouraging the private sector to take up mechanized farming in the area, a policy that continued after the Sudan gained independence in 1956. Under the new approach, the government established several state farms to demonstrate production methods and to conduct research.

The private sector response was positive, and by 1960 mechanized farming had spread into other areas of the cracking clay zone in the eastern and central Sudan. In general, the post-independence governments favored capitalist expansion in mechanized rainfed farming, and loans with easy conditions have been provided through the Sudan Agricultural Bank.[143] The owners of the schemes have therefore tended mostly to be traders and civil and military bureaucrats,[144] while traditional farmers have become seasonal agricultural laborers.[145]

**Table 3:** Decline in annual average rainfall at selected stations.

| *Station* | *Average annual rainfall (mm)* | | *Decline %* |
|---|---|---|---|
| | *1901–65* | *1965–80* | |
| Karima | 42 | 24 | 42 |
| Kassala | 338 | 256 | 24 |
| Wad Medani | 386 | 306 | 21 |
| Kosti | 514 | 437 | 15 |
| Kadugli | 778 | 670 | 14 |
| Juba | 1,025 | 943 | 8 |
| Wau | 1,166 | 1,050 | 10 |
| Yambio | 1,482 | 1,425 | 4 |

*Source:* Sudan Meteorological Department. D. Knott and R. Hewett, 'Water resources planning in the Sudan', in *The Nile, Sharing a Scarce Resource: An historical and technical review of water management and economical and legal issues*, Cambridge 1994, 207.

By 1968 more than 750,000 ha were being cultivated. The average agricultural production growth rate declined, however, from 2.9 per cent in the period between 1965 and 1980, to 0.8 per cent in the period between 1980 and 1987. The Ten Year Plan of Economic and Social Development for 1961/62–1970/71 emphasized only the modern irrigated and mechanized farming sectors, and, although the plan was abandoned in 1964, the emphasis on large-scale irrigation and mechanized farming continued.[146] According to official sources, in 1985–86, of the country's 200 million *feddans* of cultivable land, only 20.9 million were actually cultivated: 2.4 million *feddans* in the irrigated sub-sector; 8.6 million in the mechanized farming sector; and 9.9 million in the traditional farming sector.[147] Despite the fact that the area cultivated in the traditional farming sector was the biggest, this sector received the least attention; the traditional rainfed sector obtained only 1–5 per cent of all formal agricultural credits in 2001, and received few other supportive services such as research and extension. For the season 2006–07 the irrigated sector received 3.5 times the amount of government assistance allotted to the rainfed sector.[148]

## IRRIGATION IN THE COMPREHENSIVE NATIONAL STRATEGY, 1992–2002

In 1979 a master plan for utilization of the Nile waters in the Sudan was prepared with the participation of international consultants. Its primary objective was to address Nile water development over the next 20 years, staged over four successive five-years plans.[149] The master plan was interrupted by the political and financial difficulties the Sudan experienced during the last years of the Numayri government, and was not pursued after the toppling of the regime in April 1985.

The present government, which overthrew the democratic regime in June 1989, declared a ten-year plan known as the Comprehensive National Strategy, 1992–2002. The plan put food security, sustained agricultural development, efficient resource utilization and yield enhancement at the top of the agenda. The main target of the plan for irrigation was the exploitation of water resources in gravity irrigation for the plains of Great Kenanah, Dinder and El Rahad and the New Halfa Scheme to increase the cultivated area from 2.8 million *feddans* to 9 million at the end of the plan, and the rehabilitation of the existing dams: Sennar; the heightening of the Roseires Dam; the building of new dams (Hamdab, Upper Atbara and Seteit) to achieve horizontal and vertical expansion of agricultural production; expansion in the use of pumps to cultivate winter crops on the White Nile and the main Nile, the maximization of benefits from valley waters by building reservoirs, and to make use of groundwater; the finding of a solution for the problem of sedimentation that has lessened the storage capacity of the existing dams; and local manufacture of pumps, canal sluices and engravers.[150] Prior to the launching of this plan, the ministry of irrigation set in the early 1990s as its main focus an increase in the country's hydroelectric production; making water available for irrigation; increasing the efficiency of existing irrigation schemes; and forecasting and preventing floods. Floods had become more frequent in the Sudan in the late 1990s; the period between 2000 and 2002 witnessed six flood disasters, two of them on the Nile and in the northern states, two in Sennar State, one in southern Kordofan, and one in West Kordofan.[151] Little was achieved by the end of this plan and little has been done to improve accessibility to food for the poor, the vulnerable and the marginalized strata of society.

The plan was accompanied by a new economic policy based on economic liberalization, privatization and the free market. Farmers were to be charged the actual cost for providing water irrigation, and system and operation maintenance. This was one factor in the decrease of cotton production.[152] Ill-conceived ideas of privatization, and lack of coordination between those who dictate the measures and those who translate them into plans, were in turn factors in the failure of the policy.

While the government is now heavily engaged in building the Merowi Dam, problems with regard to the rehabilitation of irrigated agriculture and existing dams remain. The relevant minister in April 2007, almost four years after the end of the National Strategic Plan, said that donors had not fulfilled the commitments made at the Oslo Donors Conference of 2005 to fund irrigation and water projects. The minister said that donors had promised $124 million to execute water and irrigation projects in rural areas, $173.5 million for urban areas, $25 million for personnel training and $67 million for equipment. Meanwhile, UNICEF has provided $25 million to execute water projects in the Sudan during 2007.[153]

It is difficult to reconcile the lack of funding for vital irrigation projects and the Sudan's position as an oil exporter since 1999. The total output of oil products increased from 2.79 million tonnes in 2003 to 3.16 million tonnes in 2004.[154] Oil and its by-products constituted 81.1 per cent of total exports in 2004, while agricultural products and livestock exports ranked far behind, followed by manufactured items such as sugar.[155] By concentrating on oil, the government is repeating the same mistake made earlier regarding cotton, and not seizing the opportunity to diversify the country's exports.

## ABUNDANCE VERSUS SHORTAGE

In 1957 the Irrigation Advisor to the Sudanese government estimated the country's total water requirements, as shown in Table 4.

These calculations exceeded the Sudan's share of the Nile waters by 3 bcm. A clear water deficit would result if the Sudan implemented its policy of extensive increase of irrigated land. These projects were not implemented, and at present the Sudan still uses less than its allowed share.[156] But the government has revealed its estimated water requirement of 32 bcm by the year 2025 for food security and

**Table 4:** Sudan: Total water requirements, 1957.

| *Region* | Feddans | $m^3$ *per* feddan | *Billion cubic metres* |
|---|---|---|---|
| Southern Sudan | 500,000 | 2,700 | 1.35 |
| White Nile pumps: Sobat–Geiger | 120,000 | 3,400 | 0.41 |
| White Nile pumps: Geiger–Khartoum | 210,000 | 4,400 | 0.92 |
| Blue Nile pumps: Roseires–Sennar | 250,000 | 3,600 | 0.90 |
| Blue Nile pumps: Sennar–Khartoum | 200,000 | 4,500 | 0.90 |
| Kenana gravity area: west | 600,000 | 4,400 | 2.64 |
| Kennana gravity area: east | 580,000 | 3,600 | 2.09 |
| Kennana gravity area: south | 40,000 | 3,100 | 0.12 |
| Gezira and Managil areas | 1,800,000 | 4,500 | 8.10 |
| Butana | 500,000 | 4,500 | 2.25 |
| Northern Sudan pumps | 700,000 | 5,330 | 3.73 |
| Total | 5,500,000 | | 23.41 |
| Less 10% for transmission losses | | | 2.34 |
| As at Aswan | | | 21.07 |

*Source:* Note by Allan, Irrigation Consultant, February 1957, 591/6/50, Sudan Archive, Durham.

other essential uses.[157] But water scarcity in the Sudan is a question of quantity and timing, as the reservoir capacity amounts to only 17 per cent of the country's share according to the 1959 agreement.[158] One commentator believes that the anticipated area for cultivation in the 1990s was huge by all standards, and that the irrigated area would be far beyond what the Sudan could irrigate with its legal share of Nile water, especially if its level of irrigation management and technology is considered.[159]

About 80 per cent of the irrigated areas were developed before the mid-1960s. All these schemes were designed and developed on the basis of a constant water factor, now considered defunct, of $400m^3$/*feddan* applied at 14-day intervals.[160] Whatever remains unused by the Sudan out of her share is used by Egypt. Needless to say, the Sudan's inability to exploit her share precludes her from asking for repayment of the 'water loan' given to Egypt according to the 1959 agreement. But the Sudan could have bargained to get this loan repaid in other forms.

## PROBLEMS IN THE GEZIRA SCHEME

As the Sudan's first and still biggest irrigated scheme, comprising 12 per cent of the country's cultivated land, the Gezira Scheme may be seen as representing the problems encountered by the entire irrigated sector. Financial, technical and institutional problems there have resulted in a serious fall in productivity and a corresponding drop in farm incomes.

Water for the Gezira Scheme comes to the field from the dam on the basis of a calculation of the total requirements of the crops at any particular time. Calculations are made on a block-by-block basis. These are transformed into indents for water. These indents are transmitted to a Divisional Engineer employed by the ministry of irrigation. The ministry is responsible for the cleaning and maintenance of all waterworks from the dam to the minor canals. There are four divisional engineers in the Gezira's Main Scheme and two in the Managil Extension. When these engineers have the total of indents for their areas, they pass on the request to the headquarters at Sennar. There the total volume of water required for the entire scheme is calculated and then released into the main canal. The water can take as long as four days to travel from the dam to the northernmost parts of the scheme. Its flow is controllable after it has left the dam by means of a system of locks, water gates and overflow channels.

During its first phase (1925–60) the Gezira Scheme ran smoothly. In the second phase (1961–80) more crops were introduced, and the night storage system collapsed. Between 1981 and 1991 water charges were introduced. Other water demands have also gradually built up, further reducing the ability of a fixed supply to meet peak demand. Domestic and animal water demands have increased. Gardening and fodder cultivation have increased both inside and around the fringes of the scheme, bringing further stresses on the supply capacity of the canal system.[161]

Conflicts between the Sudan Gezira Board (SGB) and the ministry of irrigation began when tenants were required to pay for irrigation services per area cultivated. The cost recovery was low and the ministry failed to carry out maintenance, especially of the lower system, because tenants did not pay the full water service charges. Problems intensified further in the 1990s with the liberalization of the economy, as the government began to pull out of financing the scheme bit by bit.[162]

Since 2000 there has been a major change in the institutional arrangement for water management. The SGB asked for full control of the whole system, disclaiming responsibility for low productivity while the most important input – water – was not under its control. On the other hand, the ministry of irrigation held the view that the irrigation system is a national asset in which the government has invested heavily and which needs very high technical expertise for proper operation and maintenance. A committee was formed in 1998. It recommended that the ministry of irrigation be responsible for organization and maintenance of the upper system and the SGB for maintenance and organization of the lower system. This recommendation was implemented in the season 1999–2000. This opened the door for the transfer of maintenance and organization of minor canals from the SGB to tenants. The training of tenants is crucial for ensuring the success of this change.

Management of water is thus not the only problem facing the scheme. Others can be summarized as a dysfunctional 'top down' management, skewed incentives for farmers and the SGB, low productivity of crops, and canal maintenance problems affecting the delivery of irrigation water. Silting and weeds in small canals disrupt the flow.

The International Water Management Institute and the SGB have carried out research on water-management practices in selected areas. The performance of all lower-level canals has been poor owing to the physical deterioration of the structure of the canals. The high costs of canal irrigation and low crop prices have made investment in new irrigation schemes increasingly unappealing. In addition, the farmers' performance has led to a reduction in crop yield; a reduction in sorghum yield has been related to water mismanagement. Much water is wasted.[163] Low productivity is related to water management at the field level as well as insufficiency of the conveyance system during the peak water requirement season in September and October.[164]

There is a definite relationship between the amount of water available at the dam and the area of crop that can be grown in the Gezira Scheme. The official irrigation pattern requires the application of 400 $m^3$ of water per *feddan* at 14-day intervals. This implies a daily crop water use of 28.5 $m^3$/*feddan*. The possible inaccuracies involved in such a rule of thumb when extended to cover the watering of almost 2 million acres (8,094 $km^2$) will obviously have considerable effect on the water economy of the entire scheme.[165] This pattern of application was worked out for a much smaller area at the inception of the scheme. Tenants have been found to care more for irrigating their sorghum crop during water shortages than their cotton.[166]

In general, problems facing irrigated agriculture as far as water supply is concerned may be summarized as discharge capacity limitations of the main canals in meeting the peak demands; the silting of the intakes of several pumping stations; silting in irrigation channels; incompetent performance of the pumping plants;

**Table 5:** Estimated water losses in 1978 from some irrigation schemes in the Sudan.

| | *% water loss* |
|---|---|
| White Nile schemes | 54 |
| New Halfa | 38 |
| Gezira/Managil | 29 |
| Average for all irrigation schemes | 30 |

*Source:* H.R.J. Davies, 'Irrigation development programs', in G.M. Craig (ed.), *The Agriculture of the Sudan*, Oxford 1999, 361.

head and tail water dispatch problems in the field channels; uneven distribution of water within the plots; major energy shortages; inadequacy and insufficiency in renewals, replacements, spares and maintenance; water mismanagement; and major water supply problems arising from frequent critical delays in agricultural inputs and operations.[167]

## THE POSITION OF THE SUDAN IN THE NILE BASIN

### *The Permanent Joint Technical Commission (PJTC)*

The 1959 Nile Waters Agreement provided for technical cooperation between the Sudan and Egypt, to be achieved through a Permanent Joint Technical Commission (PJTC)[168]: 'In order to ensure the technical cooperation between the Governments of the two Republics, to continue the research and study necessary for the Nile control projects and the increase of its yield and to continue the hydrological survey of its upper reaches, the two Republics agree that immediately after the signing of this Agreement a Permanent Joint Technical Commission shall be formed of an equal number of members from both parties.'[169] The Commission was formed in 1960, holds four meetings every year and publishes annual reports. This institution has continued to function almost without interruption through various political crises.[170]

The PJTC is responsible for discharge measurements and gauge readings, and new gauging stations were established along the Nile and its tributaries. The present Sudanese network of hydrological stations includes 130 stage-measuring stations, 137 stage and discharge measuring stations, and eight stations that measure stage and discharge plus sediment load.[171]

Part of the PJTC's work includes monitoring the problem of water hyacinths by giving estimates of water hyacinth distribution, gauging the severity of the problem, identifying areas requiring management action, and assessing the efficacy of action. Combined efforts between the PJTC and the ministry of agriculture have been undertaken to execute programs to combat the problem.[172] However, little has been achieved and the security situation in the south has stopped the work many times.

Another area of the PJTC's mandate that has not been systematically followed up is the widening of its own membership to include other riparian states. In 1979 the seventh meeting of the Joint Supreme Ministerial Committee of Integration between Egypt and the Sudan discussed the promotion of the PJTC to include all Nile basin countries. The committee recommended that coordination of this issue should take place between the ministries of irrigation and foreign affairs in both countries.[173] Even if this had been undertaken, however, expansion could hardly have succeeded because the other countries are basically opposed to the 1959 agreement itself.

### *Integration between Egypt and the Sudan*

Relations between Egypt and the Sudan can be said to be good over water resources but at times tense over other issues. Though historically the Sudan was considered as subordinate to Egypt, a situation aggravated by British policy in the Nile valley, Egypt now has no alternative to maintaining good relations with the Sudan. The Sudan also has much to gain from cooperation with Ethiopia; dams on the Blue Nile would regulate the flow of the river and would reduce the danger of floods for the Sudan, as well as reduce the sedimentation problem in Sudanese reservoirs.

In addition to cooperation through the PJTC, attempts at political and economic integration between Egypt and the Sudan have been made during the period of Numayri's military rule of 1969–85.[174] The Egyptians saw in integration a chance to achieve the unity that they had hoped for during the period of condominium rule, a goal frustrated by the Sudan's choice of complete independence in 1956. The main element of the Egyptians' interest in unity with the Sudan is the Nile waters, a fact manifested in Egyptian complaints about their minor share in the Sudanese administration under the Condominium.[175] On his part, Numayri saw in Egypt an ally that would come to his rescue against internal troubles. An Integration Agreement was signed in February 1974. Until Numayri's downfall in April 1985, the Sudan was a close and dependable ally of Egypt.

Lavish and unrealistic economic projects toward integration between the Sudan and Egypt were launched without tangible results.[176] The failure of an experiment of integration between the Sudan's Northern Province and Egypt's Aswan Province is a good example. A Common Defense Agreement was signed in July 1976 following an attempted coup against Numayri. Another charter of economic integration was signed in November 1981.[177] This charter attributed the earlier lack of achievement in integration to a shortage of funds. A phased-in plan was agreed upon and also came to nothing.

Relations deteriorated after the military coup of 1989. Tensions between the two countries mounted, particularly after the June 1995 attempt on the life of the Egyptian President Hosni Mubarak in Addis Ababa, for which the Sudan was blamed. The issue of Nile water has been raised several times during these and other conflicts.[178] Nevertheless, attempts at closer cooperation continued. A Four Freedoms agreement was signed in 2004, giving Sudanese and Egyptians freedom of movement, residence, work and ownership in either country. Some Sudanese believe that Egypt benefits more from this agreement because of the Sudan's abundant resources. In any case, this agreement has not been implemented yet, and obstacles to execution remain, on both sides. Cairo fears a mass exodus of refugees from Darfur. It also seems that opening the Sudanese border would facilitate the movement of terrorist groups in Egypt.[179]

### *The Sudan and cooperation in the Nile basin*

The Sudan and Egypt did not take into consideration the other riparian countries when they negotiated the 1959 Nile Waters Agreement. They turned deaf ears to demands from Britain and Ethiopia to be included in any negotiations. Inviting Ethiopia in particular would pose a real danger when considering the division of shares, as it supplies the Nile with more than 85 per cent of its water. In August 1959, when negotiations were about to reopen between the Sudan and Egypt, Britain addressed a diplomatic note to Egypt, Ethiopia, the Sudan and Belgium about the needs of the East African countries. This pointed out that when the investigations preceding the 1929 Nile Waters Agreement were made, the need for water for irrigated agriculture in East Africa had not yet been appreciated.[180] The Sudan and Egypt both ignored this note and the British and Ethiopians were not persistent in following up.

This attitude persisted during the abortive negotiations of 1954–56. The Sudanese negotiators and their Egyptian counterparts ignored a British note in November 1955 calling attention to the interests of the British East African territories, and reserving the right to negotiate on their behalf with the Egyptian and Sudanese at the appropriate time for a share of the Nile waters.[181]

The 1959 Nile Waters Agreement is the source of the present conflict among the Nile basin countries. The preamble of the agreement provocatively referred to full utilization and full control of the Nile by the Sudan and Egypt, ignoring the interests of the other riparian states. However, article 5, entitled 'General Provisions', states that since the riparian countries, other than the two republics, claimed a share in the Nile waters, the signatories agreed that they should jointly consider and reach one unified view regarding those claims; if that consideration resulted in allotting Nile water to one or another of the riparian states, the accepted amount would be deducted from the shares of Egypt and the Sudan in equal parts, as calculated at Aswan. The bilateral technical commission would make the necessary arrangements with the respective states, ensuring that their water consumption would not exceed the amounts agreed upon. This wording gives the impression that the Sudan and Egypt had sole say in allocating the waters of the Nile, even though both of them are water recipients.

The 1959 agreement neglected the idea of treating the Nile valley as a hydrological unit. A study of the Nile published in 1958 observed that the ultimate development of any river could be planned properly only if its entire basin, including the catchments of all of its important tributaries, was treated as an organic whole. The Nile is no exception to this principle, and an admirable series of hydrological records provides the essential data.[182] But the Egyptian government favored a Nile conservation system under its control, and this was its major achievement in the 1959 Nile Waters Agreement.

The 1959 agreement provided no alternative basis for all-basin cooperation. It binds the Sudan and Egypt to the PJTC and they became one negotiating bloc vis-à-vis the other riparian countries. In relation to Ethiopia, the agreement tied the Sudan's hands without taking into consideration that an expansion of irrigated agriculture in the Sudan and the problem of silting in Sudanese dams could best be pursued in cooperation with Ethiopia.

Both the Sudan and Egypt nonetheless came together with the independent states of East Africa in the Hydromet project established in 1967. The initial members were the Sudan, Egypt, Kenya, Tanzania and Uganda. Rwanda and Burundi joined in 1970.[183] Ethiopia joined as an observer in 1971. The project was supported by the UN agencies, which regarded it as an important step in regional cooperation and integrated river-basin development. The objective of the project was to collect and analyze hydro-meteorological data of the catchments of the lakes in order to study the water balance of the Upper Nile.[184] It undertook a series of surveys and provided basic training for technicians and technocrats in several of the member countries, but failed to develop an effective basin-wide cooperative arrangement. It has, however, served as a significant forum for discussions regarding the creation of a viable mechanism for regional cooperation within the basin.[185]

Before Hydromet, the PJTC had advocated the technical cooperation of the Sudan, Egypt, Uganda, Tanzania and Kenya. The Sudanese and Egyptian governments agreed in 1962 to start technical, unofficial discussions with the governments of East Africa with regard to their demands from the Nile waters and other common technical affairs.[186] Conversations took place in October 1962 in Cairo with representatives of Uganda, and in March 1963 with representatives of East African countries.[187]

During meetings of the technical committee of Hydromet in December 1978, representatives of the Sudan and Egypt made a proposal for establishment of a Nile basin commission to cooperate in the natural planning of conservation, development and allocation of water resources in the basin.[188] The proposal was a step forward from the stand of the Sudan and Egypt that had regarded development in the Upper Nile to be the sole responsibility of the PJTC, but it was not implemented. The Sudan and Egypt continued to press for a Nile basin commission, however, and in March 1981, at a meeting of the irrigation ministers of the basin states in Khartoum, President Numayri argued without success for an agreement for the development of the Nile basin as a whole.[189]

Meanwhile, the Sudan was a member of Undugu,[190] a group including the governments of the Sudan, Egypt, Uganda, the Central African Republic, Zaire, and (later) Rwanda and Burundi. It was established to coordinate policies on many issues, among them and most importantly the Nile waters. Kenya, Tanzania and Ethiopia have not joined the group.[191] The Sudan was also a member of the Technical Committee for the Promotion of the Development and Environmental Protection of the Nile Basin (TECCONILE), formed in 1993.

In a game dominated by three players – Egypt, Ethiopia and the Sudan – the Sudan is the 'master of the middle', occupying the vital central position. But that position is not entirely advantageous. Egypt has a longstanding interest in developing projects in the White Nile basin and counts on the Sudan's cooperation for the realization of these objectives. The Sudan, with its tremendous potential for irrigated agriculture, is interested in exploiting this potential on a much greater scale; its priority is to pursue water-development projects on the Blue Nile, which would mean cooperating with Ethiopia.[192] If optimal water development were the overriding objective of Sudanese and Ethiopian policy-makers, they would have achieved close technical cooperation and joint projects decades ago.[193]

The first time the Sudan acted against Egypt's interests in this regard was when she signed with Ethiopia a statement of principles in 1991 over equitable use of the Blue Nile and Atbara. Nothing tangible developed from this statement, which was made at a time of deteriorating relations between the Sudan and Egypt. Nevertheless, it has been taken seriously by Egypt, which is apprehensive about ties between the Sudan and Ethiopia. Instead of making an issue out of it, however, Egypt adopted the policy of cooperation between the countries of the 'Eastern Nile'. The Sudan could have a dominant role in such a relationship if sound irrigation policy is adopted so as to exhaust her share in the Nile waters. This has yet to be done. The Sudan, with abundant water resources and good lands that could relatively easily be brought under irrigation, has tremendous potential to become a major agricultural producer and exporter.[194]

Negotiations with Ethiopia showed that the Sudan has a strong hand in relations with Egypt. The ultimate nightmare for Egypt might be for Ethiopia and the Sudan to overcome their differences and examine coolly their shared interests in joint development of their watershed in the Blue Nile, Atbara and Sobat basins.[195] In a recent interview with the *Sahafa* newspaper, the Sudanese director of the technical unit in the ministry of irrigation mentioned that the Sudan had no alternative to cooperation with Ethiopia in water projects in order to overcome the problem of sedimentation.[196]

Finally, the Sudan is a member of the Nile Basin Initiative formed in 1999. The NBI is meant to function only until an all-basin framework agreement is signed. But the Sudan shares Egypt's view of maintaining the status quo defined by the 1959 agreement: in the September 2004 Draft Framework Agreement the Sudan and Egypt entered their reservation that 'The principles and framework are without prejudice to existing agreements'.[197] As one expert puts it: 'Notwithstanding the ups and downs in other areas of Egyptian-Sudanese relations, the two nations have joined forces in all matters concerning the status quo with regard to the Nile waters. This is also evident in the ongoing Nile Basin Initiative process. Throughout this process Egypt and Sudan strongly uphold the "existing agreements". The seven upstream nations, however, insist that a new agreement be reached to supersede the already existing agreements, which do not take into consideration the rights and interests of all the riparian nations.'[198]

# 10

# Egypt and Collective Action Mechanisms in the Nile Basin

**Hosam E. Rabie Elemam**

This chapter takes as a starting point a disagreement with the dominant research tradition about the Nile basin's future. First, it is acknowledged that a combination of factors, including increasing water scarcity, diminishing water quality, rapid population growth and unilateral water development, has led academics and policy-makers alike to warn of impending conflict over shared water resources. A review of the literature on worldwide water cooperation indicates that in the early 1990s a debate[1] raged over the likelihood that water scarcity would lead to armed conflict, with the Nile basin frequently referred to as a case in point.

Despite all warnings, there has been no such conflict. On the contrary, efforts to cooperate on utilization of water resources have been growing noticeably, though the potential for conflict remains. Political conditions in the Nile basin countries have provided a window of opportunity for further progress in the cooperative development of the shared waters. This attitude should be strongly supported at all levels. Bearing in mind that states invoke cooperation and collective action, however, as the ideal way to achieve their common goals through agreed methods, this chapter will address the actual water situation in Egypt, in terms of the National Water Plan and its implications for cooperation with the other basin states. Attention will be paid to the mutual benefits that may be gained through the collective action process, and to the performance of past and current collective mechanisms in the basin.

## EGYPT'S CURRENT WATER SITUATION

Egypt has been listed among the ten countries threatened by want of water by the year 2025 owing to a rapidly increasing population.[2] Current indicators confirm that Egypt is facing increasing water needs, demanded not only by population growth but also by increased urbanization, higher standards of living and by an agricultural policy emphasizing expanded production in order to meet increasing demand.[3] Meanwhile, the per capita share of water is continuously declining: the present share is below 1,000 cm/capita/year, a figure that, according to international

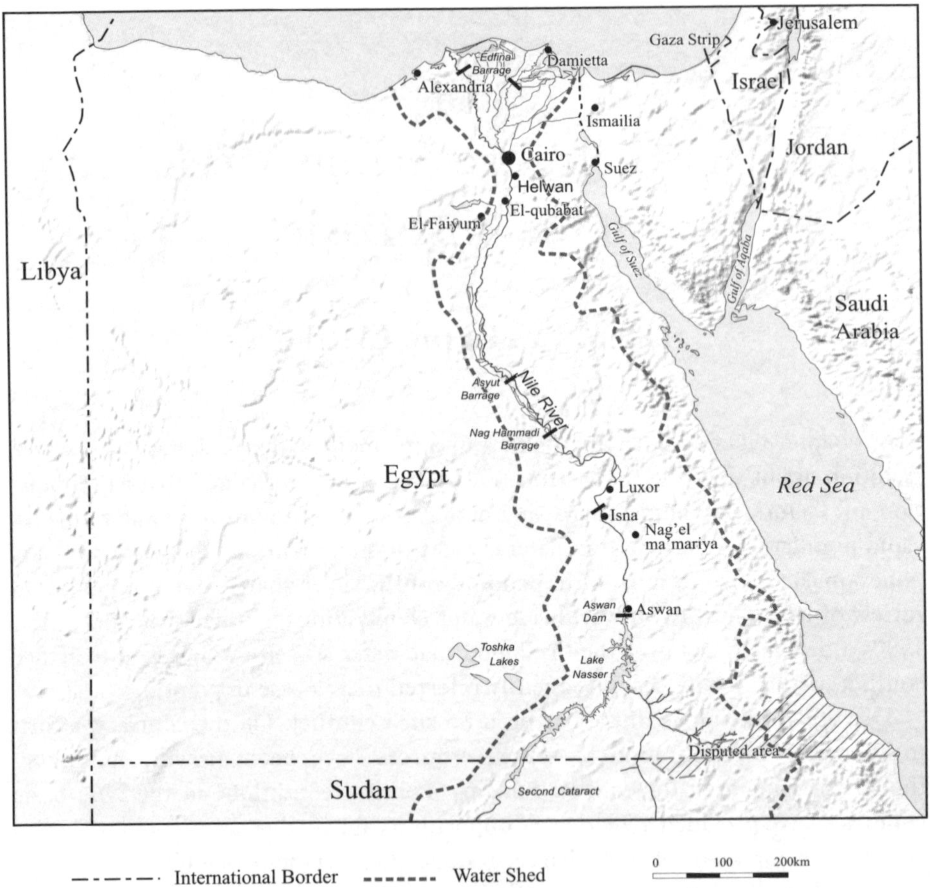


**Map 1:** Egypt.

standards, equals the water poverty limit for a nation; this value might drop to 500 cm/capita/year by 2025, which would indicate 'water scarcity'. In terms of water quality, the available data indicate rapid degradation in surface and groundwater.[4]

In the near future, water available from both conventional and non-conventional sources will be insufficient to satisfy increasing demand. Efficient and effective use of all water resources therefore requires formulation and implementation of appropriate water sector policies. This will in turn require a major shift from the classical paradigm used in water resource planning and management to a different and innovative one.

Egypt developed the National Water Resources Plan (NWRP) in January 2005, in keeping with current global thinking on how to face the dual challenge posed by current water-resource problems and the Integrated Water Resources Management (IWRM) approach. Assuming that Egypt's water future was uncertain, the plan developed various scenarios in order to devise the best strategy.[5] The holistic approach of the IWRM[6] requires water managers to look beyond the physical aspects of the water system, and to take into account the users and uses of water and the institutions involved – in other words, considering users and institutions as part of the water resources system.

We will therefore explore Egypt's available water resources, ways of improving their use, and the possibility of additional resources, in order to answer three questions. Are the current available water resources able to meet increased demand? Are there possibilities for additional resources and, if so, to what extent? And, in light of the answers to those questions, what are the implications for regional cooperation as a starting point for further research?

## THE NILE

Egypt relies completely on the water storage of Lake Nasser to sustain its annual share of Nile water, fixed at 55.5 billion cubic meters (bcm) annually under terms of the 1959 Nile Waters Agreement with the Sudan. The river in Egypt, where there is practically no rainfall, is the backbone of all activities, as has been realized since ancient times.[7] Egypt receives about 98 per cent of its fresh water from the Nile, which satisfies more than 95 per cent of the country's various water requirements. In comparison with four other critical regions of the world (China, South Asia, Southeast Asia and West Africa), the Nile basin has by far the greatest scarcity of water.[8]

Although it is considered one of the major rivers in the world, and taking into account its huge annual flow, it may be said that the Nile has not yet undergone extensive development. A major feature of the Nile basin is the swamps of the *sudd*, where much water is lost through evaporation: it is estimated that the outflow from the swamps is only about half the inflow. The large volume of evaporation from these swamps gave rise to the proposal in 1904 for construction of the Jonglei Canal that would bypass the swamps and reduce these losses. Construction began only in the 1970s, however, and was abandoned because of civil war and political instability in the Sudan.

With no significant rainfall except in a narrow strip along the Mediterranean coast, agriculture depends on irrigation from the Nile. The requirements of the agricultural sector represent the largest component of the total water demand in Egypt, consuming more than 85 per cent of available water resources. In 1861 two barrages were constructed at the apex of the Delta across the Damietta and Rosetta branches to provide water for perennial irrigation. Other barrages were built for further development and expansion of the cultivated area using perennial irrigation: the original Aswan dam, which was completed in 1902 with a storage volume of about 1 bcm, was heightened in 1934 to increase its storage capacity to 5 bcm, and a series of barrages were added between Aswan and the Mediterranean.[9] Storage was unable to meet increasing demand, and in 1971 the idea of over-year storage was finally realized with the completion of the High Dam at Aswan, the first of a series of projects for Nile control and development.

Horizontal expansion into the desert is a must, on the one hand to open new lands for agriculture, and on the other hand to accommodate the growing population. Egypt is implementing two national land reclamation projects: the North Sinai Development Project, 2,510 $km^2$ (620,000 acres); and the 'Toshka' South of Egypt Development Project. The latter aims to add 2,185 $km^2$ (540,000 acres) of arable land through irrigation from a new canal; create a base for the development of the surrounding area; build sustainable communities to accommodate some 3 million people by 2017; establish livestock, poultry and fish breeding projects to meet local and export demands; construct or upgrade transportation systems; and promote tourism.[10]

### *Water quality*

Management of water quality, control of water pollution, and environmental protection are major factors in preserving living conditions for the future. Water quality depends directly on water quantity. In Egypt, the quality of Nile water is affected by agricultural drainage, which contains salts, nutrients, pesticides, herbicides and industrial and municipal effluents from all the towns and villages of upper Egypt that drain either directly or indirectly into the river.

Although the quality of groundwater in the Nile system is still generally fairly good, pollution in some shallow groundwater bodies has reduced its suitability for raw drinking water; at present about 20 per cent of the groundwater in the Nile aquifer does not meet the standard for drinking water production. Especially in the fringes of the Nile valley and delta, where there is no protective clay cap, groundwater is highly vulnerable to pollution.[11]

### *Population growth*

Demand for water in Egypt is ever-increasing as a result of population growth and increased economic activities, as well as of improving standards of living. The population has grown from 38 million in 1977 to 66 million in 2002 and is expected to reach 83 million by 2017. Moreover, the population is concentrated in the Nile valley and the delta: 97 per cent of the population lives on only 4 per cent of the total land area. In terms of water resources, the situation is critical, with less than 900 cm available per capita/year for all purposes today, which will decrease to 670 cm/capita/year by 2017 if no additional water is made available.

In 1966, Egypt established a national family planning program, which aimed at reducing fertility and thus population growth. In 1995 a national population policy was articulated that recognized four interrelated dimensions of Egypt's population: growth, spatial distribution, characteristics and structure. Since population growth is the main reason that Egypt has reached its limits with respect to the availability of water from the Nile, further substantial growth in the population, in particular in rural areas, will be very difficult to cope with.

## DEVELOPMENT OF ADDITIONAL RESOURCES

### *Nile water*

Regional cooperation among riparian countries promises the most scope for increasing the Nile water supply. In the meantime, plans should be made and measures taken to develop other potential resources such as groundwater, rainfall and desalination.

The NWRP has emphasized that options for increasing the amount of Nile water are mainly related to upstream water conservation schemes in the White Nile basin, where a large amount of water is lost through evaporation. Reclaiming swampy areas on a smaller scale in different parts of the basin may result in less pronounced environmental impacts, while at the same time creating opportunities for agricultural development.

On the national level, enhancing the profitability of every cubic meter of water is a must. In this respect, several measures may to be taken: for instance, saving water through modern means of irrigation such as dripping and sprinkling; reduction of water wasted in the irrigation network by covering the sub-canals, at least, and fighting water weeds; proper management of river supplies; and cost/benefit analysis of yields in proportion to consumption.

### *Groundwater*

The major groundwater systems in Egypt are the Nile aquifer, the Nubian sandstone aquifer, the fissured carbonate aquifer, the Moghra aquifer, the coastal aquifer and the hard rock aquifer. Use of this water is very limited, and it subsequently contributes very little to the total water supply of Egypt. The Nile aquifer is the most important (about 87 per cent of the country's total groundwater abstraction), but since this aquifer is recharged by infiltration of excess irrigation water, and since the source of this irrigation water in Nile water is released at Aswan, the water in this aquifer is not considered a separate resource. This yield is therefore a reservoir in the Nile river system, with about 7.5 bcm/yr of rechargeable live storage.[12] The total available storage of the Nile aquifer has been estimated at about 500 bcm, but the maximum renewable amount (the aquifer's safe yield) at only 7.5 bcm. The existing rate of groundwater abstraction in the valley and delta regions is about 4.5 bcm/year, which is thus still below the potential safe yield of the aquifer.[13]

Groundwater occurs also in the western desert, in the Nubian sandstone aquifer that extends below the vast area of the New Valley governorate and the region east of Owaynat. The total volume of fresh water stored in that aquifer has been the subject of many studies and probably exceeds 150,000 bcm. This value is of mainly academic interest, however, since large-scale development is unviable owing to the great depth of the groundwater table (up to 2,000 m). In addition, this aquifer is generally non-renewable, and use of its water depends on pumping costs and depletion rate versus the potential long-term economic return. Groundwater in Sinai is mainly encountered in three water-bearing formations: the shallow aquifers of the northern peninsula; valley aquifers; and deep aquifers. Although most of the shallow aquifers are renewable, only 10–20 per cent of the deep aquifers are renewable by rainfall and flash floods.

### *Rainfall and flash floods*

Rainfall in Egypt is very scarce except along a narrow band of the northern Mediterranean coast, where an insignificant rain-fed agriculture is practiced. Rainfall occurs in winter in the form of scattered showers with a total amount that may reach 1.5 bcm/year. This amount cannot be considered a reliable source of water owing to its spatial and temporal variability.

Flash floods from brief heavy storms are a source of environmental damage, especially in the Red Sea area and southern Sinai. Many studies have been made to determine possible measures to avoid hazards caused by flash floods. Mechanisms have also been developed to harvest flash-flood water. This water could be used to meet part of the country's water requirements or it could be used to recharge the shallow groundwater aquifers. It is estimated that about 1 bcm of water on average can be utilized annually.[14]

### *Desalination of seawater*

Desalination of seawater has been given low priority because the cost of treating seawater is high compared with that of other sources, even unconventional ones such as drainage reuse. The average cost of desalination of one cubic meter of seawater is 3–7 Egyptian pounds. In spite of this, it is sometimes feasible to use this method to provide domestic water, especially in remote areas where the cost of constructing pipelines to transfer Nile water is relatively high. Desalinized water is therefore being used now to provide domestic supply for some locations along the Mediterranean and the Red Sea coasts, as well as in Sinai.[15]

The future use of such resources for other purposes (agriculture and industry) will largely depend on the rate of improvement in the technologies used for desalination and the cost of needed power. Nevertheless, brackish groundwater with a salinity of about 10,000 parts per million can be desalinated at a reasonable cost for potential use in agriculture. The amount of desalinated water used in Egypt now is in the order of 0.03 bcm/year.[16]

### *Unconventional resources*

Other sources can be used to meet part of Egypt's water requirements. These unconventional sources cannot be considered independent and consequently cannot be counted as part of Egypt's freshwater resources.

### *Reuse of agricultural drainage*

The amount of water that returns to drains from irrigated land is relatively high (about 25–30 per cent). This drainage flow comes from different sources: tail end and seepage losses from canals; surface runoff from irrigated fields; and deep percolation from irrigated fields (partially required for leaching salt). Of course, none of these sources is independent of the Nile; the first two are considered to be fresh water of relatively good quality. Agricultural drainage in the southern part of Egypt returns directly to the Nile, where it is mixed automatically with fresh Nile water that can be used for different purposes downstream. The total amount of such direct reuse was estimated at 4.07 bcm/year in 1995–96. In addition, it is estimated that 0.65 bcm/year of drainage water is pumped to the Ibrahimia and Bahr Youssef canals for further reuse. Another 0.235 bcm/year of drainage water is reused in the Fayoum, while about 0.65 bcm/year from the Fayoum is drained to Lake Qarun. Moreover, pumping stations lift about 0.60 bcm/year of drainage to the Rosetta branch of the Nile just downstream from the delta barrages for further downstream reuse.[17]

Drainage water in the delta region is emptied into the sea and the northern lakes via pump stations. The amount of drainage water pumped to the sea was estimated in 1995–96 at 12.41 bcm. This has decreased and will continue to decrease in the future to the extent that drainage water is developed and reused.[18]

### *Reuse of treated wastewater*

Increasing demands for domestic water owing to population growth, improvement in living standards and the growing use of water in the industrial sector will of course increase the total amount of wastewater available for reuse. Wastewater treatment could thus become an important source of water and should be considered in any new water resource development policy. However, proper attention must be paid to the issues associated with such reuse. These include public health and environmental hazards, as well as technical, institutional, sustainability and even socio-cultural challenges.[19]

There are few wastewater treatment plants in large cities where wastewater collection networks exist. About 1 bcm of primary treated water is used in irrigation in specific locations outside the greater Cairo region. It is expected that in the near future this volume of treated wastewater will reach 2 bcm.[20]

### *Domestic and municipal*

Municipal water demand is small compared to agricultural water demand, but the relation between public health and the quality of municipal water creates a priority for this use over all other uses. Municipal water demand includes the water supply for major urban areas (216 cities) and rural villages (some 4,525), and is estimated at about 5.5 bcm/year. Some of that water comes from the Nile system and some from groundwater sources. A small portion of the water diverted (about 1 bcm) is actually consumed, while the remainder returns to the system in the form of polluted sewage water.[21]

## COOPERATION WITH BASIN STATES

Our point of departure here is that international water resources actually tend to induce cooperation. In the largest quantitative study of water conflict and cooperation, researchers at Oregon State University found that cooperative interactions between riparian states since 1948 had outnumbered conflict interactions by more than two-to-one. Over the same period, the historical record documents 37 incidents of conflict over water,[22] while approximately 295 international water agreements were negotiated and signed.[23]

Cooperation in the Nile basin has become essential. For downstream countries, conserving water and harvesting water wasted in swamps, for instance, might be seen as the most important issue. For upstream countries, modern facilities and technical expertise would enable them to use their available water resources more efficiently. Egypt has an important role to play as a major downstream state and as an important provider of the technical abilities that have accrued through unrivalled experience in the relevant fields.

### *Egyptian technical experience*

The Ministry of Water Resources and Irrigation (MWRI) is the main agency responsible for water resources in Egypt. It controls and manages all freshwater resources and is in charge of construction, supervision, operation and maintenance of all irrigation structures and drainage networks. The ministry is also responsible for meeting the freshwater needs of all other sectors.

The long history of the ministry – it was founded in 1836[24] – has resulted in a highly sophisticated water structure in Egypt, with a variety of facilities, experiences and human resources in various government departments and economic sectors. With respect to technical expertise and capacity building, the National Water Research Center (NWRC)[25] and its 12 research institutes are mandated to implement a comprehensive research plan serving continuing projects and national development plans in general.[26] The NWRC also has a Strategic Research Unit, a Central Laboratory for Environmental Quality Monitoring, a Geographic Information System, an Information/Documentation Center, and a highly specialized Central Library. With its wide experience the NWRC provides technical consultation for all governmental agencies and the private sector, and also plays a remarkable role in capacity building through professional training for water resource specialists in Egypt and other countries.[27]

The NWRC has carried out applied research and technology transfer to address the diverse problems that face the water sector. Research results and studies related to integrated water management have helped, among other things, in identifying the quantity and quality of drainage water in Egypt that could be reused safely for irrigation; in evaluation and adoption of improved irrigation systems for old and newly reclaimed areas; and in the hydraulic design and stability of the new Esna barrage. In addressing such problems a multidisciplinary approach has been adopted to ensure comprehensiveness and successful implementation. Conversely, local and foreign universities and companies have been consulted where relevant and needed. To support the decision-making and management processes, new technology has been widely adopted along with mathematical models and decision support systems.

On the international level, NWRC acts as the coordinating unit for the African Water Resources Network and is a member of other international water resource networks. NWRC has also established relations with several international organizations and bodies. Programs involve technology transfer, training, employee exchange, conferences and seminars, field trips and similar activities. For example, the Regional Center For Training and Water Studies (RCTWS) was established to provide teaching and training possibilities and recreational facilities.[28] The center has developed systematic procedures to promote professional capabilities for participants from Africa and the Arab and Mediterranean countries in the fields of irrigation, drainage, water resources development and management.

During a joint meeting of the Nile basin countries' ministers of water resources with donor countries and agencies in Geneva in June 2001, Egypt was selected to host the Nile Basin Training and Capacity Building Shared Vision program for the basin countries. And the RCTWS was chosen to host that project's regional office.

### *Egypt's activities in the basin*

The most famous recent example of Egyptian activity in the Nile basin is the Uganda–Egypt aquatic weed control project (Ug–Eg AWCP). This project had its origin at the end of 1997, when heavy rainfalls in the equatorial regions increased the level of Lake Victoria by as much as 1.5 m, equivalent to more than 100 bcm of runoff. The additional water volume led to a high discharge into the Victoria Nile and consequently high levels in Lake Kyoga, where large quantities of water hyacinth and floating mats of papyrus caused huge blockages at the lake outlet. This led in turn to the inundation of all the communities along the shores of Lake Kyoga, and to outbreaks of epidemic diseases such as typhoid and cholera. The Ugandan government was under pressure to find a quick solution to the problem and invited all countries and donors to offer help.[29]

Egypt responded to this invitation and in August 1998 delegated a team of experts from the ministry of water resources and irrigation to evaluate the situation and make recommendations. The experts found that the blockage of the lake's outlet had been caused not only by water hyacinths, but also by an accumulation of densely vegetated floating masses of giant papyrus plants. The government of Egypt allocated $13.9 million as a grant to Uganda, of which $8.3 million would supply equipment and $5.6 million would pay for three years of machine operation. The project activities included a training program for Ugandan engineers and technicians and environmental studies of Lake Kyoga.[30]

This project has enhanced the environmental, social, economic and health conditions of people living along the shores of Lake Kyoga. Rehabilitation helped fishermen to bring their fresh fish to market, resulting in increasing production and subsequent enhancement of the general regional economy; fishing is Uganda's second source of national income, on which more than 5 million people depend directly or indirectly. Removal of floating aquatic weeds in Lakes Victoria, Kyoga and Albert also resulted in much improved navigation, and curtailed the spread of disease. For future reference, a detailed manual was prepared for operating the project equipment, with full descriptions of all machines, their technical specifications, and methods of operating each unit for maximum efficiency. Finally, removing the floating weeds led to the establishment of experimental farms along the shores of the lake, using the project equipment.[31]

Several other cooperative projects are of interest. In 1996, Egypt granted $4.2 million to Kenya to excavate 100 groundwater wells in arid and semi-arid zones; discussions are under way between the two governments with regard to the implementation of another phase of this project.[32] In terms of capacity building, the International Postgraduate Diploma on shared water resources was established in the department of irrigation and hydraulics of Cairo University, to train water resource professionals from the Nile basin countries. The course covers relevant engineering, political, geographical, socio-economic and environmental aspects of the subject.

### *Collective action mechanisms in the basin*

The historical record indicates an increased likelihood of conflict in basins lacking institutions that can accommodate changing political, hydrologic or other conditions; where international water institutions exist, relations among riparian states become more cooperative than in basins without treaties or other cooperative management mechanisms.

If a cooperative management mechanism is to be adopted by riparian states to provide opportunities for mutual benefit, some tools are needed. These include water conservation programs, appropriate education and public awareness, and watershed and flood management. Environmental diplomacy can be effective in facilitating dialog, but changes in institutional and public attitude and behavior should be considered a cornerstone by making water an issue for everyone, and by establishing mechanisms to support consensus building on shared objectives for sustainable development.[33]

A great strength of cooperative water regimes is that, once established, they tend to be resilient even in conflict situations. It may be, however, that investment in enforcement mechanisms is necessary to enhance interaction and ensure fruitful and stable cooperation; while interaction can be a path toward establishing such enforcement mechanism, it cannot take its place.[34]

In Africa, the idea of collective action mechanisms was launched at the meeting of the heads of state and government in Lagos on 28–29 April 1980. In terms of water resources, recommendations were adopted in line with the Action Plan of the United Nations Water Conference of March 1977. Those concentrated on institutional strengthening as a problem that could be approached on three levels: the national (where National Water Committees should be established in countries that had not yet done so); the sub-regional level (where existing sub-regional organizations such as river and lake basin commissions should be strengthened); and the regional level (in the form of an Inter-governmental Committee on Water for the African Region).[35]

In the Nile basin, however, the collective action process had started much earlier. Members of the Permanent Joint Technical Commission for the Nile waters (PJTC) met representatives of Tanganyika, Uganda and Kenya in October 1961 to discuss the establishment of a project for the hydro-meteorological survey of the catchments of Lakes Victoria, Kyoga and Albert. Following an initial study in 1963, five of the riparian countries (Egypt, Kenya, the Sudan, Tanzania and Uganda) requested the cooperation of the UNDP in a hydro-meteorological survey to study the water balance of Lakes Victoria and Albert. The participating governments signed the plan of operation in May 1967, and the project officially commenced work in August.

Hydromet was a highly successful example of technical cooperation among eight countries. Since the potential control and regulation of the Nile has a direct bearing on the economic development of all the riparian countries, it was evident to these countries that a high priority must be placed on the collection of hydro-meteorological data and an investigation of the meteorology, hydrology and hydraulics of the Upper Nile basin. The long-term objectives of Hydromet were to assist the participating governments in the planning of conservation and development of the water resources of the Upper Nile, and to provide the groundwork for inter-governmental cooperation on storage, regulation and water use.

As work in the Upper Nile basin progressed, the participating countries and governments of Rwanda, Burundi and the Congo (Zaire) requested UN cooperation to extend the project area to include the portion of the Lake Victoria catchment within Rwanda and Burundi, and that of Lake Albert in the Congo. By the end of 1971 the government of Ethiopia had joined the project with observer status. A considerable amount of hydro-meteorological data was collected, with photography, ground surveys and hydrographic surveys. The training of staff from participating countries was also a major activity of the project.[36]

Egypt and the Sudan, motivated by the 1980 Lagos action plan, meanwhile began a diplomatic offensive for the establishment of basin-wide cooperation comprising all riparian countries. Those efforts resulted in a meeting in Kinshasa on 3–4 September 1984, and the creation of the group known as Undugu,[37] consisting of Egypt, the Sudan, Uganda, Zaire and the Central African Republic; Ethiopia, Kenya and Tanzania never joined the group, but Rwanda, Burundi and Tanzania assumed observer status at this or subsequent meetings: Rwanda became a member in 1988, Burundi in 1990. Although Ethiopia never joined, it supported the work and even submitted a framework of cooperation for the basin states for the Undugu meeting held in Addis Ababa in May 1992.

The objectives of Undugu were to create a mechanism for exchanging information and points of view on common issues in the basin, and to focus the attention of the people of the basin on the importance of the Nile water resources and, consequently, to achieve socio-economic development.

After repeated invitations, the UNDP sent a preliminary mission and later a fact-finding mission to the basin states to study possibilities of development. The outcome of these missions was the most important achievement of Undugu.[38] The preliminary mission visited the region in February–March 1989 and presented a comprehensive report for a long-term development and cooperation plan. The fact-finding mission visited in May–June and evaluated available water resources, hydropower production and water demand.[39]

A number of international meetings took place in response to both political changes within the basin states and a growing awareness that future development options would require more strategic and multi-sector thinking. In December 1992 the water resources ministers of the Congo, Egypt, Rwanda, the Sudan, Tanzania and Uganda met in Kampala and agreed to fold Hydromet into a Committee for Promotion of Development and Environmental Protection of the Nile Basin (TECCONILE), in order to coordinate the common interests and objectives of the basin countries. The other four riparian countries participated as observers and eventually became members (with the exception of Eritrea). A council of ministers of water affairs was formed, with a technical group acting as a steering committee.

TECCONILE began to function in 1993. At the third ministerial meeting, at Arusha in February 1995, the project was extended for another three years (to December 1998), with tasks that included establishment of a data base, studying national water-use plans and shared projects, and preparation of the cooperative and institutional framework.

Under the umbrella of TECCONILE, the Nile 2002 conference series was initiated in February 1992. The first meeting was arranged a year later at Aswan, and meetings were held annually thereafter in one or another of the basin states. The original purpose of the Nile 2002 conferences was to provide an informal mechanism for riparian dialog and the exchange of views between countries and with the international community.

### *The Nile Basin Initiative*

At the second Nile Technical Advisory committee meeting held at Arusha in September 1998, an agreement was reached on a plan of action and policy guidelines for the establishment of the Nile Basin Initiative. In September 1999 the NBI secretariat, which superseded the disbanded TECCONILE, was officially opened at Entebbe. The NBI aimed to develop water resources of the Nile basin in a sustainable and equitable way to ensure prosperity, security and peace for all its peoples; to ensure efficient water management and the optimal use of resources; to ensure cooperation and joint action between the riparian countries for their mutual benefit; and to eradicate poverty and promote economic integration. The NBI for the first time brought together all the riparian countries, and has taken significant steps to implement a concrete strategy for sustainable use and development of this shared resource for the benefit of all basin states.

The question of whether collective action in the Nile basin should focus on a sub-basin approach or on basin-wide cooperation has been the subject of much discussion. Support for the sub-basin model assumes limits to what could be achieved on a basin-wide level without first laying a firm sub-regional foundation. Thus a sub-basin model would be more appropriate and effective because it would defuse existing tensions without jeopardizing basin-wide cooperation. It might be more useful and productive to establish a regional regime through intermediate steps, undertaking such policy initiatives as any riparian state would wish to adopt even without international cooperation.[40]

Others argue – more persuasively, in our view – that attainment of basin-wide cooperation is more likely to be achieved now than later. They envisage establishment of sub-basin arrangements within a wider basin framework, as in the case of the subsidiary action programs of the NBI. In this view, what has made the NBI process and its achievements possible so far is the fact that the riparian states have managed to reach agreement on a basin-wide set of principles that legitimizes particular projects. In other words, it was not possible to promote 'action on the ground' or sub-basin activities without a basin-wide framework of principles.

## THE SHARED VISION PROGRAM

The primary purpose of the Shared Vision Program is to create an enabling environment for cooperative management and development in the Nile basin through a limited, but effective, set of basin-wide activities and projects. These projects address the major water-related sectors and themes deemed critical by the Nile riparian states to ensure an integrated and comprehensive approach to water resources development and management, and serve as catalysts for broader socio-economic development. The projects are the Nile Trans-boundary Environmental Action program, which is hosted by the Sudan; the Nile Basin Regional Power

Trade (Tanzania); the Efficient Water Use for Agricultural Production (Kenya); the Water Resources Planning and Management (Ethiopia); the Confidence Building and Stakeholder Involvement, and the Socio-economic Development and Benefit Sharing projects, hosted by the NBI secretary at Entebbe; and the Applied Training Project in Egypt.[41]

## THE SUBSIDIARY ACTION PROGRAM (SAP)

The success of the NBI, by far, lies in one of its institutional innovations, namely the application of the principle of subsidiarity, or management of the basin on the lowest appropriate level. This has led to institutional division into an Eastern Nile Subsidiary Action Program (ENSAP)[42] comprising Ethiopia, the Sudan and Egypt (and Eritrea too, were it to formalize its participation), and the Nile Equatorial Lakes Subsidiary Action Program (NELSAP)[43] comprising Kenya, Uganda, Tanzania, the Congo (Kinshasa), Rwanda and Burundi, as well as Egypt and the Sudan. The basic rationale is that by reducing decision-making complexity the process of cooperation can be facilitated.

SAP activities encompass many areas: river regulation; water harvesting and conservation; hydropower generation; irrigated food production; watershed management and soil erosion control; reduction of evaporation losses from swamps; fisheries development; transport and navigation development; eco-tourism development; weed control; waste-water treatment, pollution control and water quality management; and water-use efficiency improvement.

## THE COOPERATIVE FRAMEWORK

One of the projects prioritized in the Action Plan aims to establish a Nile Basin Cooperative Framework. This project was initiated in 1996 with UNDP funding, to enable the riparian countries to establish a forum for a process of legal and institutional dialog aimed at reaching agreement on core legal principles and institutional arrangements. A draft cooperative framework was produced in early 2000, encompassing general principles, rights and obligations, and an institutional structure. Negotiations over the draft moved the riparian states toward important compromises, but some issues remain to be resolved and the process is continuing.[44]

## POSSIBLE PROBLEMS

A prominent commentator has argued that while Egypt has tried to build an agenda around 'process' issues, Ethiopia has stressed principles. To be sure, Egypt as the main beneficiary of the status quo has proposed processes by which the needs of all riparian countries can be identified and expertise and resources can be mustered to meet them. To critics, the focus on process suggests problems arising from the two-state regime constructed on the basis of the 1959 Nile Waters Agreement between Egypt and the Sudan. For Egypt, in this view, cooperation means sharing in basin-wide efforts to enhance the flow of the Nile and its tributaries, so that any needs of the upper basin states can be met from the incremental gains from enhanced flow. Should Egypt adopt the same logic enshrined in the 1959 agreement, by which the Sudan received two-thirds of incremental gains?[45]

Considering the potential of the Nile basin and total annual rainfall estimated at 1,686 bcm per year, we may estimate that only about 5 per cent of the total amount is currently in use. Thus, the 'process' approach may be the most logical for cooperation among the basin states. The issue of how to share benefits should not be a critical problem, especially if the basin states agree on the factors of equitable and reasonable utilization mentioned in the United Nations Convention of 1997 (taking into account that those factors are subject to addition or deletion, in terms of the specific characteristics of each basin).

Moreover, the notion of allocating incremental gains according to the logic enshrined in the 1959 agreement is vague. First, the expected allocation of incremental gains today would be among ten countries; that is, the needs of riparian states will differ, and some countries may not ask for water but for hydropower. So efforts related to benefit sharing should be concentrated on the real needs of each party.

We argue that it is no longer valid to say that Egypt is the main beneficiary of the status quo. Even the Egyptian National Water Plan itself accepts that the current system is unable to meet future needs. The status quo, as mentioned in this plan, includes not only the 55.5 bcm Nile quota, but also all the available water resources in Egypt.

A second question arises from discussion of collective action in the utilization of shared freshwater: to what extent does existing law contribute to optimal and equitable utilization? That standard, adopted in the 1997 UN convention, implies attaining maximum possible benefits for all watercourse states and achieving the greatest possible satisfaction of all their needs, while minimizing the detriment to, or unmet needs of, each.[46]

In terms of international law, cooperation is considered a basic principle. It is worth mentioning that the arena of international law includes a wide variety of international instruments calling for cooperation between parties with regard to utilization of international watercourses.[47] The importance of cooperation in relation to the utilization of international watercourses and other common natural resources has been emphasized repeatedly in declarations and resolutions adopted by intergovernmental organizations, conferences and meetings.[48] The European Water Charter of 1967, for instance, held that water knows no frontiers. As a common resource it demands international cooperation;[49] optimal and equitable utilization of shared water basins cannot be achieved unilaterally, and collective action mechanisms should be targeted to reach it.

Two famous initiatives adopted the model of cooperation in the utilization of common water resources and in the field of environmental protection: the Declaration of the Stockholm Conference on the Human Environment in 1972; and the Mar del Plata Action Plan,[50] adopted by the United Nations Water Conference in 1977. The latter contains a number of recommendations relating to regional and international cooperation with regard to the use and development of international watercourses. Recommendation 90, for example, provides that cooperation between states in the case of international watercourses, 'in accordance with the Charter of the United Nations and principles of international law, must be exercised on the basis of the equality, sovereignty and territorial integrity of all States, and taking due account of the principle expressed, inter alia, in principle 21 of the Declaration of the United Nations Conference on the Human Environment'.

An instrument expressly recognizing the importance of cooperation between states as to the effectiveness of procedural and other rules concerning international watercourses is the Rules on Water Pollution in an International Drainage Basin, adopted by the International Law Association in 1982.[51] Article 4 of the Rules state: 'In order to give effect to the provisions of these articles, States shall cooperate with the other States concerned.'

The UN Convention of 1997 on the Law of the Non-navigational Uses of International Watercourses adopted the principle as a general obligation. Article 8 indicates the basis of cooperation by referring to the most fundamental principles upon which cooperation between watercourse states is founded, such as sovereignty, equality, territorial integrity, mutual benefit and good faith. The objectives of cooperation were not raised in detail; the Commission came to the conclusion that a general formulation would be more appropriate, especially in view of the wide diversity of international watercourses and the uses thereof, and the needs of watercourse states.

Although these legal instruments are included under the umbrella of international law, in some cases states cannot reach the desired level of cooperation. Indeed, it has been argued that international law regarding utilization of natural resources shared by two or more states is a controversial area partly because of the value of natural resources in national policy, and partly because states often invoke legal rules and principles only when these support their own interests.[52] This raises the question of whether cooperation is limited to declarations and conventions, or can succeed in dealing with competing national interests.

If states turned only to established principles of international law, upstream countries would tend to adopt and apply the principle of absolute territorial sovereignty (the Harmon Doctrine), while downstream states defended their right of territorial integrity, invoking the 'no harm principle' that would limit the activities of upstream states.

There are, in fact, many legal principles and rules that can be applied to support both positions. How are differences to be resolved? It should be mentioned that most international water-related disputes have been resolved by negotiation rather than litigation.[53]

Negotiation offers parties an opportunity to exchange information about their domestic constraints and explore their differences in valuation, preferences, risk aversion, or other dimensions of a dispute. Through negotiation, parties will therefore be in a better position than third parties (such as judges or arbitrators) to reach an efficient, equitable and stable outcome. Another benefit of direct negotiation is the potential for establishing formal and informal channels for further communication, exchange of information and coordination of policies. These channels can build confidence in good faith, and establish lasting relations beyond the signing of an agreement. Third-party settlement, on the other hand, may leave the parties in an adversarial frame of mind that is not conducive to further cooperation. Moreover, negotiation experience creates an impetus to formalize interaction through institutions for the joint management of a disputed watercourse; litigation cannot impose long-term cooperation.[54]

If parties choose to negotiate rather than litigate a dispute, it should be noted that international law provides only legal frameworks, thus allowing the negotiating parties great latitude in determining and presenting their cases. Accordingly, they are able, for instance, to raise or ignore factors related to equitable utilization, or to modify periods of prior notification of planned measures, and so on, in line with the special characteristics of the waterway in question, in our case the Nile basin. In terms of the 1997 UN Convention, basic principles such as equitable and reasonable utilization, the obligation not to cause significant harm, regular exchange of data and information, provision of information concerning planned measures, and protection and preservation of ecosystems are open to detailed interpretation.

## THE 1959 NILE WATERS AGREEMENT AND INTERNATIONAL LAW

By far the most controversial issue in the Nile basin is the existing Nile Waters Agreement between Egypt and the Sudan. With the independence of the upstream countries, that and other agreements were denounced as invalid because they had been negotiated on behalf of – or without reference to – African dependencies, and were thus not binding on successor governments.

This position completely contradicts the rules and provisions of international law, in particular the principle of Succession of Treaties, which is governed by the 1978 Vienna convention on the succession of states in respect of treaties.[55] While independence may allow repudiation of existing treaties – under the *tabula rasa* or clean slate principle – an important exception to this principle, in terms of the Vienna convention, applies. Article 11 specifies that a succession of States does not affect a 'boundary established by a treaty or obligations and rights established by a treaty and relating to the regime of a boundary'. Moreover, article 12/2 stipulates that a succession of States does not affect 'obligations relating to the use of any territory, or to restrictions upon its use, established by a treaty for the benefit of a group of States or of all States and considered as attaching to that territory'; or 'rights established by a treaty for the benefit of a group of States or of all States and relating to the use of any territory, or to restrictions upon its use, and considered as attaching to that territory'.

Article 62 of the Vienna convention on the law of treaties[56] provides that newly independent states can repudiate existing treaties only because of a fundamental change of circumstances. As one observer has put it: 'Occasionally states seek to displace and explain away obvious contractual undertakings they have entered into[;] like businessmen who sometimes regret having entered into a contract when prices have gone up or down, states seek legal formulae to rid themselves of an obligation.'[57] This is the ideal refuge for the crafty statesman who wishes to prove that a state finds itself in circumstances different from those prevailing when the treaty was entered into. The stipulation that a change should have been fundamental does little to relieve the hazard of this escape mechanism since the state that wishes to denounce a treaty is apparently solely to decide what is fundamental. It has been suggested that this doctrine of *rebus sic stantibus*, dangerous as it is to the stability of international legal relations, should be replaced by a doctrine of continuous consent applicable for certain treaties, namely those that restrict territorial sovereignty in the area of a state.[58]

International law does not provide for suspension or denunciation of a treaty because it does not contain any provision for its own termination. The International Court of Justice, ruling in the famous Gabcikovo–Nagymaros dispute,[59] concluded that, regarding the determination of treaties, insofar as the treaty does not contain either provisions of determination, or an indication of the parties' intention to admit denunciation or withdrawal, it could be terminated only on the limited grounds enumerated in Articles 60–62 of the Vienna convention of 1969, which are declaratory of customary law.[60] In addition, the court explained that a distinction must be made between, on the one hand, a determination under the law of treaties of whether a treaty is in force and has been properly suspended or denounced, and, on the other hand, an evaluation under the law of state responsibility of the extent to which the suspension or denunciation of a treaty in violation of the law of treaties engages the responsibility of the state concerned. The law of state responsibility emphasizes the effects of an unlawful suspension or denunciation. Moreover, the court confirmed, regarding the succession of states in respect of treaties (by 12 votes to three), that Slovakia had became a party to the treaty as successor to Czechoslovakia upon the dissolution of the latter state on 1 January 1993, on the basis of the particular nature of the treaty. The treaty had established a territorial and navigational regime and created rights and obligations attaching to the relevant parts of the Danube, leaving it unaffected by a succession of states.[61]

This is the attitude of international law regarding the issue of existing agreements. Regarding the Nile basin states, then, the question is: if the parties chose to institute proceedings in The Hague, would the decision satisfy any of them? And would it help them to meet their needs or wishes?

The Nile water in Egypt is not only fully used, but it is also not enough to meet future needs. Since this water use dates back millennia, a distinction must be made as to whether the 1959 Nile Waters Agreement established any additional rights for Egypt, or simply stated its existing uses and rights in this water.

Establishment of the current institutional structures, under the umbrella of the NBI, creates an opportunity for the riparian states to rethink their positions. International law has recognized, and the International Court of Justice has ruled in favor of, the validity of agreements of the 1959 kind, and the principle that no party can terminate those agreements unilaterally, or even implement any provisional solutions along the river without a state of necessity. The Court has defined such necessity as imminent and grave peril.

Increasing water demand means that the status quo of the Nile water is unsustainable. All the riparian states, through the NBI, have begun to recognize the need for a comprehensive regime to regulate the Nile. Since downstream countries could be harmed by any upstream measures that might affect the quantity or quality of water or both, prior notification and consultation on such planned measures is a necessity. What, we might ask, would be the future of the Nile basin without the Nile Basin Initiative?

# 11
# Some Conceptual Issues Regarding the Study of Inter-state Relationships in River Basins

**Terje Tvedt**

The issue that will define the future of the River Nile and thus the region as a whole remains: under what circumstances would sovereign states – some water rich and other water poor, some upstream powers and other downstream powers – voluntarily agree to manage their shared water resources for the greater good of all the states and all the inhabitants of the basin?

The problem of the Nile is also related to the general question of how to share common resources in an optimal way. This question of global relevance has given impetus to an extensive literature and a number of influential analytical models for understanding collective-action problems in river basins all over the world.[1] We will discuss some of these models from a Nile basin perspective, and show that they neglect a crucial issue; they overlook the ways in which the specific physical characteristics of river basins frame human action, and how important it is to understand the relationship between water systems, river control, and social and economic development and development patterns.

## DOGMAS OF WAR, PEACE AND INTERNATIONAL WATERS

Since the 1980s – when Egypt's foreign minister, Boutros Boutros-Ghali, predicted that the next war in the Middle East would be fought over water, and the vice-president of the World Bank, Ismail Serageldin, said that 'the wars of the next century' would 'be about water' – a number of scenarios have foreseen water wars as a feature of geopolitical power struggles. Water scarcity is bound to lead to war, it is said, because water is absolutely essential to human life.[2]

The general 'theory' of 'water war' can be criticized on two grounds. Firstly, it can be described as a mechanical misunderstanding of the relations between resource scarcity and conflict, and between climatic change (predictions) and the form of social and political adaptations. Secondly, very few examples can substantiate the hypothesis and quite a number of cases can falsify it. But although the alarmist 'water war' dogma must be rejected, there have indeed been water wars in the past, including in the Nile basin, and armed conflict over how to control and share the Nile cannot be ruled out.

A competing theory has recently become popular: that instead of leading to war, water will fuel greater interdependence among states, and that 'water and river basins are [therefore] pathways to peace'.[3] Coming together to manage shared water resources will prevent conflict and build trust even in otherwise conflict-ridden areas. This reassuring argument refers to 'historical lessons': interstate war is unlikely, it is said, because for thousands of years there have in fact been no wars specifically over water resources, and the evidence shows that water interdependence does not lead to war. The real problem, according to this way of reasoning, is that the provocative rhetoric that politicians aim at their own constituencies can antagonize their neighbours.

There are at least three fundamental problems with such generalizations. They evoke 'historical lessons' that for historical–philosophical reasons cannot be drawn and that are therefore no guide for how states and others will act in the future. The past cannot in this way predict the future, especially at a time when unprecedented technological ability to control water has coincided with uncertainty about future climate and water landscapes, thus fuelling distrust and competition, and also underlining the need for cooperation. Problematically imprecise definitions of 'cooperation' and 'conflict' must be questioned when, if only owing to cost, initiatives to cooperate will always outnumber instances of war. But although water as a 'pathway to peace' is a simplistic slogan, the challenges of optimal water planning have brought some of the world's most implacable enemies to the negotiation table and led to agreements and institutions that survived strained relations.

For the Nile countries both options have been available and continue to be available, and the Nile issue will never be settled once and for all: the Nile waters might become a pathway to peace or a currency of war, or both, at different historical junctures.

## THE 'TRAGEDY OF THE COMMONS'

What guidance can general theories of optimal resource management give us for future developments? One very famous model invoked to explain behaviour and solve problems related to the management of shared resources is the one developed by Garret Hardin, often condensed as the 'tragedy of the commons'.[4] This theory is exemplified by a certain type of physical space: a pasture shared by herders, each of whom wishes to maximize his yield; each additional animal has both a positive and negative effect as the herder gets a higher return, but the pasture is degraded. By 'the remorseless working of things', the actions of self-interested individuals do not promote the public good. Hardin writes: 'Therein is the tragedy. Each man is locked into a system that [causes] him to increase his herd without a limit – in a world that is limited … Freedom in a commons brings ruin to all.'[5] Based on this simple idea, grand models of human action and social development have been formulated. To this very influential paradigm two main objections may, however, be raised: Hardin's general argument of the 'remorseless working of things' is not working the same way in relation to rivers and their waters as in relation to pastures.[6]

A pasture, as a physical space in nature and thus also as a resource, is fundamentally different from a river. The perceived model pasture is ecologically quite uniform, as are the economic activities, adaptation mechanisms and strategic choices of the actors sharing it. Most rivers, and even more so long rivers – in particular a river such as the Nile, which traverses three climatic zones from tropical Africa through the Sahara to the Mediterranean, one-tenth of the continent – are ecologically extremely varied, and lend themselves to differing strategic choices and economic adaptations at various points along their courses. Importantly, in such physical and social circumstances these need not conflict with those of other users, or the character of the conflict and the potential for cooperation and collective action might be different.[7]

The theoretical configuration associated with the pasture ecology will therefore not be reproduced in the context of a river basin, owing to its physiography, topography and the consequent unequal structural position of actors in relation to the resource. One riparian can use or even control the river to maximize yield without negatively impacting other users or the river itself. That Egypt, for example, exploited the river for thousands of years had no effect on the areas that make up present-day Kenya or Tanzania, and the country's extensive exploitation even since the 1970s has been of negligible importance for the lack of Nile development in many of these countries.[8] In river basins, moreover, pursuit of self-interest upstream can *benefit* downstream users. A hydroelectric dam, for example, might in certain hydrological settings protect downriver areas against flood and reduce silting, when that is considered a problem. The Roseires Dam in the Sudan would have had very negative effects in Egypt had it been built during those millennia when seasonal flood irrigation dominated Egyptian agriculture, but after 1971, when the Aswan Dam was put into operation, Roseires' benefits outweighed by far the disadvantages (it has trapped Blue Nile silt, thus protecting the Aswan reservoir against siltation) in Egypt. If Ethiopia decides to build more dams on one of the Blue Nile tributaries, the flow of the Nile might be reduced in Egypt, but it will also reduce the amount of silt in the river that is currently threatening the downstream reservoirs with destruction. These might be examples of how the natural domain and rights domain of resources alter with time, technological change, and the circumstances of stakeholders. Rivers thus fulfil different demands or needs at different places, and it is this natural character that can encourage cooperative action as rational individual behaviour.

Rivers, it should furthermore be recalled, unlike the pasture in Hardin's model, change (in some cases also dramatically) and will always do so, not only as a result of human interference but also as a consequence of changes in rainfall patterns, atmospheric pressure and so forth. Permanent insecurity and endless fluctuation guarantee that an individual (or state), acting rationally to maximize yield, will try to cooperate to achieve common control of the river because these changes, the scale of the uncertainty and the magnitude of the task make it impossible to do so alone. This physical world thus introduces social science theory and management models to a set of variables that should not be overlooked.

The 'tragedy of the commons' model fails to take account of power relations reflecting and stemming from what always has to be different positions within a physical location such as a river basin. Problematic in the case of the hypothetical pasture, this omission of power is unrealistic when considering large river basins and the management of watercourses traversing several countries and climatic zones. In such enormous physical spaces, people have developed a wide variety of resource adaptations. Actors share the same resource – the river – but most often they tend to conceive of themselves as living in different 'water worlds'. The classic dilemma of a dominant individual incentive that creates a suboptimal social equilibrium will therefore not naturally emerge in river basins, where moreover (and paradoxically only in a literal sense) the resource both varies widely from place to place and is still always the same.

Societies along the Nile are neither equally capable of harming their common resource nor equally likely to suffer the consequences of others' behaviour, not only because some live upstream and others downstream, but also because individual action need not negatively affect other actors (although this of course may happen, and very deliberately so). Differing technological capabilities and other social factors, as well as physical location along the river, affect ability to participate in collective action. In a river basin the distribution of both benefits and costs is decidedly biased, and it will continue to be so as long as people live where they live because these places will be influenced by the river's physiography, hydrology, topography and longitudial profile. Trans-boundary watercourses do not constitute common pool resources that can be exploited jointly and simultaneously, and thus the 'tragedy of the commons' proposition is unsuitable.

Application of Hardin's model in river basins will tend to downplay the possibility and reality of sincere cooperative efforts and cooperative opportunities, opportunities made possible and continuously reproduced by the river due to its physical nature. It might also tend to downplay the extent to which river waters may be used to establish an enforced and willed 'tragedy of the commons'. Individual actors' power over river water may exert much more political power than individual actors' power over pastures, due to the enormous importance of river waters in many societies (not only where the resource is scarce) and because of their physical character. The model will also tend to support the conventional wisdom that water resources, such as the pasture, must be managed as one resource, or on a basin-wide scale, devaluing the importance of the fact that spatial discrepancy between benefits and costs of cooperation at the basin scale may be so great that other scales should be advanced in order to offset this discrepancy.

## COLLECTIVE ACTION THEORIES

The theory of collective action holds that when rational individual behaviour and companies' profit-seeking fail to provide public goods, the common or shared interest of a group might enable collective action.[9] A number of formal theories and models treat collectivities rather than individuals as units of analysis.[10] Mancur Ohlson's famous model of the 'rational' individual questioned this willingness to cooperate. He argued that for all types of social activity, when people can benefit from cooperation without contributing to it, they will do so. Since, by Ohlson's definition, a 'collective good' is one that, if provided to one member of a group, cannot be withheld from any other,[11] it follows that collective action is 'irrational', and the bigger the group, the more irrational collective action becomes. Therefore, what needs explanation within this way of thinking is collective action, not collective inaction, which will be regarded as natural. This premise has been important in theorizing about why some 'public goods' need to be provided coercively through the political system. There is of course a very voluminous literature discussing Ohlson's theories. Our modest point here is only to situate this model of social action within the geographical and physical constraints and potentials of a river basin – far from the perfect market mechanisms.

When analysing collective action in river basins in general and in the Nile basin in particular these abstract and generalized models show their limitations. One reason is that because, throughout history, the Nile's physical character has continually created widely different possibilities for human adaptation and action along its course, the physical context cannot and should not be compared to the abstract context or idea of a market. If the ten riparian states are regarded as a group, the issue of the 'free rider' does not arise since benefits and costs are not allocated by market forces, but are partially affected by factors that individuals can neither create nor control, including the physical character of the river, the history of the river and river control works in each country, and the existence and relative importance of other water resources that can be exploited by the same actors. A lack of collective action among actors in such contexts is thus very natural, but cannot be explained by social facts alone or by this way of reasoning about the free-rider argument: that one or some of the riparian countries' are benefiting without contributing.[12] It is possible to analyse the problem of collective action in the Nile basin as if the physical and historical context and scale do not matter, but it is not analytically advisable to do so.

Theories about problems of collective action all share the idea that excluding non-contributors to a collective benefit is a major cost. They differ, however, over the cost or difficulty of devising physical or institutional means to exclude others.

But what the river basin context makes clear is that some of these differences stem from the physical 'water world' itself. General models of collective action derived from abstract notions of the market cannot explain or accommodate the conceptual differences and conflicts that develop in a river basin, which are based not only on changeable social variables but also on fixed geographical positions, in addition to the fact that the 'collective good' is most often exploited sequentially. Behavioural economics has focused attention on the importance of perceptions and perceived risks as critical factors, but not in the context of large-scale river basins. These differences can be compared metaphorically to the philosophy of the Cubist school in art, which holds that observers see the same object differently because they occupy different positions relative to it – from the right, the left, from below or above, and so forth – each standpoint providing a different but perfectly valid view. But in the Nile basin they cannot choose to occupy new positions relative to the Nile, because people live where they live. The Nile, as other objects in nature, exists independently of any theories about it, and continues to influence people's lives in different ways whether they are aware of it or not. The time sequence in which a problem of action is located must also be important: countries in a large river basin develop differently, and at different times their needs for water will vary considerably, as will their technological and economic capacities to control or use the water resource that is the object of collective action.

Empirical studies show how the competing conceptual constructions adopted by various 'stakeholders' reflect the distinct perspectives that different state elites and leading politicians occupy within physical – and socio-historical – locations along the Nile.[13] Such empirical descriptions of differences in approach and conceptualization help us to comprehend the obstacles to resource optimization in river basins. The Nile basin's intrinsically complex structural attributes have impeded and will continue to impede the search for a cooperative water-use regime. Managerial blueprints or misplaced resentment of 'free riders' will confuse the real issues, obstacles and opportunities for both collective action and cooperation (which is not the same thing).

## THE 'COMMON PROPERTY RESOURCE' MODEL

'Common property resource' has become a very influential term in the international debate about managing resources, including river basins. The model is predicated on certain relations between the spatial domains of resources and their users. Like the 'Rowland–Ostrom Framework', this has been suggested as a model for effective, sustainable trans-boundary water management.[14]

The term 'common property resource' was popularized by Elinor Ostrom in 1990 to denote natural resources used by many individuals in common.[15] Her point of departure was that such common property resources have long been overexploited and misused by individuals acting in their own interests. The conventional solution to such sustainability problems has been government regulation or privatization. Ostrom argued a third way: durable cooperative institutions organized and governed by resource users themselves. The central question that arises is therefore how a group of principals who are in an interdependent situation organize and govern themselves to obtain continuing joint benefits when all face temptations to free-ride, shirk, or otherwise act opportunistically.

The 'design principles' needed for a stable 'common property resource' arrangement and to foster contingent self-commitment of members,[16] according to this model, are clearly defined boundaries; congruence between appropriation and provision rules, and local conditions; arrangements for most appropriators to take part in the decision-making process; effective monitors accountable to the appropriators; graduated sanctions for appropriators who do not respect the rules; conflict-resolution mechanisms that are cheap and easily accessible; government recognition of rights to organize; and, in cases of larger common property resources, multiple layers of enterprises, with small, local organizations at their base.

There are problems with this concept's applicability to river basins in general and to the Nile basin in particular. The Nile, like other large basins, has unclear physical boundaries, and these are often thus areas of social and political contention. If, for example, upstream countries develop rain-based agriculture within the watershed instead of irrigated agriculture, to what extent should the amount of rainfall falling in this part of the watershed be considered part of the available sharable water in the Nile basin as a whole? A more controversial question is whether the waters of the Nile can and ought to be pumped outside the river basin itself. In upstream countries, Egypt has been accused of doing this by bringing Nile water to Sinai and the New Valley project, and Egypt has objected to Tanzania pumping water from Lake Victoria to Shinianga. And in Ethiopia, the authorities have for a long time discussed plans for sending water in tunnels to dry, drought-prone areas. Because of the physical character of water, and the fact that the boundaries of a river basin are always in flux, this whole issue of 'clearly defined boundaries' creates a divisive element. Furthermore, excluding users is extremely difficult, if not impossible, because of the special challenges related to water control, and in any case water transfer from one basin to another is already common.[17]

Ostrom's design principles pose other problems. Although the congruence issue is at the very heart of the whole Nile problem, correspondence between appropriation and provision rules and local conditions is difficult to achieve. Local conditions cannot be reduced to social or economic issues alone, although such differences may be extreme in large river basins, but must take account also of physical differences in the watercourse itself and of its environment. The decision-making process must be complicated when ten countries share a resource, but have different interests in the process or its outcome. Effective monitoring is also problematic; hydrological data are regarded in some cases as state secrets, and instead of forming a basis for cooperation are part of the problem that cooperation should solve. Cheap and effective systems of conflict resolution will be difficult to establish and maintain, owing to geographical distance, political atmosphere and cultural differences. In suggesting that the Rowland–Ostrom model be applied globally, the author states: 'The most critical and difficult part of the first step is for people who share a common pool water resource to agree that a crisis situation exists'. But on a river such as the Nile, almost 7,000 km long, this is most unlikely, unless the crisis is defined upstream and has clear and immediate downstream consequences (for the waters take many months to flow from Lake Victoria to Egypt).

Such models disregard the spatial and physical dimension that any solution to water management problems, and their social consequences, must address. People will interpret them very differently, while such general concepts create the impression that there exists a kind of blueprint for solving complicated issues. Such models also make the recent history of the Nile basin unintelligible. An alternative framework is needed with a much greater emphasis on how to understand collective action theoretically and conceptually, and on what has been focused on in this book: how patterns of action will vary according to physical characteristics and the technical, social, economic, environmental and institutional history of this particular river basin.

This book has described some of these conditions prevailing in the Nile basin, highlighting some structural and historical contexts for current Nile diplomacy and policies. These contexts are also in a flux, like the river itself. It has been supposed, for example, that the costliness of separate storage facilities will encourage countries to cooperate.[18] This argument rests on assumptions about the power and almost monopoly of the World Bank in financing such projects. The World Bank's requirement for funding has been that other stakeholders in the basin should not object to the project in question. Foreign investments for the development of the Nile waters have therefore been difficult to achieve, not least since the downstream riparian states have maintained the right to veto projects envisaged by upstream states. States in the basin have therefore tried to bypass the World Bank, and time and again they have succeeded. Separate storage facilities financed by institutions other than the World Bank have been quite common from the very 'beginning' of the post-colonial period. The High Dam at Aswan that the World Bank, under the leadership of Eugene Black, strongly wished to fund in the early 1950s was eventually financed by Moscow.[19] Building a dam, the highest in Africa, on the Tekezze (Atbara), with financial and technical support mainly from China, is making Ethiopia more independent from other basin states than it was before China entered the African theatre. Tanzania's project to pump water from Lake Victoria has been financed from other sources than those coming from the World Bank, as has the biggest dam on the Nile, at Merowe in the Sudan. Since there is no international agreement yet to which potential donors or loan-institutions all subscribe, hopes that prohibitive costs might induce cooperation seem naïve and overlook the new opportunities created by the new economic actors on the world scene.

It has also been argued that cooperation is bound to come about once the Nile basin experiences a crisis serious enough to force the issue.[20] The problem with this prophecy is imagining circumstances that would affect all the countries of the Nile basin in the same way or in a way that the main actors would regard as a crisis, and thus bring them together in joint action. Models based on assumptions that Egypt will or must benefit from future dams upstream (i.e. that in reality there are no conflicting interests when it comes to dam-building upstream) are wrong: whether it does or not depends on the type of dams that are built and their purposes. Dams are forms of structural power that reflect interests; here as elsewhere, control of nature implies power over other human beings.

## CRITICAL 'MASS THEORY' AND THE NILE BASIN INITIATIVE

Critical mass theory may be relevant in highlighting and understanding aspects of contemporary history of Nile cooperation, and especially the role and potentials of basin-wide institutions. According to this way of thinking, what matters is not that everyone benefits, but that there exists a 'critical mass' of highly interested and resourceful people who provide or decide to provide collective benefits for others. Collective action may rest on an initial event of 'joint commitment' to which each participant makes a contribution. The Nile Basin Initiative (NBI), formally launched in February 1999 by the Council of Ministers of Water Affairs of the Nile basin states, the subsequent establishment of the Nile Basin Secretariat at Entebbe in Uganda in 2002, and the Nile Basin Discourse (being a network of civil society organizations from the ten countries of the Nile Basin supported by international donors) may be interpreted in such a way. The NBI is defined as a 'partnership initiated and led by the riparian states of the Nile River through the Council of Ministers of Water Affairs of the Nile Basin states' (Nile Council of Ministers). Its aim is to develop the river 'in a cooperative manner, share substantial socioeconomic benefits, and promote regional peace and security'. The Nile Basin Initiative was based on what, when considered in the long term, can be seen as a 'revolutionary' idea: that the river was a shared treasure of all the basin states, and on NBI's operational design about 'sharing benefits' authorized piecemeal execution of selected water control projects that were jointly sanctioned by the collaborating states. The Subsidiary Action Programs (SAPs) of the NBI should implement joint investment programmes supported by donors with the aim of 'contributing to poverty eradication, promotion of economic development and to reverse environmental degradation in the basin'. In the Nile equatorial lakes subregion – comprising Kenya, Uganda, Tanzania, DR Congo, Rwanda and Burundi – projects such as the Lakes Edward and Albert Fisheries Pilot Project, and the Mara, Sio–Malaba–Malakisi and Kagera Integrated River Basin Management, have been developed. Eastern Nile Subsidiary Action Program (ENSAP) focuses on the Nile and its tributaries within the Eastern Nile countries of Egypt, Ethiopia and the Sudan, and encompasses the sub-basins of Baro/Akobo/Sobat, portions of the White Nile, Abay/Blue Nile, Tekezze/Settit/Atbara, and the Main Nile. The Eastern Nile Technical Regional Office, based in Addis Ababa, is the implementing arm of ENSAP. There can be no doubt that as a result of exchange of ideas and propositions in a context of institutionalized collaboration over time, actors involved in the NBI process have to a certain extent become jointly committed, for example, to discuss water-sharing agreements or common water projects, and are thereby obliged, so to speak, to act as if they were a single person. An account of collective action from this perspective has the merit of explaining the fact that those working to develop a river together understand that each can demand corrective action of another, acting in ways perceived detrimental to their stated common goal. This means that, at the core of collective action, a 'collective intentionality' may be required, because this prior commitment might reproduce contexts of action that benefits further cooperation.

The NBI has been criticized for a lack of results, and because the institution in itself gives the impression that there has been more cooperation than has actually been the case. This assessment is understandable, especially given the inflated expectations that were propagated by some actors when it was established. But the NBI has served as a catalyst for cooperation, and has some quite remarkable achievements to its credit. Despite many opportunities for disputes and despite its general inability so far to deliver projects that really make a difference, the NBI has managed to build more trust among the riparian countries. The Nile sharing issue has not been settled, but the NBI has reintroduced a framework for basin-wide planning under the control of the basin states themselves. There are now institutional frameworks for regional cooperation. Through the NBI, states in the Nile basin may obtain foreign aid otherwise unavailable to them, and an institutional arrangement has been created whereby third parties have been given an opportunity to mediate disputes and help riparian countries to sustain negotiating processes. A multinational group of water experts and communication experts, each coming from one of the Nile basin countries, has been established by the NBI to advocate cooperation and national restraint in exploiting the Nile. The NBI and the way it has operated may be seen as an institutional recognition that the 'status quo' is unsustainable for all the parties to the Nile, and a comprehensive new regime is needed to regulate and control the river in an optimal way.

There is no simple institutional solution to the collective action problem in the Nile basin, and no ready-made model that can be copied. This book has highlighted multiple modes of conflict and cooperation surrounding the use and management of the Nile, along with the temporal and spatial scales and the dialectic character of societal development and the physical character of the river system – all factors that are central to an understanding of the relationship between the actors within the basin and of the history of the whole region.

# Notes

## 1 ABOUT THE IMPORTANCE OF STUDYING THE MODERN HISTORY OF THE COUNTRIES OF THE NILE BASIN IN A NILE PERSPECTIVE

1 According to recent statistics from the United Nations the territory of 145 countries falls within international basins: 33 are located almost entirely within such basins. The UN has discerned 263 international river basins, which cover 45.3 per cent of the planet's land surface and are home to 40 per cent of its population.

2 See L. Teclaff, *The River Basin in History and Law*, The Hague 1967.

3 In Terje Tvedt, *The Nile. An annotated bibliography*, 2nd ed., London and New York 2004, there is a quite comprehensive overview of books, reports and articles dealing with the river published up to the year 2000 (more than 3,000 titles). In Terje Tvedt with Eirik Hovden, *A Bibliography on the River Nile*: *Nile plans and reports 1960–2006*, III, Bergen 2008, projects, plans and reports on all major projects on the Nile are listed. Finally, in Terje Tvedt and Eirik Hovden, *A Bibliography on the River Nile: Literature on the River Nile published 2000–2006*, II, Bergen 2008, books, articles and reports published between 2000 and 2008 are registered.

4 See Terje Tvedt, *The River Nile in the Age of the British. Political ecology and the quest for economic power*, London and New York 2004, and Cairo 2006; Robert Collins, *The Nile*, New Haven and London 2002.

5 The roles of Egypt, the Sudan and Ethiopia in Nile development have been covered by many authors. Egyptian–Ethiopian relations have received most attention. See e.g. Yacob Arsano, 'Predicaments of cooperation in the Nile basin', in Katsuyoshi Fukui *et al.* (eds), *Ethiopia in Broader Perspective: Papers of the XIIIth International Conference of Ethiopian Studies*, Kyoto 1997, 29–47; Collins, *The Nile*, 213–17; Daniel Kendie, 'Egypt and the hydro-politics of the Blue Nile River', *North African Studies* 6:1, 1999, 141–69; Michael T. Klare, *Resource Wars*: *The new landscape of global conflict*, New York 2001, 153–4; Christopher Kukk and David Deese, 'At the water's edge: Regional conflict and cooperation over fresh water', *UCLA Journal of International Law and Foreign Affairs* 21:1, 1996, 21–64; Ashok Swain, 'Ethiopia, the Sudan and Egypt. The Nile River dispute', *Journal of Modern African Studies* 35:4, 1997, 675–94; Tesfaye Tafesse, *The Nile Question: Hydropolitics, legal wrangling, modus vivendi and perspective*, London 2001; John Waterbury, *Hydropolitics of the Nile valley*, Syracuse 1979; and *idem*, *The Nile Basin: National determinants of collective action*, New Haven 2002. Waterbury's books present interesting analysis of Nile basin developments, but omit detailed empirical study of Kenya, Tanzania, Burundi and Rwanda.

6 This historical overview is based on Tvedt, *The River Nile*. This book is based on reading of a large number of hitherto unused sources and documents, and it presents alternative explanations of the whole imperial project and of central historical events in the basin during the period between 1882 and the late 1950s. The book also

analyses how these regional developments were linked to events of world historic importance.

7 The term 'river war' is from Winston Churchill, a war correspondent in 1898, whose history of the Sudan campaign is *The River War. An account of the reconquest of the Soudan*, London 1899.

8 Crucial structural aspects of the Nile's hydrology that influenced British strategy included the fact that it has no tributaries in Egypt itself; that for rational water planning in Egypt precise knowledge of upstream hydrology was absolutely crucial; that the Blue Nile contributed around 80 per cent of water measured at Aswan, but contained so much silt that damming it for the benefit of Egyptian summer agriculture was impractical at the time; that the White Nile was the most important tributary during the summer season; that the Central African lakes were valuable reservoirs; but also that the White Nile tributaries while running through the southern parts of the Sudan lost about half their waters in an immense swampy and flat landscape. The British government sent their best hydraulic experts in the wake of the advancing flotilla and army to report on river conditions and potentials for river control. See especially the reports by William Garstin, all referred to in Tvedt, *The River Nile*.

9 This story and interpretation of the occupation of the Sudan is put in a much wider context in Tvedt, *The River Nile.*

10 The Under-Secretary of the Ministry of Public Works in Egypt in the early 1890s, Colin Scott-Moncrieff, used this expression in Colin Scott-Moncrieff, The Nile, *Proceedings* 14, Royal Institution of Great Britain, 25 January 1895, 405-418, p. 410.

11 See R. Robinson and J. Gallagher, *Africa and the Victorians: The official mind of imperialism*, London 1961; and the later edition, R. Robinson and J. Gallagher (with Alice Denny), *Africa and the Victorians: The official mind of imperialism*, London 1981. The idea that the Sudan could be compared to Afghanistan as a buffer state between competing imperial powers was a cornerstone in this book's story of imperial expansionism in Africa as a whole. This misrepresentation of Sudan's strategic position has been repeated in many recent studies of the history of the region, also in books specifically dealing with the Nile: e.g. Waterbury, *The Nile Basin*.

12 Credit is due to Sir William Garstin, whose 'Report … upon the basin of the Upper Nile', *Foreign Office Blue Book, Egypt, No. 2*, 1904, was seminal.

13 See Tafesse, *The Nile Question*, p. 62, for this assessment.

14 The very close relationship between politicians and water engineers was established by Cromer in the 1880s. This relationship was developed and maintained, which meant that the water experts of the 1940s and 1950s also had a very good relationship with the British political strategists; this holds for Allen in the Sudan, Hurst in Egypt and especially Hawks in Uganda (see Tvedt, *The River Nile* for a description of this relationship).

15 See Tvedt, *The River Nile* for a detailed reconstruction of this dramatic history.

16 Reconstructing the history of Nile policies and Nile projects since the colonial era poses methodological difficulties. The National Archives in London have a very large collection of relevant official and secret documents, and most planning reports can be found in one place, but there is no complete archive of post-colonial plans. Since the diplomatic correspondence of the independent Nile states is unavailable, collected published and grey planning reports must suffice.

17 See Hosam E. Rabie Eleman, Egypt and Collective Mechanisms in the Nile Basin, in Terje Tvedt (ed), *The River Nile in the Post-Colonial Age*, London, 2010, 217–235.

18 Frederick W. Frey and Thomas Naff, 'Water: An emerging issue in the Middle East', *The Annals of the American Academy of Political and Social Science* 482:1, 1985, 65–84, p. 78.

19 See Fadwa Taha, The History of the Nile Waters in the Sudan, in Terje Tvedt (ed), *The River Nile in the Post-Colonial Age*, London, 2010, 179–216.

20 Prime Minister Meles Zenawi, interview on the televised documentary, 'The future of water, part III: The water lords', 2008 (producer Panopticon AS).

21 For a discussion of the Ethiopian position, see Y. Arsano, 'Challenges of effective cooperation in the Nile Basin', paper presented at the conference, *Green Wars? Environment between Conflict & Cooperation in the Middle East and North Africa*, organized by the Heinrich Boll Foundation, Beirut, 2–3 November 2007, Gefinor Rotana Hotel.

22 See Pascal Nkurunziza, Burundi and the Nile, Resource Management and National Development in Terje Tvedt (ed), *The River Nile in the Post-Colonial Age*, London, 2010, 13–29; Robert Balgira, Rwanda and the Nile: Water Plans and their Implementation, in Tvedt, *Ibid.*, 31–56; Honest Prosper Ngowi, Unlocking Growth and Development Potential: The Nile Basin Approach in Tanzania, in Tvedt, *Ibid.*, 57–71; Raphael M. Tshimanga, The Congo Nile: Water Use, Policies and Challenges, in Tvedt, *Ibid.*, 73–91; Mary C. Mwiandi, The Nile Waters and the Socio-Economic Development of Kenya, in Tvedt, *Ibid.*, 93–124; James Mulira, Independent Uganda and the Nile: Hydrolelectric Projects and Plans, in Tvedt, *Ibid.*, 125–160; and Yacob Arsano, Institutional Development and Water Management in the Ethiopian Nile Basin, in Tvedt, *Ibid.*, 161–178. For a more overall overview of literature and reports on Nile development, see the bibliographies cited in note 3.

23 *Ibid.*

## 2 BURUNDI AND THE NILE: WATER RESOURCE MANAGEMENT AND NATIONAL DEVELOPMENT

1 Rashid Mbaziira, Nsubuga Senfuma and Rachael McDonnell, 'Institutional development in the Nile equatorial lakes sub-basin: learning from the experience of the Kagera Basin Organization', University of Oxford, 2006, retrieved from http://www.iwmi.cgiar.org/Africa/files/RIPARWIN/05/EARBM_Papers/Theme4/Rashid%20Mbaziira.doc.

2 Burundi tourism, August 2009, retrieved from http://musemakweli.ifrance.com/musemakweli/ touribu.html.

3 'The Anglo-Belgian Agreement of 12 May 1934', *English Historical Review* 57, 226. Text of Trusteeship Agreement as Approved by The General Assembly of the United Nations, 13th December, 1946, retreived from: http://www.kituochakatiba.co.ug/TanganyikaTrusteeshipAgreement1946.pdf.

4 FAO, *Enquête Aquastat, 2005: l'irrigation en Afrique en chiffres; le rapport sur l'eau 29, 2005*, retrieved from www.fao.org/ag/agl/aglw/aquastat/countries/burundi/printfra1.stm.

5 Gouvernement du Burundi, Ministère de l'energie et des mines, *Plan directeur national de l'eau, Rapport de base, Volet III: Ressources en eau.V. Planning the future*, Vienne 1998, 99.

6 D.Y. Mohamoda, 'Nile Basin cooperation, a review of the literature', *Current African Issues* 26, Uppsala, 2003.

7 Ministère de l'Energie et des Mines, *Enquêtes nationales sur la couverture en eau et assainissement. Rapport DGHER*, Bujumbura, 2001.

8 Ministère de l'Aménagement du Territoire, Politique Sectorielle du Ministère de l'Aménagement du Territoire, de l'Environnement et du Tourisme, *Document ministériel*, April 2006.

9 Missions Economiques and MINE-DGTPE, *L'Energie et l'eau au Burundi, fiche de synthèse*, 8 March 2005, retrieved from www.pefac.net/pdf/101540.pdf.

10 Gouvernement de Burundi, *Planning the future*, 99.

11 *Ibid.*

12 S.L.M.A. Mekouar, *Projet de loi sur les marais au Burundi: étude juridique de la FAO en ligne*, August 2001, retrieved from www.fao.org/Legal/default.htm.

13 *Ibid.*

14 'Le milieu naturel', 2007, retrieved from www.netpress.bi/geo/Mil.htm.

15 Mekouar, 'Projet de loi sur les marais au Burundi'.

16 Gouvernement de Burundi, *Planning the future*, 99.

17 Ministère de l'Aménagement du Territoire, *Document ministériel.*

18 M.K. Gishinge, *Analyse contextuelle en matière de gestion integrée des ressources en eau (GIRE) au Burundi. Rapport final*, PROTOS, March 2006, retrieved from http://www.protos.be/protosh2o/copy_of_Links-met-bestanden/Rapport_GIRE_Burundi_finaldef.pdf.

19 S. Nkokoni, *Regional development planning of the Kagera river basin in eastern Afrika under the Kagera basin organization (KBO). A case study of hydropower development planning and related environmental impacts*, Bergen 1983.

20 Mbaziira and McDonnell, 'Institutional development'.

21 Mohamoda, 'Nile Basin cooperation'.

22 Mohamoda, 'Nile Basin cooperation'.

23 US Department of State, Press release, 'New research predicts dire water, food shortages for sub-Saharan Africa', Washington DC 2003.

24 S.R.A. O'Reilly *et al.*, 'Climate change decreases aquatic ecosystem productivity of Lake Tanganyika, Africa', *Nature* 424:14, August 2003, 766–8.

25 Simon I. Hay *et al.*, 'Hot topic or hot air? Climate change and malaria resurgence in East African highlands', *Trends in Parasitology* 18:12, December 2002, 530–4.

26 P.P. Rudahirwa, 'Baisse continue de la production électrique aux centrales de Rusizi I et Rusizi II. Bukavu. Echo de Goma et d'ailleurs', 2005, retrieved from www.pole-institute.org/site_web/echos/echogoma13.htm.

27 E. Sinarinzi, 'Programme d'action national d'adaptation des ressources en eau aux changements climatiques. Secteur des ressources en eau. Projet UNDP & Gouvernement du Burundi', 2006, retrieved from ww.napa-pana.org/private/modules/knowledgebox/io/file.php.

28 Institut Géographique du Burundi, *Evolution climatique actuelle et développement des scénarios de changements climatiques à l'horizon 2050*, Bujumbura, 2001.

29 E. Sinarinzi, 'Convention Cadre des Nations Unies sur les Changements Climatiques: première communication nationale', Bujumbura, August 2001.

30 E. Sinarinzi, 'Programme d'action national'.

31 Gouvernement du Burundi, Ministère de l'Aménagement du Territoire de l'Environnement et du Tourisme, *National adaptation plan of action to climate change* (PANA), Bujumbura, 2007.

## 3 RWANDA AND THE NILE: WATER PLANS AND THEIR IMPLEMENTATION

1 L.H. Gann and Peter Duignan, *The Rulers of German Africa, 1884–1914*, Stanford 1997, 5.

2 Steven Schoenherr, *The Versailles Treaty. June 28, 1919*, February 2004, retrieved from http://history.sandiego.edu/gen/text/versaillestreaty/vercontents.html.

3 H.H. Codere, *The Biography of an African Society, Rwanda 1900–1960*, Tervuren 1973; Sarah W. Freedman and Timothy Longman, *Rwanda*, Berkeley 2006.

4 Georges Mortehan, 'L'agriculture au Rwanda-Urundi, notes techniques', *Bulletin agricole du Congo Belge*, 1921.

5 Helmut Strizek, 'La situation des droits de l'homme au Rwanda. La vie après le génocide', in *Secteur 'Droits de l'Homme'*, Aachen 2003; Jan Vansina, *Antecedents to Modern Rwanda*, Kampala 2004.

6 République Rwandaise, 'National water resources management policy', July 1998, in *M.d.l.A.E.E.e.D. Rural*, Kigali 1998.

7 Leigh Kemp, *The Albertine Rift*, January 2007, retrieved from http://www.go2africa.com/africa-travel-articles/the-albertine-rift.

8 BBC News report, *Team reaches Nile's 'true source'*, January 2007, retrieved from http://news.bbc.co.uk/2/hi/uk_news/england/hampshire/4864782.stm. Also available as an article named *Rwanda is Nile's 'true source'*, April 2006, on http://www.rwandagateway.org/article.php3?id_article=1761.

9 SHER Conseil, *Etude hydrologique du bassin versant de la Kamiranzovu*, Kigali 1986 (unpublished extract).

10 TBW Ingénieurs Conseils, *Plan directeur et système de gestion des ressources en eau*, Kigali 1992.

11 J. Czekanowski, *Forschungen im Nil-Kongo-Zwischenebeit. Ethnographie, Uele-Ituri-Nil-Lander. Zwischenseengebiet*, vol. 2, Leipzig 1917, 135–6.

12 Republic of Rwanda, *Third General Census of Population and Housing of Rwanda – August 2002*, Kigali 2003.

13 République Rwandaise, *Rwanda: stratégie de réduction de la pauvreté, 2002*, Kigali 2002.

14 République Rwandaise, *Indicateurs de développement du Rwanda, 2004*, Kigali 2004.

15 République Rwandaise, *Enquête démographique et de santé du Rwanda, 2005*, Kigali 2005.

16 *Ibid.*

17 République Rwandaise, *Indicateurs de développement du Rwanda, 1999*, Kigali 1999.

18 République Rwandaise, *Indicateurs de développement du Rwanda, 2002*, Kigali 2002.

19 République Rwandaise, *Indicateurs de développement du Rwanda, 2004*.

20 FEWS NET, World Food Program, reports covering the period from 2005 to 19 February 2007.

21 République Rwandaise, *Direction des statistiques, 2005*, Kigali 2005; Rwanda development programme, National Bank Report, 2004, Kigali 2004.

22 République Rwandaise. Ministère des Infrastractures, *Document des stratégies sectorielles du Ministère des infrastructures 2005–2010*, Kigali 2004.

23 SNC-Lavallin and HydroQuebec International, *Strategic/sectoral, Social and Environmental Assessment of Power Development Options in Burundi, Rwanda and Western Tanzania, Draft report no 1.* The report has been commissioned by the World Bank, Rwanda 2004.

24 Kigali Institute of Science and Technology, *UNEP-GEF pilot project on reducing the vulnerabilty of the energy sector to the impacts of climate change in Rwanda*, Kigali 2006.

25 Théophile S. Ruberangeyo, *Etat des lieux du processus de formulation de la politique de l'eau au Rwanda et évaluation des besoins pour son amélioration et sa mise en œuvre*, Kigali 2005.

26 République Rwandaise, *Constitution du Rwanda 4 June 2003*, Kigali 2003.

27 Groupement BRL Ingénierie, Projet National de Gestion des Ressources en Eau, *Legislation sur l'eau et institutions*, Kigali 2005.

28 Congo Belge, 'Extraits de "Codes et Lois du Congo Belge"', ed. E.d.C.e.L.d.C. Belge, *Matières sociales et économiques, 9eme partie législation économique*, Tome III, 31 March 1959.

29 The AIDR (Association International de Développement Rural) was created after independence. Three founders were members of the colonial government in Rwanda. The association operated from about 1966 to 1982, mainly in the water sector. See Gregor Stangherlin, *Les organisations non gouvernementales de coopération au développement en Belgique. Origine, Cadre Juridique, Cofinancement et Enjeux*, Brussels 2001.

30 The Régie de production et de distribution d'eau et d'électricité was created in 1936 and replaced by Electrogaz in 1976.

31 République Rwandaise, *Indicateurs de Plan Quinquennal de Développement Economique, Social et Culturel, Synthèse du Plan 1972–1976*, Kigali 1976.

32 République Rwandaise, Secrétariat d'Etat au Plan National de Développement, *Rapport d'exécution du I° Plan Quinquennal de Développement Economique et Social et Culturel 1966–1970*, Kigali 1971.

33 Peter H. Gleick, *The Human Right to Water*, Oakland CA 1999.

34 United Nations High Commissioner for Human Rights, *Declaration on the Right to Development*, Geneva 1986.

35 Law 18/99 of 30 August 1999 ended Electrogaz's monopoly because it had operated at a loss and did not satisfy public demand, in particular because of high technical deficits.

36 République Rwandaise, *Etude institutionnelle de la gestion des adductions d'eau en mileu rural*, Kigali 1985.

37 World Bank, *The World Bank report, 1994: Infrastructure and development*, The World Bank, Oxford University Press, March 2007, retrieved from http://www-wds.worldbank.org/external/default/WDSContentServer/WDSP/IB/1994/06/01/000009265_3970716142907/Rendered/PDF/multi0page.pdf.

38 *Ibid.*

39 Gérard Prunier, *The Rwanda Crisis: History of genocide*, Columbia University Press, New York 1995.

40 Food and Agriculture Organization of the United Nations. *Mise en valeur du Bassin du Nil pour la production agricole: Suivi, prevision et simulation. Region Afrique: Burundi, Egypte, Ethiopie, Kenya, Ouganda, Republique Unie de Tanzanie, Rwanda, Soudan, Zaire. Compte rendu final*, Rome 1992, 15.

41 SHER, 'Projet de rehabilitation des adductions d'eau en milieu rural (Rwanda) 1993–1995', *Eau potable et assainissement*, Kigali 1995; *idem*, 'Identification du projet transition institutionalle du secteur de l'alimentation en eau potable en milieu rural', *Eau potable et assainissement*, Kigali 1998.

42 Government of Rwanda: United Nations Development Programme, *Country cooperation, 2004–2008. Project profiles*, Kigali 2004.

43 Rwanda Utilities Regulatory Agency, Law 39/2001 of 13 September 2001, Kigali 2007.

44 Rwanda Environment Management Authority, v. 2003, 2007.

45 SHER, 'Projet de gestion nationale des ressources en eau, etude sur la connaissance et gestion des données sur l'eau (Rwanda), 2004–2005', *Eau potable et assainissement*, Kigali 2005.

46 Norconsult and Electrowatt, Kagera river basin development. Phase II. Sectoral studies, 1975.

47 SHER Conseil, *Etude sur la Connaissance et Gestion des données sur l'eau (2005). Projet de Gestion Nationale des Ressources en Eau – Etude sur la Connaissance et Gestion des données sur l'eau (Rwanda)*. The study was commissioned by the World Bank, Kigali 2005.

48 Republic of Rwanda, *The Study on Rural Water Supply Project in the Eastern Region (Phase III). January 1992*, Kigali 1992.

49 Igip, *Etude complémentaire de faisabilité de l'alimentation en eau de la ville de Kigali à partir des ressources en eau de la région de Ruhengeri*, Kigali 2003.

50 République Rwandaise, *Inventaire national de l'alimentation en eau potable en milieu rural au Rwanda, édition 2001*, Kigali 2001.

51 Expreco, *Master Plan for Food Production and Water Use in Umutara*, Kigali 2002.

52 Robert Baligira, *Process of Formation and Chemical Composition of the Phreatics Water (in Kibungo Province), in case of applying hydrogeochemical methods for mines survey and appreciation [of] their quality for drinking use*, PhD, Moscow 1993.

53 *Ibid.*

54 Foraky Africa Rwanda, *Forages d'exploitation de 100 m de profondeur équipe en piézomètre 5 pouces avec essai de pompage*, Kigali 2004; *Forages d'exploitation de Nzove a 62.00 m*, Kigali 2004; *Carottage de 15.00 m et de 15.30 m de profondeur pres de la riviere Nyabaringo a Nzove*, Kigali 2004.

55 Daniel C. Clay and Laurence A. Lewis, 'Land use, soil loss, and sustainable agriculture in Rwanda', *Human ecology* 18:2, 1990.

56 République Rwandaise, *Rapport d'exécution du I° plan quinquennal.*

57 Sogreah Ingénieurs Conseils, *Etude du plan directeur d'aménagement et de mise en valeur des vallées de la Nyabarongo et de l'Akanyaru*, Kigali 1988.

58 Vincent Ngarambe, 'Plan stratégique de transformation de l'agriculture au Rwanda', *Gestion et utilisation de l'eau et des sols*, Kigali 2004.

59 Lux-Development is a Luxembourg cooperative project started in December 2004. The project finances the Bugesera Rural Economic Development Project for a five-year implementation period.

60 John Bayigamba, 'Lux-development dedicated to end famine', *NewTimes*, Rwanda 24 March 2007.

61 Organisation des Nations Unies pour l'Alimentation et l'Agriculture, *Aquastat, Système d'information de la FAO sur l'eau et l'agriculture (Rwanda), 2005*, March 2007, retrieved from http://www.fao.org/nr/water/aquastat/countries/rwanda/indexfra.stm.

62 Republic of Rwanda, *2020 Vision*, Kigali, November 2002.

63 Harindintwali Révérien, *Marshlands Master Plan: Sélection de 48,000 ha de marais aménageables echéance (2010–2012)*, Kigali 2006.

64 Vincent de Paul Kabalisa, *Analyse contextuelle en matière de gestion intégrée des ressources en eau au Rwanda*, Kigali 2006.

65 Ministry of Lands, Forestry, Water and Mines, *Initial National Communication on Climate Change*, Kigali 2005, 187.

66 Norconsult and Statnett, *Opportunities for Power Trade in the Nile River, Scoping Study, Draft data report*, Rwanda 2000.

67 A.T. Félicien and R. Bikumu, *La problématique du déficit énergétique dans la sousrégion des Grands Lacs africains*, Goma 2005.

68 Prof. James Mulira, private communication, May 2007.

69 République Rwandaise, *Deuxième plan quinquennal de développement economique, social et culturel 1977–1981*, 1976.

70 République Rwandaise, *Projet d'étude d'alimentation en eau potable de la région orientale*, Kigali 1985.

71 République Rwandaise, 'Connaissance et gestion des données sur l'eau au Rwanda, 2004–2005', in *Eau Potable et assainissement*, Kigali 2005.

72 Robert Baligira, *Examen critique et enquête de l'état de la gestion des données de la qualité de l'eau au Rwanda ainsi que les recommandations pour son amélioration*, Kigali 2007.

73 P. James Money, *Report on the Assessment of the Impact and Sustainability of Community Based Water and Environmental Sanitation Facilities*, Kigali 2004.

74 Electrogaz, *Rapport de l'analyse de la qualité de l'eau effectué par Electrogaz 1997–2006*, Kigali 2006.

75 République Rwandaise, *Atelier sur la gestion intégrée des ressources en eau au Rwanda*, Kigali 2002.

76 SHER, 'Etude sur la connaissance et gestion des données sur l'eau'.

77 Organisation des Nations Unies pour l'Alimentation et l'Agriculture (FAO), *Rapport au Gouvernement du Rwanda sur la Pisciculture et le développement des pêches*, March 2007, retrieved from http://www.fao.org/docrep/005/E5927F/E5927F00.htm.

78 Terje Tvedt, *The River Nile in the Age of the British*, London and New York 2004.

79 Yosef Yacob, *Tough talk over a defunct treaty: the case of the 1929 Nile Waters Agreement*, January 2007, retrieved from http://www.tigrai.org/News/Articles2004/TheNileByYacob.html.

80 Owen McIntyre, Environmental protection of international watercourses under international Law, *Agreement between the United Kingdom and Belgium Regarding Water Rights on the Boundary between Tanganyika and Ruanda-Urundi*, November 1934, London, retrieved from http://ocid.nacse.org/tfdd/tfdddocs/45ENG.htm, August 2009.

81 Information and Public Relations Office, Belgium Congo and Ruanda-Urundi, Brussels 1960.

82 S. Nkonoki, *Regional development planning of the Kagera river basin in eastern Afrika under the Kagera basin organization (KBO). A case study of hydropower development planning and related environmental impacts*, Bergen 1983.

83 United Nations Development Program, 'Hydrometeorological survey of the catchments of Lakes Victoria, Kyoga and Albert', *Hydrological Studies of Selected River Basins*, Geneva 1974, 355.

84 P.P. Howell, *East Africa's Water Requirement: The Equatorial Nile Project and the Niles Waters Agreement of 1929*, London 1994.

85 Norconsult and Electrowatt, *Rapport Technique, Volume 7 – Aménagement de la Rivière Kagera, Phase II – Burundi, Rwanda, République Unie de la Tanzanie – Etude sectorielle, Hydrologie*, Rwanda 1975.

86 John Waterbury, *The Nile Basin: National determinants of collective action*, New Haven and London 2001.

87 The first decree related to the Kagera Basin Development Organization (KBO) dates to 25 November 1975.

88 Charles Kazooba, 'Rwanda, Burundi join EAC in March', *News Times*, 13 February 2006.

89 Nile Basin Initiative, *Nile Council of Ministers, Policy Guidelines for the Nile River Basin Strategic Action Program*, Entebbe 1999.

## 4 UNLOCKING ECONOMIC GROWTH AND DEVELOPMENT POTENTIAL: THE NILE BASIN APPROACH IN TANZANIA

1 The author acknowledges the support of the Nile Basin Research Program (NBRP) at the University of Bergen, and the assistance of researchers and staff there, especially Professor Terje Tvedt; of participants in the 8th International Conference of the International Academy of African Business and Development (IAABD) at London Metropolitan University; and of Professor Ashok Swan, who reviewed an early version of this work.

2 T. Tvedt, *The River Nile and its Economic, Political, Social and Cultural Role: An annotated bibliography*, Bergen 2000, vii.

3 In 1964, Tanganyika merged with Zanzibar to form Tanzania. To avoid confusion we use the latter name throughout this essay.

4 T. Tvedt and E. Hovden, *Development of the Nile 1960–2006. Plans and reports: a partially annotated bibliography*, Bergen 2006.

5 Tvedt, *The River Nile.*

6 For details of Tanzania's and other countries' positions on the 1929 agreement, see e.g. C. Okidi, 'Legal and policy regime of Lake Victoria and the Nile basin', paper presented to the Institute for Development Studies, University of Nairobi, March 1980; Tvedt, *The River Nile*; and *idem*, *The River Nile in the Age of the British: Political ecology and the quest for economic power*, London and New York 2004.

7 C. Nyamurunda, Southern Africa Documentation and Cooperation Centre, communication of 9 February 2004.

8 Nile River Basin Commission, *Draft Agreement on the Nile River Basin Cooperation Framework*, Kampala 2004.
9 *The East African*, 1–7 January 2007.
10 *The Guardian*, 12 December 2006.
11 *Taarifa ya Utekelezaji wa Mradi wa Maji Kutoka Ziwa Victoria Hadi Miji ya Kahama na Shinyanga*, Dar Es Salaam 2007.
12 Southern Africa Documentation and Cooperation Centre, *Victoria Water Started*, February 2004.
13 *Financial Times*, 31 January 2001.
14 *The Express*, 30 August–5 September 2001.
15 These include United Nations Development Program and World Meteorological Organization, *Hydrometeorological Survey of the Catchments of Lake Victoria*, Unpublished Report 1974.
16 The Swahili word means 'brotherhood'.
17 Y. Arsano, *Ethiopia and the Nile: Dilemmas of national and regional hydro politics*, PhD, Zurich 2004.
18 United Republic of Tanzania, *National Strategy for Growth and Reduction of Poverty*, Dar Es Salaam 2005, 115.
19 See Tvedt and Hovden, *Development of the Nile*.
20 Tractebel Energy Engineering, *Studies on technical and economic justification of the interconnection of networks linked to Rusumo Falls Hydro Power Plant, vol. 1: text, final edition, vol. 2: tables, figures and appendices. Burundi, Rwanda, Tanzania, Uganda*, Unpublished Report 1995.
21 Tractebel Energy Engineering, *Institutional and tariff studies for Rusumo Falls Hydroelectric Power Station. Phase 1, Summary*, Unpublished Report 1997.
22 KBO, *Development programme for the Kagera basin. Final report, vol. 3: Energy*, Unpublished Report 1982.
23 KBO, *Environmental impact studies of the Rusumo Falls Hydroelectric Power project*, Unpublished Report 1993.
24 Tractionel Electobel Engineering, *Rusumo Falls Hydroelectric Scheme*, Unpublished Report 1988.
25 Norconsult and Electrowatt, *Kagera river basin development, phase II: pre-feasibility studies, Kagera river hydropower developments, Rusumo Falls Hydropower Project, Kishanda Valley Hydropower Project, Kakono Dam Hydropower Project*, Unpublished Report 1976.

## 5 THE CONGO NILE: WATER USE, POLICIES AND CHALLENGES

1 'Letter from Mr H.M. Stanley, on his journey from the Albert Nyanza to the southern side of Victoria Nyanza', *Proceedings of the Royal Geographical Society and Monthly Record of Geography*, New Monthly Series 11:12, December 1889, 720–6.
2 For metrics and other data this study relies on R.H. Hughes and J.S. Hughes, *Zaire*, Tresaith 1987. A still useful general work is H. Wellington Wack, *The Story of the Congo Free State*, New York, London 1905.
3 Hughes and Hughes, *Zaire*.

4 Maher Abaza, Africa–Europe electrical interconnection and prospects of worldwide interconnections, CIGRE keynote address, Paris 1994.

5 This includes inter-basin water transfer between the Congo and Lake Chad, and the Congo river and the Nile: personal communication with the DRC minister of the environment, November 2007.

6 R.G. Taylor, N.L. Rose, A.W. Mackay, V. Panizzo, L. Mileham, I. Ssemmanda, C. Tindimugaya, B. Nakileza, A. Muwanga and J. Hau, *Climate Change and the Aquatic Ecosystems of the Rwenzori Mountains, Uganda, Research Report No. 113*, University College London 2007.

7 Hughes and Hughes, *Zaire*.

8 K.J. Sene, 'Theoretical estimates for the influence of Lake Victoria on flows in the upper White Nile', *Hydrological Sciences: Journal des sciences hydrologiques* 45:1, February 2000.

9 T. Tvedt, *The River Nile and its Economic, Political, Social and Cultural Role: An annotated bibliography*, Bergen 2000; *idem*, *The River Nile in the Age of the British: Political ecology and the quest for economic power*, London and New York, 2004; J.V. Sutcliffe and Y.P. Parks, *The Hydrology of the Nile*, IAHS special publication 5, Wallingford 1999.

10 H.E. Hurst *et al.*, *The Nile Basin*, Cairo 1946, cited in Sutcliffe and Parks, *The Hydrology of the Nile*.

11 Sutcliffe and Parks, *The Hydrology of the Nile*; R.O. Collins, *The Waters of the Nile: Hydropolitics and the Jonglei Canal, 1900–1988*, Oxford 1990.

12 Taylor *et al.*, *Climate Change*.

13 Gouvernement de la République Démocratique du Congo, National Assessment of 1994, *Monographie de la province orientale*, Kinshasa 1998.

14 National Assessment of 1994, PNSAR 1998.

15 FAO, UNDP/FAO Regional Project for Inland Fisheries Planning (IFIP), *Report of the Technical Consultation on Management of the Fisheries of Lakes Edward and Mobutu*, Kampala 1990.

16 E.V. De Giessen, *Peace Park Amid Violence? A report on environmental security in the Virunga-Bwindi region*, The Hague 2005.

17 Land ownership is governed by law No. 73-021 of 20 July 1973, as amended by law No. 80-008 of 18 July 1980, which makes the state the sole owner of land. This law provides for prior agreement by a customary representative landowner before legal agreement is reached with the government. Numerous legislative texts regulating the use of natural resources in the DRC are subordinated to this legal framework. A legislative text of 12 July 1932 prescribes measures for permitting fishery exploitation, licensing, indigenous rights and the duties of concessionaires. It is supplemented by a legislative text of 21 April 1937 regulating and prescribing measures for fishing activities and fishing materials, and by Ministerial Act 047/ CAB/ MIN/ ECNT/94 of 18 February 1994, modifying and complementing Act 042/ CAB/ MIN/ ECNT/92 of 6 April 1992 on the exploitation and export of aquarium fish. A legislative text of 6 May 1952 on the concession and administration of inland waters and legislative text No. 74-009 of 10 July on the territorial line of seawater and measures concern their use, protection and exploitation. Numerous ordinances deal with water resource protection and management. An Ordinance Law of July 1914 dealt with pollution and

contamination. Ordinance Law No. 52/443 of 24 December 1952 provided protection for sources, aquifers, streams and rivers, provided for wastewater, and further regulated water rights and ownership. Ordinance Law No. 74/569 of 3 December 1958 regulated irrigation in order to protect public sanitation. Ordinance Law No. 69/041 of 22 August 1969 concerned nature conservation. Ordinance Law No. 71-079 of 26 March 1971 defined the role of the government in rainfall water and wastewater networks. Ordinance Law No. 77-022 of 22 February 1977 defined the duties and responsibilities of the services and directorate of the Department of Environment, Nature Conservation, Waters and Forests. Ministerial Act No. 0072/CAB/ENER/94 of 16 November 1994 described measures and regulations for hydroelectric power plants. Ordinance Law No. 77-019 of 22 February 1977 regulated REGIDESO's charges. Ordinance Law No. 81-013 of 2 April 1981 includes general legislation for mining and hydrocarbons. Ordinance Law No. 91-348 of 27 December 1991 fixed the rates and modalities of taxes and charges for the Ministry of Energy. Ministerial Act No. E/SG/O/0133 C2/93 of 7 March 1993 established conditions for obtaining groundwater and surface water permits, conditions for carrying out drilling of aquifers, and exploitation of groundwater. See K.P. Mbalanda *et al.*, *Les textes législatifs et juridiques en matière environnementales de la République Démocratique du Congo*, Kinshasa 2006.

18 Mbalanda *et al.*, *Les textes*, provides a comprehensive list.

19 Yacob Arsano, *Ethiopia and the Nile: Dilemmas of national and regional hydropolitics*, PhD, Zurich 2004; Metawie Abdel Fattah, 'History of co-operation in the Nile basin', *International Journal of Water Resources Development* 20:1, 2004, 47–63.

20 Arsano, *Ethiopia*.

21 T. Majzoub, *L'Éthiope, le Nil et le droit international public*, Congres International de Kaslik-Liban, 1998.

22 Digitalcongo.net 3.0: Interview with President Kabila, 14 September 2007, retrieved on 14 March 2008.

23 République démocratique du Congo, Ministère de l'énergie, *Électrification du Nord-Kivu, Centrale hydroélectrique de Semliki, Études de préfaisabilité, Version provisoire*, March 2005, cited in NBI-NELSAP, *Strategic/Sectoral, social and environmental assessment of power development options in the Nile Equatorial Lakes Region, Stage II, revised final report*, vol. 2, 2005.

24 F. Corsi, Projet Régional PNUD/FAO pour la Planification des Pêches Continentales (PPEC), *Evaluation des pêcheries zaïroises des lacs Idi Amin/Edouard et Mobutu Sese Seko*, Bujumbura, Burundi 1990.

25 DECIDI is composed of four main associations: the Union des Associations Culturelles et de Développement de l'Ituri (UNADI), the Association pour le Développement des Peuples Forestiers (ADPF), Initiatives pour le Développement d'Aru (IDA/ISEAV-ARU), and the Syndicat d'Initiatives paysannes et Associations Culturelles des Ressortissants d'Aru a Bunia (SIP/LIBARU).

26 Projet Agriculture, Travaux Publics et Energie pour le District de l'Ituri, retrieved 14 March 2008 from www.societecivile.cd/membre/decidi.

27 'Congo-Kinshasa, Les 5 chantiers de Kabila exigent des milliards de dollars US. *Le Phare* (Kinshasa) le 17 septembre 2007: Energie pour tous', retrieved 15 March 2008 from http://fr.allafrica.com/stories.

28 Interview with President Kabila, 14 September 2007, retrieved 14 March 2008 from http://digitalcongo.net.
29 Taylor *et al.*, *Climate Change*.
30 Dominic Johnson, 'Shifting sands: Oil exploration in the Rift valley and the Congo conflict', 2003, retrieved from www.pole-institute.org.

## 6 THE NILE WATERS AND THE SOCIO-ECONOMIC DEVELOPMENT OF WESTERN KENYA

1 Patricia Kameri-Mbote, 'From Conflict to Cooperation in the Management of Transboundary Waters: The Nile experience', in Marc Berthhold (ed.), *Linking Environment and Security: Conflict prevention and peacemaking in East and Horn of Africa*, Washington DC 2004, 16–17.
2 *Ibid.*, 19.
3 FAO Corporate Documentary Repository, 'Irrigation potential in Africa: a basin approach', 6, retrieved 19 March 2007 from www.fao.org/docrep/W4347E/w4347e0k.htm.
4 Eric O. Odada, 'Mitigation of environmental problems in Lake Victoria, East Africa: causal chain and policy options analysis', *OMBIO* 33:1–2, February 2004, 13.
5 George S. Ongweny, 'Water resources of Lake Victoria drainage basin in Kenya', in C.O. Okidi (ed.), *Natural Resources and the Development of the Lake Victoria Basin of Kenya*, Nairobi 1979, 74.
6 Bruce Berman and John Lonsdale, *Unhappy Valley: State and Class Conflict in Kenya and Africa*, London 1992, 50.
7 John O. Oucho, 'Population and its implication for resource development in the Lake Victoria basin', in Okidi, *Natural Resources*, 11.
8 Eric O. Odada *et al.*, *Lake Victoria Basin Environmental Outlook: Environment for development*, Nairobi 2006, 4.
9 Odada, 'Mitigation', 13.
10 Tvedt, Terje, *The Nile in the Age of the British: Political Ecology and the quest for Economic Power*, London and New York 2004, 34.
11 *Ibid.*
12 Bethwell A. Ogot, 'British administration in the Central Nyanza District of Kenya, 1900–60', *Journal of African History*, 4:2, 1963, 249.
13 Caroline Elkins, *Imperial Reckoning: The untold story of Britain's gulag in Kenya*, New York 2005, 9–10.
14 Johnston was Special Commissioner for the Uganda Protectorate between 1899 and 1901.
15 George Bernnet, 'Settlers and politics of Kenya', in Vincent Harlow and E.M. Chilver (eds), *History of East Africa*, Nairobi 1982, 265.
16 *Ibid.*, 293.
17 Council of Ministers, *Individual Applications for Use of the Nile Waters*, CO 822/2202, Public Record Office (PRO), London.
18 John Oyaro Oucho, 'Rural-rural migration field in Kenya: The case of Kericho Tea Estates complex in a regional setting', *Geografiska Annaler*, series B, human geography 66:2,1984, 123–34.

19 C.C. Wringley, 'Kenya: the patterns of economic life, 1902–45', in Harlow and Chilver, *History of East Africa*, 222–3.
20 *Ibid.*, 275.
21 Edwin B. Olouch, 'Farm resources use and household differentiation: A case study of smallholder farm households in Kisumu District, Western Kenya', MA, Agricultural University of Norway 1990, 2-4.
22 B.A. Ogot and W.R. Ochieng (eds), *Decolonization and Independence in Kenya*, London 1995, 64.
23 *Ibid.*, 145.
24 'East African Claim for a share in the Nile Waters: Text of a British Note Dated 11th August, 1959', sent to UAR, Ethiopia, the Sudan and Belgium, Allan Papers, 591/9/36, Sudan Archive, Durham.
25 John Waterbury, *The Nile: National determinants of collective action*, New Haven and London 2001, 153.
26 W.H. Luce, Political Advisor to the Governor-General of the Sudan, to T.E. Bromley, African Department, Foreign Office, 5 May 1955, Allan Papers, 591/5/40, Sudan Archive, Durham.
27 'East African claim for a share in Nile waters', British note to UAR, Ethiopia, the Sudan, Belgium, 11 August 1959, 591/9/36, Sudan Archive, Durham.
28 Brief for Secretary of State's Visit, 'East Africa and the Use of Nile waters', *Individual Applications for Use of Nile Waters* (1959–61) CO 822/2202, PRO, London.
29 *Ibid.*
30 D.G. Reid to Richard Turnbull, 25 November 1959, Individual application for use of Nile waters, CO 822/2202, PRO, London.
31 Permanent Joint Technical Commission for Nile Waters, First annual report, 1960/61.
32 *Ibid.*
33 Individual applications for use of Nile waters, 1959–61, CO 822/2202, PRO.
34 Waterbury, *The Nile Basin*, 4.
35 Governor of Kenya to Macleod, Secretary of State for Colonies, 13 August 1960, CO 822/2202, PRO.
36 Council of Ministers Meetings: Kenya's water requirement from Lake Victoria drainage system, 17 October 1962, CO 822/2204, PRO, London.
37 Minutes of Meeting of Ministers for Kenya, Uganda and Tanganyika, 21 June 1962, CO 822/2204.
38 Secretary of State, Foreign Office to Sir Ralph Stevenson, Cairo, 21 April 1955, Allan Papers, 591/5/38JE, 1423/39, No. 99, Sudan Archive, Durham.
39 Extract from 54th (Routine) Meeting: Kenya Council of Ministers, 14 November 1962, CO 822/881, PRO.
40 Malcolm MacDonald to Secretary of State for Colonies, 15 March 1963, CO 822/3186, PRO.
41 Yosef Yacob, 'Tough talk over a defunct treaty: the case of the 1929 Nile Waters Agreement', retrieved 30 January 2007 from www.ethiopianforum.com/articles/Nile-Treaty.html.
42 Arthur Okoth-Owiro, *The Nile Treaty: State succession and international treaty commitments: a case study of the Nile water treaties*, Occasional Paper, East Africa, 9, 2004, 17–18.

43 Ashok Swain, 'The Nile River basin: Too many cooks, too little broth', *SAIS Review* 22:2, 2002, 4–5.
44 Okidi, *Natural Resources*, 1.
45 Joseph L. Awange and Obiero Ong'ang'a, *Lake Victoria: Ecology, resources, environment*, New York 2006, 80.
46 R.J.M. Swynnerton, *A Plan to Intensify the Development of African Agriculture in Kenya*, Kenya Department of Agriculture, Nairobi 1954, 40.
47 Ongweny, 'Water resources', 74.
48 *Ibid.*
49 Isaack O. Nyambok and Colins Y.O. Owayo, 'Mineral resources of the Lake Victoria basin', in Okidi, *Natural Resources*, 162–7.
50 V.A.O. Odenyo and V. D'Costa, 'Soils of the Lake Victoria basin', in Okidi, *Natural Resources*, 175–96.
51 Ogot and Ochieng, *Decolonization*, 84.
52 H.W.O. Okoth-Ogendo, 'Public interest and private benefit in land and policy: An enquiry into some societal and land use issues in the Lake Victoria', in Okidi, *Natural Resources*, 306.
53 Awange and Ong'ang'a, *Lake Victoria*, 81.
54 *Ibid.*, 82.
55 C.O. Okidi, 'Irrigation activities and institutions in Kenya LVB', in George M. Riugu and Mandivamba Rukiuni (eds), *Irrigation Policy in Kenya and Zimbabwe*, Nairobi 1987, 271.
56 Joseph Mbugua, 'Water resource use efficiency in L. Victoria', in Riugu and Rukuni, *Irrigation Policy*, 42.
57 Sadik Toksoz, 'An accelerated irrigation and land reclamation program in Kenya: Dimension and issues', Development Discussion Paper 14, Harvard Institute for International Development, March 1981, 17.
58 Okidi, 'Irrigation activities', 112.
59 Okidi, 'Irrigation activities', 272.
60 Okidi, 'Irrigation activities', 107.
61 Rudo Niemeijer, 'Commercialization of Rice and nutrition: a case from West Kenya' in Joachim Braun and Eileen Kennedy (eds.) *Agricultural Commercialization, Economic Development, and Nutrition*, Baltimore and London 1994, pp. 264–275, 268.
62 *Ibid.*, 240.
63 Steve Moses Nang'endo, 'The web of poverty: Women and sugarcane farming in Bokoli Location, Bungoma District of Kenya', in *Gender Issues Research Report Series* 4, January 1998, 39–40.
64 D.A. Obara, 'Cotton production and marketing in the Lake Victoria basin of Kenya', in Okidi, *Natural Resources*, 277–80.
65 *Ibid.*, 284.
66 Eric O. Odada *et al.*, 'Mitigation of environmental problems in Lake Victoria, East Africa: Causal chain and policy options analyses', *AMBIO* 33:1–2, February 2004, 19.
67 Korinna Horta, 'The IFC, forests and the PanAfrican Paper Mill in Kenya', *Environmental Defence*, 14 May 2001.
68 Crispin Bokea and Moses Ikiara, *The Macroeconomy of the Export Fishing Industry in Lake Victoria*, Nairobi 2000, 2.

69 *The Times*, 4 January 1928, MAF 209/798, PRO, London.
70 Local Native Council, Minutes and Agenda held on 30–31 May 1932, 3. PC/NZA/2/2/, Kenya National Archives (KNA).
71 Denham to Secretary of State for Colonies, 14 August 1924, Kenya Colony: Development of Inland Fisheries in Lake Victoria, MAF 209/798, PRO, London.
72 Criminal Jurisdiction for Central Kavirondo Native Tribunal, 1937, CO 1018/125, PRO, London.
73 P.C. Kongere, 'Production and socio-economic aspects of the fisheries in Lake Victoria basin', in Okidi, *Natural Resources*, 410.
74 E.B. Worthington, 'Geography and the development of East Africa', *The Geographical Journal* 116:1/3, July–September 1950, 30–1.
75 E.B. Worthington, 'The lakes of Kenya and Uganda', *The Geographical Journal* 79:4, April 1932, 275.
76 *The Times*, 4 January 1928.
77 Worthington, 'The lakes', 293.
78 *Ibid.*, 290–1.
79 Richard E. Dent and E.B. Worthington, 'The introduction of fishes to East African lakes', CO 533/420, PRO, London.
80 Regan to Secretary of State of Colonies, 3 February 1932, CO 533/420, PRO, London.
81 Colonial fisheries, CO 533/449/7, PRO, London.
82 *Ibid.*
83 P.C. Kongere, 'Production and socio-economic aspects of fisheries in Lake Victoria, Kenya', in Okidi, *Natural Resources*, 420.
84 Bokea and Ikiara, *The Macroeconomy*, 3.
85 *Ibid.*, 7.
86 Awange and Ong'ang'a, *Lake Victoria*, 72.
87 Odada *et al.*, 'Mitigation', 15.
88 John S. Balirwa *et al.*, 'Biodiversity and fisheries sustainability in the Lake Victoria basin: An unexpected marriage?', *Bioscience* 53:8, August 2003.
89 M. Medard *et al.*, 'Women and gender participation in fisheries sector in Lake Victoria', retrieved 26 January 2007 from http://www.worldfishcenter.org/Pubs/Wif/wifglobal/wifg_africa_victoria.pdf.
90 *Ibid.*, 156.
91 M. Medard *et al.*, 'Women and gender', 155.
92 C.O. Okidi, 'History of the Nile and Lake Victoria basin through treaties', in P.P. Howell and J.A. Allan, *The Nile: Sharing a scarce resource: a historical and technical review of water management and of economical and legal issues*, Cambridge 1994, 326–7.
93 Awange and Ong'ang'a, *Lake Victoria*, 147.
94 Awange and Ong'ang'a, *Lake Victoria*, 79.

## 7 INDEPENDENT UGANDA AND THE NILE: HYDROELECTRIC PROJECTS AND PLANS

1 Uganda, Physical Setting, retrieved 2 May 2007 from www.photius.com/countries/uganda/geography/index.html.

2 S. Johnson *et al.*, 'Lake Victoria basin: Experience and Lessons learned,' COWI, 2002. See also Lake Victoria catchment area, retrieved 3 October 2007, from http://www.eac.int/lvdp/basin.htm.

3 For an analysis of Uganda's history in a Nile perspective, see T. Tvedt, 'Political Ecology and the Quest for Economic Power', in *The River Nile in the Age of the British*, London and New York 2004, Cairo 2006.

4 *Ibid.*, 27.

5 *Ibid.*, 32.

6 *Ibid.*, 27.

7 Odd Erik Berlid, *Hydro-politics on the Nile: Negotiating the Owen Falls Dam*, MA Thesis, Department of History, University of Bergen, 2001.

8 Press communiqué: 'Equatorial Nile', 10 January 1953, Public Records Office, London.

9 Tvedt, *The River Nile*.

10 J. Waterbury, *The Nile Basin National Determinants of Collective Action*, New Haven and London, 2002.

11 Tvedt, *The River Nile*, 224.

12 J. Kamau, 'Can EA Win the Nile War?', *The Nation*, Kenya, 28 March 2002.

13 Tvedt, *The River* Nile, 223–4.

14 *Ibid.*, 38.

15 For a more detailed description of this story, see *ibid.*

16 Foreign Office minute 1956, cited in *ibid.*, 300.

17 Tvedt, *The River Nile*, 309.

18 Ministry of Water and Environment, Uganda, Uganda National Water Development Report – 2005, retrieved 10 May 2007 from www.unesco.org/water/wwap/wwdr2/case_studies/uganda/pdf/8_water_energy.pdf.

19 Ministry of Energy and Mineral Development, Uganda, retrieved 11 May 2007 from www.energyandminerals.go.ug/EnergyPolicy.pdf.

20 Lana Saravic *et al.*, 'Energy and Development', *The Impact of Hydroelectric Dam Construction on the Environment*, Norwegian University of Science and Technology-Makerere University Summer Course, retrieved 5 June 2007 from http://folk.ntnu.no/lekangso/kurs2005/project-work/Group%20B_report.pdf.

21 E. Biryabarema, 'Heavy rains boost power generation', *The Monitor*, Kampala, 21 September 2007, 1.

22 T. McConnell, 'UN Accuses Uganda of Draining Lake Victoria', *The Independent*, London, 9 February 2006.

23 D. Kull, 'Connections Between Recent Water Level Drops in Lake Victoria, Dam Operations and Drought', *AfricaFocus Bulletin*, retrieved 5 May 2007 from www.irn.org/programs/nile/pdf/060208vic.pdf.

24 Angelo Opio *et al.*, 'The Real Culprits in Lake Victoria Water Drop', *The Nation*, Kenya, 3 March 2006.

25 I. Kasita, 'More Power Expected as Water Levels Rise', *The New Vision*, Kampala, 9 February 2007.

26 Tristan McConnell, 'UN Accuses Uganda'.

27 R. Okeyo, 'EALA MPs Accuse Uganda Over Receding Lake Victoria', *The Link*, 21 July 2006.

28 *Ibid.*

29 B. Among, 'Lake Victoria Levels Rise', *Daily Monitor*, Kampala, 26 June 2007.

30 Saravic *et al.*, 'Energy and Development'.

31 International Rivers Network (IRN), 'Uganda Harnessing Power for the People', Nairobi, 12 March 2003.

32 IRN, Report by Prayas Energy Group, 'The Bujagali Power Purchase Agreement – an Independent Review', 20 November 2002.

33 Frank Muramuzi of NAPE recommended that the WB should investigate the poor quality of the bank's assistance to the Uganda government during the negotiation of the Bujagali project and that there should be a participatory review of the bank's anti-corruption policies: IRN Report, 20 November 2002.

34 J. Tumusiime, 'Hydro challenges to the Victoria Nile at Bujagali Falls', *World View Magazine* online 15:4, Fall 2002.

35 Environmental News Service (ENS), 'Power Giant AES Withdraws From Uganda Dam Project', 17 October 2006.

36 S. Lilley, 'AES Backs Out of Bujagali Dam Project', *Corp Watch*, 28 August 2003.

37 'Bujagali Energy Ltd. Project description', International Finance Corporation (IFC) Project No. 24408, 20 December 2006.

38 W. Wafula *et al.*, 'Electricity Prices to Drop by 50%', *The Monitor*, 22 August 2007; F. Masiga, 'ADB Boss Calls for More Energy Solutions', *Daily Monitor*, Kampala, 6 October 2007; and 'World Bank Approves Funding for Hydropower', *IRIN Humanitarian News and Analysis*, Monday 3 December 2007.

39 IFC Summary Proposal, 20 December 2006; see also Economic and Financial Evaluation Study by Independent Consulting Power Planning Associates (UK) Commissioned by IFC February 2007.

40 IFC 2007, 'Bujagali Hydro Project: Frequently Asked Questions', www.worldbank.org/Bujagali; see also R.J. Burnside, International Ltd Canada, July 2006.

41 F. Muramuzi *et al.* of NAPE Uganda, 'NGO Letter to World Bank President', Bank Information Centre (BIC), 10 January 2004.

42 Wafula *et al.*, 'Electricity Prices to Drop by 50%'.

43 F. Muramuzi *et al.*, Letter to J. O'Connor, World Bank Country Representative Kampala, Uganda', IRN, 2 November 2006.

44 I. Kasita, 'World Bank Flags Off Bujagali Dam Project', *The New Vision*, 8 April 2007.

45 Aryamanya Mugisha, 'Nema Clears Bujagali Dam Construction', *The New Vision*, Kampala, 25 April 2007.

46 I. Kasita, 'World Bank Flags Off Bujagali Dam Project'.

47 F. Masiga, 'The World Bank Approves Funding Bujagali Project', *The New Vision*, 28 April 2007.

48 Kasita, 'World Bank Flags Off Bujagali Dam Project'.

49 G. Rulekere, 'Power Crisis Hits Harder in Uganda', U.G. Pulse, *Emboozi* newspaper, 30 October 2006.

50 M. Karugaba, 'Power Shortage Causes Ugs 432 b Annual Loss', *New Vision*, 28 April 2007, 46.

51 IRN, 'Managing Social Risks: Bujagali Hydropower Project Case Study', April 2006.

52 J. Karuhanga, 'Uganda Power Shortage Impacting Revenue Collections', Brogger News Network (BNN), 31 October 2006.

53 M. Kabagambe, quoted in S. D'Ujanga, State Minister for Energy, 'Government to Loan Bujagali Builders Shs.176 b to Kick-start Project', *The New Vision*, 21 March 2007.
54 'Uganda: Electrical Power – Overview', *MBendi Information for Africa*, 11 May 2006.
55 'Uganda Energy Policy', retrieved 14 May 2007 from www.energyandminerals.go.ug/EnergyPolicy.pdf.
56 FAO-AGL, 'Gateway to Land and Water Information: Uganda National Report', last updated 4 March 2004.
57 International Development Research Centre (IDRC), Canada, 'Egypt Sends Water Experts to Uganda', 1 February 2005.
58 East African Community Secretariat Arusha, Tanzania, 'Lake Victoria Basin Commission – Special Report', April 2006, retrieved 27 November 2007 from www.fao.org/ag/agL/swlwpnr/reports/y_sf/z_ug/ug.htm.
59 'Fish Industry in Uganda', retrieved 13 May 2007 from www.ugandainvest.com/fishing.pdf.
60 D. Nakaweesi, 'Uganda Fish Exports Earns UGS 255 Billion in 2005', *Daily Monitor*, Kampala, 4 February 2006.
61 Phillip Nabyama, 'East African: Countries Close to Marketing Region's Tourism As Bloc', *East African Business Week*, Kampala, 19 November 2007.
62 From Uganda Ministry of Tourism, Trade and Industry, Kampala, Uganda.
63 From Jovino Akaki, Tourism State Minister, Ministry of Tourism, Trade and Industry, Uganda, 2006.
64 K.J. Kelley, 'New Surveys Show East Africa May Have Huge Oil Deposits', *The East African*, 10 May 2004.
65 D. Johnson, 'Shifting Sands and Oil Exploration in the Rift Valley and the Congo Conflict', retrieved 29 November 2007 from www.pole-institute.org/documents/heritage05.pdf.
66 M. Olupot, 'New Oil Discovery Worth Sh15,000b', *The New Vision*, Kampala, 24 June 2007.
67 For a discussion of the historical context of this agreement, see Tvedt, *The River Nile*, 144.
68 'International Agreements on Use of River Nile Water', *Biosafety News* 43, August 2003.
69 Kefyalew Mekonnen, 'The Defects and Effects of Past Treaties and Agreements on the Nile River Waters', retrieved 2 June 2007 from http://ethiopians.com/abay/engine.html.
70 Sir William Gowers, Governor of Uganda, criticized the Agreement in a dispatch to the Secretary of State, Uganda Protectorate, No. 603, Public Relations Office, London.
71 Waterbury, *The Nile Basin*, 153; F. Webber, 10 March 1960, Public Relations Office, London.
72 Y. Yacob, 'Tough Talk over a Defunct Treaty: The Case of the 1929 Nile Waters Agreement', retrieved 12 June 2007 from www.ethiopianforum.com/articles/Nile-Treaty.htm.
73 IRIN, 'The Nile: Water Conflicts', *Science in Africa*, May 2003.

74 'K. Museveni of Uganda calls on Egypt to stop egoism towards the Nile', *ArabNews.com*, 4 February 2004.

75 Ministry of Foreign Affairs Kampala Uganda, Press Release, Foreign Policy Overview, 3 March 2007.

76 Waterbury, *The Nile Basin*, 150–1.

77 J. Ntambirweki, 'Colonial Treaties and Legal Regime of the Nile Valley: Rethinking the Legal Framework into the Twenty First Century, with Special Emphasis on Uganda's Interests', *The Uganda Journal* 43:1, December 1996, and 44:2, December 1997.

78 *Ibid.*, *The Uganda Journal* 44:2.

79 Waterbury, *The Nile Basin*, 159.

80 C. Onyango-Obbo, 'Africa's Struggle for Nile Water Grows Turbulent', Reuters, Tanzania, 9 February 2004.

81 G. Kateihwe from Uganda, 'How should the Nile waters be shared?', BBC, Focus on Africa, 1705 GMT, 13 March 2004.

82 Musoke from Uganda, BBC, Focus on Africa.

83 John Kamau, 'Can EA Win the Nile War?', *The Nation*, 28 March 2002.

84 C. Hanley, 'Africa States Work to Share Nile River Water', *Black American web.com*, December 2005.

85 *The East African*, Nairobi, 19 December 2006; *Pan African News Wire*, 1 January 2007.

86 A. Okoth-Owiro, *The Nile Treaty, State Succession and International Treaty Commitments: A Case of the Nile Water Treaties*, Nairobi 2004.

87 *Ibid.*, 28.

88 A. Swain, 'The Nile Basin Initiative: Too Many Cooks, Too Little Broth', *SAIS Review* 22, Summer–Fall 2002.

## 8 INSTITUTIONAL DEVELOPMENT AND WATER MANAGEMENT IN THE ETHIOPIAN NILE BASIN

1 Government of Ethiopia, 'Resettlement and rural development in Ethiopia', Milan 2002.

2 Aberra Mekonnen and Deksios Tarekegn, 'Water resources of Ethiopia', paper presented at the National Workshop on Water Resources Utilization: Water Supply, Agriculture and Hydroelectric Power, Addis Ababa 2001; Mehret Debebe, *Power Sector Reform Direction*, Addis Ababa 2002, 7.

3 Bairu Tafla, 'The father of rivers: the Nile in Ethiopian literature', in Haggai Erlich and Israel Gorshoni (eds), *The Nile: Histories, cultures, myths*, Boulder and London 2000, 154–6.

4 Damaka Mattaferia, 'The Nile river, need for equitable use of its waters', *Ethiopian Register*, July 1994, 46–57; August 1994, 41–9; September 1994, 51–3; November 1994, 39–41; December 1994, 40–6.

5 *Digest of Old Ethiopian judgments*, nd, III, 303, 305; X, 1226, 1228.

6 *Ethiopia Observer* 2:2, 1958, 93.

7 John Waterbury, *The Nile Basin: National determinants of collective action*, New Haven and London 2002, 64.

8 Zewde Gebre Selassie, 'The Nile River Question: In a new era of cooperation among riparian states', paper presented at the Nile 2002 Conference, Addis Ababa, 26–29 June 2000.

9 Zewde Gebre Selassie, 'The Nile Question: 1955–1964, the Ethiopian perspective', paper presented at Tel Aviv University, May 1997.

10 Robert Collins, *The Waters of the Nile: Hydropolitics and the Jonglei Canal, 1900–1988*, Princeton 1990, 166.

11 Lahmeyer Consulting Engineers, *Gilgel Abbai Scheme*, Addis Ababa 1962.

12 Collins, *The Waters*, 277; Paolo Dieci and Claudio Viezzoli, *Resettlement and Rural Development in Ethiopia*, Milan 1992, 61; Robin Clarke, *Water: The international crisis*, London 1991, 161.

13 Tesfaye Tafesse, *The Nile Question: Hydropolitics, legal wrangling, modus vivendi and perspectives*, Hamburg 2001, 49; Dieci and Viezzoli, *Resettlement*, 343–5.

14 *The Ethiopian Herald*, 20 April 2001.

15 Jon Martin Trolldalen, *International Environmental Conflict Resolution: The role of the United Nations*, Washington DC 1992, 79.

16 Thomas Naff, 'Conflict and water use in the Middle East', in P. Rogers and P. Lydon (eds), *Water in the Arab World*, Cambridge MA 1994, 253–84.

17 Panel of Experts, Nile Basin Cooperative Framework, Draft of the Cooperative Framework, Revision 2, Entebbe, 10 December 1999, 5.

## 9 THE HISTORY OF THE NILE WATERS IN THE SUDAN

1 Terje Tvedt, *The River Nile in the Age of the British: Political ecology and the quest for economic power*, London and New York 2004, 113.

2 According to the first census, conducted in 1956, the total was 10,262,536. Of the estimated 40.2 million in 2006, 30.75 million are in the north and about 9.46 million in the south. See FAO/WEP Crop and Food Supply Assessment Mission to Sudan, 1 February 2007, retrieved from www.fao.org/giews.

3 Saghayroon El Zein, 'The water resources of the Nile for agricultural development in the Sudan', in Mohammed Obeid (ed.), *Aquatic Weeds in the Sudan*, Khartoum 1975, 2.

4 D. Knott and R. Hewett, 'Water resources planning in the Sudan', in P.P. Howell and J.A. Allan (eds), *The Nile, Sharing a Scarce Resource: An historical and technical review of water management and economical and legal issues*, Cambridge 1994, 207.

5 *Ibid.*, 205.

6 Report by Sir Alexander Gibb and Partners, *Estimation of Irrigable Areas in the Sudan 1951–1953*, 1954, v.

7 1 *feddan* = 0.42 ha or 0.0042 km$^2$ or 1.038 acres.

8 Gibb and Partners, *Estimation of Irrigable*, iv.

9 *Ibid.*, 60.

10 *Ibid.*, 61.

11 John Garang de Mabior, *Identifying, Selecting, and Implementing Rural Development Strategies for Socio-economic Development in the Jonglei Projects Area, Southern Region, Sudan*, PhD, Iowa State University 1981, 28.

12 G.W. Grabham, *Water Supplies in the Anglo-Egyptian Sudan*, Khartoum, Sudan Government 1934, 9.
13 Tvedt, *The River Nile*, 110.
14 *Ibid.*, 112.
15 Republic of the Sudan, *Sennar Buri Project*, Sennar Hydro Electric Power Station, inauguration of stage two, 2 November 1962.
16 Tony Barnett, *The Gezira Scheme: An illusion of development*, London 1977, 159.
17 Tvedt, *The River Nile*, 93.
18 T.M. Ali, *The Cultivation of Hunger: State and agriculture in the Sudan*, Khartoum 1989, 30.
19 J.D. Tothill (ed.), *Agriculture in the Sudan*, Oxford 1948, 600.
20 Arthur Gaitskell, 'The Sudan Gezira Scheme', *African Affairs*, 1952, 51:205, 311–12.
21 Tothill, *Agriculture in the Sudan*, 604–7.
22 *Ibid.*, 608.
23 *Ibid.*, 209.
24 Characteristics of the *saqiya* system have been discussed in Anders Bjorkelo, *Prelude to the Mahdiyya: Peasants and traders in the Shendi region, 1821–1885*, Cambridge 1989, 64–5.
25 Tothill, *Agriculture in the Sudan*, 597.
26 Tvedt, *The River Nile*, 97.
27 *Ibid.*, 155.
28 Tothill, *Agriculture in the Sudan*, 790.
29 *Ibid.*, 792.
30 Abbas Hidaytalla Abdalla, 'Water supply factors in Sudan', in A.B. Zahlan (ed.), *Agricultural Sector of Sudan: Policy & systems studies*, London 1986, 316–17.
31 Ali, *The Cultivation of Hunger*, 42.
32 Tothill, *Agriculture in the Sudan*, 749.
33 Ministry of Irrigation and Hydro-electric Power, *Irrigation by Pumps from the Nile in the Sudan*, Khartoum 1955, 7–8.
34 Abbas Hidaytalla, 'Water supply factors in Sudan', 314.
35 *Ibid.*, 714.
36 Tvedt, *The River Nile*, 251.
37 G.R. Warburg, 'Sources of the Nile and Egyptian-Sudanese relations', *Middle Eastern Studies* 43:3, May 2007, 477.
38 *Al-Sudan min 13 February 1841 ila 12 February 1953*, Egyptian Government, 1953; 2nd edited version, Fadwa Taha, 'The Sudan from 12 February 1841 to 12 February 1953', 2006, 52–53. See also M.W. Daly, *Empire on the Nile: the Anglo-Egyptian Sudan, 1898–1934*, Cambridge 1986.
39 For the full text of this ultimatum and other relevant documents, see Government of Egypt, *Al-Sudan min 13 February 1841 ila 12 February 1953*, (ed.) Fadwa Taha, Khartoum 2006, 52–3.
40 The commission was composed of J.J. Canter Cremers, Chairman; R.M. MacGregor, British Delegate; and Abdel Hamid Sulayman, Egyptian delegate.
41 *Exchange of notes between His Majesty's government in the United Kingdom and the Egyptian government in regard to the use of the waters of the River Nile for irrigation purposes, Cairo, May 7, 1929, Treaty Series No. 17* (1929) Cmd. 3348, 6.

42 Tvedt, *The River Nile*, 145. See the same reference for further details about the 1929 Nile Waters Agreement.

43 Tothill, *Agriculture in the Sudan*, 596–7.

44 A detailed account of Sudanese–Egyptian negotiations on the Nile waters is in Fadwa Taha, *The Sudanese Role in Negotiating and Reaching a new Nile Waters Agreement, 1949–1959*, a modified version of a seminar paper presented on 16 March 2007 to the Nile Basin Research Programme, University of Bergen. See also: I.H. Abdalla, 'The 1959 Nile Waters Agreement in Sudanese-Egyptian relations', *Middle Eastern Studies* 7:3, 1971, 329–41.

45 C.D.O. Farran, 'The Nile Waters Question in international law', *Sudan Notes & Records*, 41, 1960, 98.

46 Proceedings of the Sudan Legislative Assembly, First Assembly 1948–52.

47 First Legislative Assembly, Daily Digests, 17 May 1951, 345–9.

48 Ministry of Irrigation and Hydro-Electric Power, *The Nile Waters Question: The case for the Sudan, the case for Egypt and the Sudan's reply*, Khartoum 1955.

49 United Kingdom Trade Commissioner, Khartoum, to African Department, Foreign Office, 6 February 1954, PRO FO 371/108514.

50 Ministry of Irrigation and Hydro-Electric Power, *The Nile Waters Question*, 3.

51 *Ibid.*

52 *Ibid.*

53 Cory, the American member of the 1920 Commission, accepted the vested rights of Egypt and the Sudan to irrigate 5,400,000 and 400,000 *feddans* consuming 40 and 1.5 bcm respectively; excess water over and above such vested rights should be divided equally between Egypt and the Sudan according to prospective cultivable lands in each country.

54 Salah Salim to Ismail El Azhari, Prime Minister of the Sudan, 25 July 1955, 591/15/52, Sudan Archive, Durham.

55 Minister of Irrigation and Hydro-Electric Power to the Minister of Works, Egypt, 20 November 1955, 591/6/7–8, Sudan Archive, Durham.

56 Morrice, Irrigation Adviser to the Minister of Irrigation and Hydro-Electric Power, 28 December 1955, 591/6/18, Sudan Archive Durham.

57 The Egyptian government had moved troops to Halayib, a border area, claiming it belonged to Egypt even though it had been defined as a Sudanese constituency in the elections of 1954.

58 Robert Collins, *The Waters of the Nile: Hydropolitics and the Jonglei Canal, 1900–1988*, Princeton 1996, 265–6.

59 W.N. Allan to Mohammed el Rashid Sid Ahmed, Ministry of Irrigation, Sudan, 10 June 1959, 591/9/3, Sudan Archive, Durham.

60 Collins, *The Waters of the Nile*, 269.

61 Beshir Abdel Rahim *et al.*, 'The waters of the Nile and the future of the Sudan', 21 August 1955, 591/5/60–67, Sudan Archive, Durham.

62 The full text of the agreement is in Permanent Joint Technical Commission for Nile Waters, *First Annual Report, 1960/1961*, 7–15.

63 Collins, *The Waters of the Nile*, 112.

64 *Ibid.*

65 Ministry of Irrigation and Hydro-electric Energy, *Control and Use of Nile Waters in the Sudan*, Khartoum 1975, 26–7.

66 First Sudanese Parliament, *Second Session of the First Parliament*, No. 1, 10 March 1954, 165.
67 M.M. Abdel Salam, 'Agricultural policy formation and administration', in Zahlan, *Agricultural Sector of Sudan*, 409.
68 Ministry of Irrigation and Hydro-electric Energy, *Control and Use of Nile Waters in the Sudan.*
69 Terje Tvedt, *The Management of Water and Irrigation: The Blue Nile*, Bergen 1991, 3.
70 *Ibid.*
71 D.J. Shaw, *The Managil South-Western Extension: An extension of the Gezira Scheme*, Bulletin 9, 1965, 10.
72 M.M. Khogali, *The Managil Extension*, Khartoum 1974.
73 Shaw, *The Managil South-Western Extension*, 8.
74 Khogali, *The Managil Extension.*
75 *Ibid.*, 12.
76 Sudan Republic, *Sudan Almanac* 1963, 65–6.
77 John R. Randell, 'Patterns of settlement in the Managil Extension to the Sudan, Gezira Scheme', *Sudan Notes & Records*, 47, 1966, 88.
78 H. Bell, *Irrigation by Gravity from the River Atbara: A preliminary report on a dam at Khashm El Girba and the area it can command*, Part 1, Khartoum n.d., 1.
79 *Ibid.*, 48.
80 Mohammed H.S. Ibrahim, 'Irrigation projects in the Sudan: the promise and the reality', *Journal of African Studies* 10:1, 1983, 3.
81 M. Shazali Osman and H.E. El Hag, 'Irrigation practices and development in the Sudan', *Sudan Notes & Records*, 55 1974, 100. See also Gunnar R. Sørbø, *Tenants and Nomads in Eastern Sudan: A study of economic adaptations in the New Halfa Scheme*, Uppsala 1985.
82 *Ibid.*, 8.
83 *Al-Raya'm*, 2 April 2007. A reporter from the newspaper went to Khashm El Girba and interviewed the dam's administrators.
84 *Ibid.* (referring to statements made by officials of the ministry of irrigation in a forum on 'Uses of water, the present and the future' organized by the Sudan Engineering Society).
85 William Garstin, *Report upon the Basin of the Upper Nile*, cmd. 2165, London 1904.
86 Administration Agreement (Roseires Irrigation Project) between the Republic of the Sudan and Kreditanstalt Fur Wiederaufbau and International Development Association and International Bank for Reconstruction and Development, 14 June 1961.
87 *Ibid.*
88 Sir Alexander Gibb and Partners, *Memorandum on Utilization of Water from Roseires Reservoir*, January 1966.
89 *Ibid.*, 3.
90 *Ibid.*, 28.
91 Ministry of Irrigation and Hydro-Electric Power, article on Roseires Dam project, November 1961.
92 Permanent Joint Technical Commission for Nile Waters (PJTC), 'Agreement between the republic of the Sudan and the United Arab Republic for the full utilization of the Nile waters', in *First Annual Report 1960–1961*, 8.

93 *Al-Raya'm*, 17 April 2007.
94 John Waterbury, *The Nile Basin: National determinants of collective action*, New Haven and London 2002, 136.
95 Ministry of Irrigation and Hydro-Electric Power, *The Nile Waters Question*, 13.
96 H. Morrice and W. Allan, *Planning for the Ultimate Hydraulic Development of the Nile Valley*, London 1959.
97 *African Research Bulletin Economic, Financial and Technical Series* 36:8, 16 August–15 September 1999.
98 J.A. Allan, 'Developing policies for harmonized Nile waters development and management', in P.P. Howell and J.J. Allan (eds), *The Nile: Sharing a scarce resource*, 387.
99 *Al-Ingaz*, 20 November 1991.
100 Waterbury, *The Nile Basin*, 137.
101 'Sad Merowi', *Monthly Scientific Magazine* 48, August/September 2007.
102 Ibrahim, 'Irrigation projects in the Sudan', 9.
103 M. El Hadi Abu Sin, *Livestock Economy and Attitude of Tenant: Rahad and Khashm El Girba projects, a comparative study*, Khartoum 1983.
104 Ibrahim, 'Irrigation projects in the Sudan', 9.
105 Democratic Republic of the Sudan, *Rahad Irrigation Project*, December 1977.
106 Ibrahim, 'Irrigation projects in the Sudan', 10.
107 R.O. Collins, *The Jonglei Canal: The past and present of a future*, Durham 1986, 11.
108 Tvedt, *The River Nile*, 68.
109 *Ibid.*
110 *Ibid.*, 229.
111 P.P. Howell (ed.), 'The Equatorial Nile project and its effects in the Sudan', *Geographical Journal* 119:1, March 1953, 33–48. All reports of the Jonglei Canal Project are kept in the PJTC library, Khartoum.
112 Garang de Mabior, *Identifying, Selecting, and Implementing Rural Development Strategies*, 44–51.
113 Collins, *The Waters of the Nile*, 122.
114 Tvedt, *The River Nile*, 233.
115 Democratic Republic of the Sudan, Ministry of Culture and Information, *Jonglei Project (Phase 1)*, 51.
116 PJTC, Annual Report 1973–74.
117 Anthony Sylvester, *Sudan under Nimeiri*, London 1977, 178.
118 PJTC Anuual Report, 1977–78.
119 *Jonglei Project (Phase 1)*, 64.
120 Garang de Mabior, *Identifying, Selecting, and Implementing Rural Development Strategies*, 53.
121 *Ibid.*, 219–20.
122 Dahilon Yassin Mohamoda, 'Nile Basin cooperation: A review of the literature', *Current African Issues* 26, 2003, 28.
123 *Sudan Tribune*, 18 April 2007.
124 *Ibid.*
125 *Al-Sudani*, 16 May 2007.
126 Ministry of Irrigation and Hydro-Electric Power, *Irrigation by Pumps*, 1.

127 Omer Mohammed Osman, 'Some economic aspects of private pump schemes', *Sudan Notes & Records*, 39, 1958, 40.

128 Mahmoud El Zain, *Environmental Scarcity, Hydropolitics and the Nile: Population concentration, water scarcity and the changing domestic and foreign politics of the Sudan*, PhD, The Hague 2007, 141.

129 *Ibid.*, 41.

130 Special Report, FAO/WFP Post Harvest Assessment Mission to Sudan, 25 May 2006.

131 Ministry of Irrigation and Hydro-Electric Power, *Irrigation by Pumps*, 30.

132 Office of the United Kingdom Trade Commissioner, Khartoum to African Department, Foreign Office, 25 October 1955, PRO, FO 371/113737.

133 'The supply and demand factors for agricultural products in the Sudan', in A.B. Zahlan (ed.), *Agricultural Sector of Sudan*, 84.

134 *Ibid.*

135 Osman Ali Osman El Makki, *Challenges and Possibilities of Drip and Canal Irrigation in Northern Sudan*, MA, Norwegian University of Life Sciences, 2006, 3.

136 Abbas Hidaytalla, *Water Supply Factors in Sudan*, 321.

137 *Ibid.*, 303.

138 E.A. Farah *et al.*, 'Sources of groundwater recharge at the confluence of the Nile, Sudan', *Environmental Geology* 39:6 April 2000, 667.

139 Ministry of Irrigation and Hydro-Electric Power, *Control and Use of the Nile Waters*, Khartoum 1975.

140 Saghayroon El Zein, 'The water resources of the Nile for agricultural development in the Sudan', in Mohammed Obeid (ed.), *Aquatic Weeds in the Sudan*, Khartoum 1975, 2.

141 Gibb and Partners, *Estimation of Irrigable Areas in the Sudan*, 48.

142 Ministry of Irrigation and Hydro-Electric Power, *Control and Use of the Nile Waters*.

143 *Ibid.*

144 Ali, *The Cultivation of Hunger*, 68.

145 *Ibid.*

146 K. Wohlmuth, 'National policies for agriculture', in G.M. Craig (ed.), *The Agriculture of the Sudan*, Oxford 1999, 438.

147 *Ibid.*

148 *Al-Rayaam*, 1 May 2007.

149 Abbas Hidaytalla, *Water Supply Factors in Sudan*, 326.

150 Government of the Sudan, *Comprehensive National Strategy 1992–2002*, vol. 1.

151 Mahmoud El Zain, *Enviromental Scarcity, Hydropolitics, and the Nile*, 163.

152 Simon A. Mason, *From Conflict to Cooperation in the Nile Basin*, PhD, Zurich, 2003, 155–7.

153 *Al-Rayaam*, 8 April 2007.

154 *Bank of Sudan 44th Annual Report*, 2004.

155 *Ibid.*

156 Statement by the Minister of Irrigation, *Al-Khartoum*, 28 March 2001.

157 Ashok Swain, 'Ethiopia, the Sudan, and Egypt: The Nile river dispute', *Journal of Modern African Studies* 35:4, 1997, 682. See also *Sudan Strategic Plan 1992–2002*, vol. 1, and Sudan country paper 1996 presented to the Conference of the Nile 2002.

158 Mason, *From Conflict to Cooperation*, 156–7.

159 Mahmoud El Zain, *Environmental Scarcity, Hydropolitics, and the Nile*, 325.
160 Abbas Hidaytalla, *Water Supply Factors in Sudan*, 317.
161 *Ibid.*, 318.
162 *Ibid.*
163 C.J. Stigter *et al.*, 'Farmers Practice in On-Farm Irrigation Management in the Gezira Scheme, Central Sudan', *Rural Environ. Eng.*, 38, 2000, 20–29.
164 A.M. Adeeb, 'Water productivity of food crops in Gezira Scheme, Sudan', *Environmentally Sound Technology in Water Resources Management*, 2006, 515–802.
165 Barnett, *The Gezira Scheme*, 106.
166 *Ibid.*
167 Abbas Hidaytalla, *Water Supply Factors in Sudan*, 315.
168 *PJTC First Report 1960–1961.*
169 *Ibid.*, 10.
170 Waterbury, *The Nile Basin*, 133.
171 *PJTC Reports.* See also Abbas Hidaytalla, *Water Supply Factors in Sudan*, 303.
172 *PJTC Report 1965–1966* and *1966–1967.*
173 Statement of the seventh meeting of the Joint Supreme Ministerial Committee of Integration between Egypt and the Sudan, 15 January 1979, in 'Integration between the Sudan and Egypt' prepared by Sudan News Agency (in Arabic), 36.
174 For more views about Unity of the Nile Valley, see Gabriel R. Warburg, *Historical Discord in the Nile Valley*, London 1992.
175 For more details, see Fadwa Taha, *The Anglo-Egyptian Dispute over the Sudan 1937–1947*, Khartoum 2006.
176 'Integration between the Sudan and Egypt', Sudan News Agency, 41–72.
177 Full text in 1/1300/4158, Miscellaneous, National Record Office, Khartoum.
178 Mohamoda, 'Nile Basin cooperation: A review of the literature', 19.
179 *Sudan Tribune*, 21 April 2007.
180 'East African claim for a share in Nile waters', 11 August 1959, sent to UAR, Ethiopia, Sudan, Belgium, 591/9/36, Sudan Archive, Durham.
181 Text of a note presented by Her Majesty's government to the Egyptian government on 22 November 1955, PRO FO371/119062.
182 Morrice and Allan, *Planning for the Ultimate Hydraulic Development of the Nile Valley.*
183 The Permanent Joint Technical Commission, *Tenth Annual Report 1969–1970*, 48.
184 Ministry of Irrigation and Hydro-Electric Power, *Control and Use of the Nile Waters*, 26.
185 Mohamoda, *Nile Basin Cooperation*, 19–20.
186 The Permanent Joint Technical Commission, *Second Annual Report 1961–1962*, Cairo 1965, 1.
187 The Permanent Joint Technical Commission, *Third Annual Report 1962–1963*, 63.
188 Collins, *The Waters of the Nile*, 288.
189 *Ibid.*, 290.
190 A Swahili word meaning 'brotherhood'.
191 Tvedt, *The Management of Water and Irrigation*, 25.
192 Mohamoda, *Nile Basin Cooperation.*
193 Waterbury, *The Nile Basin*, 137.
194 *Ibid.*, 171–2.

195 *Ibid.*, 149.

196 *Al-Sahafa*, 7 July 2007.

197 Draft Agreement on the Nile River Basin Cooperative Framework, Negotiation Committee, Entebbe, 20–24 September 2004.

198 Yacob Arsano, *Ethiopia and the Nile Dilemmas of National and Regional Hydropolitics*, unpublished Ph.D dissertation, University of Zurich 2004.

## 10 EGYPT AND COLLECTIVE ACTION MECHANISMS IN THE NILE BASIN

1 Asit K. Biswas, 'Water for sustainable development: A global perspective', *Wasteland News* 7:2, 1991, 18–21; *idem*, 'Management of international water resources: some recent developments', in *idem* (ed.), *International Water of the Middle East: From Euphrates Tigris to Nile*, Bombay and New York 1994; John Bullock and Adel Darwish (eds), *Water Wars: Coming conflicts in the Middle East*, London, 1993; Marq De Villiers, *Water Wars: Is the world's water running out?* London 2001, 235–48; M. Falkenmark, 'The massive water scarcity now threatening Africa: Why isn't it being addressed?', *Ambio* 18:2, 1989, 112–18; P. Gleick, *Water in Crisis: A guide to the world's freshwater resources*, New York 1993; Thomas Homer Dixon, 'The ingenuity gap: Can poor countries adapt to resources scarcity?', *Population and Development* 21:3, 1995, 587–612; *idem*, 'Environmental scarcity, mass violence and the limits to ingenuity', *Current History* 95, 1996, 359–65; Jacques Leslei, 'Running dry: What happens when the world no longer has enough fresh water?', *Harper's*, 1 July 2000; Sandra Postel, *Last Oasis: Facing water scarcity*, Washington DC and New York 1997; J.R. Starr, 'Water wars', *Foreign Policy* 82, Spring 1991, 17–36.

2 Hussein I. Abdel-Shafy and Raouf O. Aly, 'Water issues in Egypt: resources, pollution and protection endeavors', *Central European Journal of Occupational and Environmental Medicine*, 8:11, 2002, 3–21.

3 M. Hvidt, 'Water resource planning in Egypt', Odense University, retrieved from http://www.netcomuk.co.uk/~jpap/hvidt.htm.

4 Bayoumi B. Attia, 'Water as a basic human right in Egypt', *Global issue papers* 11, supplement 1, September 2004.

5 MWRI, *National water resources plan for Egypt, planning sector, 2017*, Cairo 2005.

6 Integrated Water Resources Management (IWRM) is defined as a process that promotes the coordinated development and management of water, land and related resources, in order to maximize the resulting economic and social welfare in an equitable manner without compromising the sustainability of vital ecosystems.

7 Control of the Nile began in about 3400 BC, and the Egyptians constructed El-Kafara Dam in Wadi Garawi as early as 2700–2600 BC.

8 Olli Varis, 'The Nile basin in a global perspective: Natural, human and social-economic resources nexus', *Water International* 25:4, 2000, 624–37.

9 The downstream barrages are Esna, Nag Hammadi, Asyut, Damietta, Rosetta, Zifta, Edfina and Farascour.

10 South valley development project, retrieved from http:/www.mwri.gov.eg/En/project_toshka%20.html.

11 'Contributions to environmental management of Egypt's groundwater resources', EMGR project, final report, October 1999. For technical definitions of pollution, see V. Novotny and H. Olem, *Water Quality: Prevention, identification and management of diffuse pollution*, New York 1994.

12 Attia, 'Water'.

13 Egyptian National Committee on Irrigation and Drainage (ENCID), 'Unknown resource', retrieved from www.mwri.gov.eg/Encid/english/irregye.htm#.

14 *Ibid.*

15 *Ibid.*; Attia, 'Water'.

16 ENCID, 'Unknown resource'.

17 *Ibid.*

18 *Ibid.*

19 *Ibid.*

20 Attia, 'Water'.

21 *Ibid.*

22 The history of international water treaties dates as far back as 2500 BC, when two Sumerian city-states crafted an agreement ending a water dispute along the Tigris. The Food and Agricultural Organization of the United Nations has documented more than 3,600 international water treaties dating from AD 805 to 1984. Although the vast majority concern navigational issues, a growing number address water as a limited and consumable resource.

23 Department of Geosciences, Oregon State University, 'Transboundary freshwater dispute database project', United States 2003.

24 The structural administration of water resources and irrigation began in 1836, when the Public Works Department was established by Mohamed Ali Pasha. After many reorganizations and name changes, its descendant is today's Ministry of Water Resources and Irrigation.

25 The irrigation authorities in Egypt have since the eighteenth century collected data on the Nile basin, and this has enabled investigators to carry out many studies related to irrigation and the means of utilizing and developing Egypt's water resources. In an attempt to keep up with worldwide developments the ministry established a Water Research Center (WRC) in 1975. Owing to its pioneering work the WRC was reorganized in 1994 as the National Water Research Center (NWRC). See Fathy El-Gamal, 'Irrigation in Egypt and [the] role of [the] National Water Research Center', retrieved from http://resources.ciheam.org/om/pdf/b31/01002074.pdf.

26 The institutes of the NWRC are the Water Management Research Institute (WMRI), Drainage Research Institute (DRI), Water Resources Research Institute (WRRI), Nile Research Institute (NRI), Hydraulic Research Institute (HRI), Channel Maintenance Research Institute (CRMI), Research Institute for Groundwater (RIGW), Construction Research Institute (CRI), Mechanical and Electrical Research Institute (MERI), Survey Research Institute (SRI), Coastal Research Institute (CoRI), and Environmental and Climate Research Institute (ECRI). Special agencies include the Strategic Research Unit (SRU), the Central Laboratory for Environmental Quality Monitoring (CLEQM), the Information/Documentation Center, the Central Library, GIS Unit, and Gender Equality Unit.

27 Fathy El-Gamal, 'Irrigation'.

28 In 1982 the Training and Manpower Development Unit (TMDU) of the MWRI was established as part of the NWRC. In 1985, TMDU expanded its scope of services with support from USAID. A milestone in the professional development program was establishment of the Training Center of the MWRI (TC-MWRI) in 1994.

29 R.R. El-Emam Hosam, 'Country experiences in promotion of environmental justice in Africa: The experience of Egypt', All Africa Environmental Law and Justice Conference, Nakuru, Kenya, 2006.

30 A joint technical committee was formed to study the environmental impact of removing the weeds from Lake Kyoga without adversely affecting fish production, water quality and quantity, wildlife and wetlands.

31 Ministry of Water Resources and Irrigation, Egypt, and Ministry of Agriculture, Animal Industry & Fisheries, Uganda, 'Uganda-Egypt aquatic weed control project', unpublished.

32 A.F. Metawie, 'History of cooperation in the Nile basin', *Water Resources Development* 20:1, 2004, 47–63.

33 International Conference on Sustainable Development for Lasting Peace: Shared Water, Shared Future, Shared Knowledge, 'Actions to promote sustainable management of trans-boundary water resources in the southeastern Europe and Mediterranean region: recommendations for action', 6–7 May, Athens 2003.

34 A. Gebeyehu, 'The role of large water reservoirs', *Ethiopian Economic Association, Second International Conference on the Ethiopian Economy June 3–5, 2004*, Addis Ababa 2004; A. Wolf, 'Conflict and cooperation along international waterways', *Water Policy* 1:2, 1998, 251–65; E. Benvenisti, 'Collective action in the utilization of shared freshwater: The challenges of international water resources law', *American Journal of International Law* 90:3, 1996.

35 Organization of African Unity (OAU), 'Lagos plan of action for the economic development of Africa, 1980–2000', Lagos 1980.

36 Metawie, 'History'.

37 *Undugu* means 'brotherhood' in Swahili.

38 B. Ghali, 'Water management in the Nile river valley', *International Policy Review* 104, April 1991.

39 M. Zahran, 'Egyptian diplomacy and cooperation among the Nile basin states', *International Policy Review* 99, January 1990.

40 J. Waterbury, *The Nile Basin: National determination of collective action*, New Haven and London 2002; Dagne *et al.*, 'Towards a Cooperative Use of the Nile: A legal perspective', *Cambridge Review of International Affairs*, vol. 12, 1999; Ashok Swain, 'The Nile river basin initiative: Too many cooks, too little broth', *SAIS Review* 22:2 Summer–Fall 2002.

41 See www.nilebasin.org/documents.

42 In May 1999, a ministerial meeting was held in Addis Ababa to consider opportunities for cooperative development within the Eastern Nile, and for the purpose of establishing a joint technical team (ENSAPT) to define and prioritize common areas of interest of the eastern Nile countries in water resources and water related projects. ENSAP is supposed to comprise actual development projects at the sub-basin level involving two or more countries. This will allow moving from planning to action,

addressing what needs to be done at the regional level and development opportunities with trans-boundary implications. The ENSAP projects are as follows: 1. Integrated Water Resources Planning and Management (Eastern Nile Planning Model/Baro-Akobo Multi-purpose Water Resources); 2. Flood and Drought Management (Flood Preparedness and Early Warning); 3. Hydropower Development and Regional Power Trade (Ethiopia–Sudan Transmission Interconnection/Eastern Nile Power Trade Investment Program); 4. Irrigation and Drainage Development; 5. Watershed Management; and 6. Eastern Nile Technical Regional Office (ENTRO). ENTRO's core activities include support for ENSAP management, to ensure effective multi-country coordination; secretariat support to ENSAPT and ENCOM; and liaison with donors and other partners. Following an inception phase (November 2003–May 2004), an agreement was signed between Great Britain and ENTRO to establish and support ENTRO's Social Development Office, which seeks to strengthen the capacity of ENTRO, ENSAPT and country teams of investment projects to integrate social development into project preparations, to establish and implement plans for working with civil society in the Eastern Nile, and to provide input on broader social development/poverty issues, as required, at either the ENTRO Office or National Focal Point level.

43 NELSAP consists of 13 projects on water use in agriculture, management and conservation of lakes and wetlands, watershed management, water hyacinth and weed control, and hydropower. The original institutional arrangement consisted of a Nile Equatorial Lakes Council of Ministers (NEL-COM) and a Technical Advisory Committee (NEL-TAC). Funding has been sought for detailed preparation and implementation of projects through the International Consortium for Cooperation on the Nile (ICCON), which was established to coordinate donor support. NELSAP projects are: Water Use in Agriculture; Sustainable Management and Conservation of Lakes and Linked Wetlands (Fisheries Project for Lake Albert and Lake Edward); Watershed Management, including development of a Framework for Cooperative Management of the Water Resources of the Mara River Basin, Kagera River Basin Integrated Water Resources Management, and Development of a Framework for Co-operative Management of the Water Resources of the Malakisi-Malaba-Sio River Basins; Water Hyacinth and Water Weed Control (Water Hyacinth Abatement in the Kagera River Basin); Hydro-power Development and Power Trade (Rusumo Falls Hydro-Electric Power Development; Transmission Interconnection between Kenya and Uganda/DRC, Burundi and Rwanda/Burundi and Rwanda/Uganda and Rwanda); and the NEL Coordination Unit.

44 Cesar A. Guvele, 'The Nile Basin Initiative and its implications in post conflict South Sudan', participatory workshop ensuring secured livelihoods, food security and economic growth through good economic governance and sustainable management of our natural resources, June 2003. Supported by the USAID-USDA PASA in collaboration with the University of Missouri, Tuskegee University and the University of Maryland Eastern Shore. Report retrieved from http://cafnr.missouri.edu/iap/sudan/doc/nile-basin.pdf.

45 Waterbury, *The Nile Basin.*

46 *Convention on the Law of the Non-navigational Uses of International Watercourses*, retrieved from www.thewaterpage.com.

47 A survey of international agreements, decisions of international courts and tribunals, declarations and resolutions adopted by intergovernmental organizations, conferences and meetings, and studies by intergovernmental and non-governmental organizations relating to the principle of cooperation is contained in the Special Rapporteur's third report, 1987 *Yearbook*, II, 24ff., document A/CN.4/406 and Add. 1 and 2, para. 43–58.
48 General Assembly resolution 3281 (XXIX) of 12 December 1974.
49 Principle XII of the European Water Charter, adopted on 28 April 1967 by the Consultative Assembly of the Council of Europe (Recommendation 493 (1967)), and on 26 May 1967 by the Committee of Ministers (resolution (67) 10); text reproduced in 1974 *Yearbook*, II, 342–3, document A/CN.4/274, para. 373.
50 *Report of the United Nations Water Conference, Mar del Plata, 14–25 March 1977* (United Nations publication, Sales No. E.77.II.A.12).
51 ILA, *Report of the Sixtieth Conference*, Montreal 1982.
52 C.O. Okidi, *Natural Resources and the Development of Lake Victoria Basin of Kenya*, Institute for Development Studies, University of Nairobi, Occasional paper 34, 1979.
53 Charles B. Bourne, 'Mediation, Conciliation and Adjudication in the Settlement of International Drainage Basin Disputes', *The Canadian yearbook of international law*, vol. 9, 1971, 114–158.
54 Benvenisti, 'Collective action'.
55 Retrieved from www.un.org/law/ilc/texts/treasucc.htm#abstract.
56 UN treaty series, *Vienna Convention on the Law of Treaties 1969.*
57 I. Detter, *International Law*, Stockholm 1993.
58 Detter, *International Law*.
59 In July 1993, Hungary and Slovakia instituted proceedings before the International Court of Justice on the basis of a special agreement signed in Brussels on 7 April. The Court held, by clear majorities, that both Hungary and Slovakia had breached their obligations under the bilateral Treaty on the Construction and Operation of the GabEikovo-Nagymaros System of Locks signed on 16 September 1977, with effect from 30 June 1998. It found unlawful Hungary's unilateral suspension and subsequent abandonment of the project contemplated by the treaty, as well as Slovakia's subsequent unilateral diversion of the Danube. Most important, the court concluded that the treaty was still in force and that its object and purpose must be carried out in good faith by both parties. P.F. Bekker, 'Gabcikovo-Nagymaros Project (Hungary/ Slovakia), Judgement', *American Journal of International Law* 92:2, April 1998, 273–8.
60 *Ibid.*
61 *Ibid.*

## 11 SOME CONCEPTUAL ISSUES REGARDING THE STUDY OF INTER-STATE RELATIONSHIPS IN RIVER BASINS

1 R. Axelrod, *The Evolution of Cooperation*, New York 1984; R.V. Bartlett, 'Comprehensive environmental decision making, can it work?', in N.J. Vig and M.E. Kraft (eds), *Environmental Policy in the 1990s: Towards a new agenda*, Washington DC

1990; H. Beach *et al.*, *Transboundary Freshwater Dispute Resolution: Theory, practice and annotated references*, New York 2000; S. Born and W. Sonzogni, 'Integrated environmental management: Strengthening the conceptualization', *Environmental Management* 19:2, 1995, 167–81; F. Correia and J. de Silva, 'International framework for the management of transboundary water resources', *Water International* 24:2, 1999, 86–94; C. Dufournaud and J. Harrington, 'Temporal and spatial distribution of benefit and costs in river-basin schemes: a cooperative game approach', *Environmental Planning A* 22, 1990, 615–28; R.B. Keiter, 'Beyond the boundary line: Constructing a law of ecosystem management', *University of Colorado Law Review* 65:2, 1994, 293–333; M. Falkenmark and J. Lundqvist, 'Focusing on the upstream/downstream interdependencies and conflicts of interests – steps and procedures towards coping with management challenges', *Analytical Summary from the Proceedings of the SIWI/WRA Seminar*, Stockholm 1999; F.W. Frey, 'The political context of conflict and cooperation over international river basins', *Water International* 18, 1993, 54–68; G. Hardin, 'The tragedy of the commons', *Science* 162, 1968, 1243–8; L. Martin, 'Heterogeneity, linkage and common problems', *Journal of Theoretical Politics* 6:4, 1994, 473–93; J.D. Priscoli, 'Epilogue: Does water integrate or separate us?', *Water International* 15:4, 1990, 236–8.

2 See e.g. J.R. Starr, 'Water wars', *Foreign Policy* 82, spring 1991, 17–36; T. Homer-Dixon, *Environment, scarcity and violence*, Princeton 1999; Jason Morrissette and Douglas A. Borer, 'Where oil and water do mix: environmental scarcity and future conflict in the Middle East and North Africa', *Parameters*, winter 2004–05, 86–101.

3 See e.g. A.T. Wolf (ed.), *Conflict Prevention and Resolution in Water Systems*, Cheltenham 2002; Anders Jägerskog, 'The Jordan river basin: explaining interstate water co-operation through regime theory', Occasional Paper 31, *Water Issues Study Group*, SOAS, London 2003; Aron T. Wolf, A. Kramer and A. Carius, *State of the World 2005 Global Security Brief #5: Water Can Be a Pathway to Peace, Not War*, 1 June 2005, World Watch Institute, www.worldwatch.org/node/79; Nils Petter Gleditch, 'Armed conflict and the environment: A critique of the literature', *Journal of Peace Research* 35:3, 1998, 381–400.

4 Hardin, 'The tragedy of the commons'.

5 *Ibid.*, 1244.

6 The extensive literature assessing Hardin's thesis includes R.M. Aggarwal and T.A. Narayan, 'Does inequality lead to greater efficiency in the use of local commons? The role of strategic investments in capacity', *Journal of Environmental Economics and Management* 47, January 2004, 163–82; and E. Ostrom *et al.* (eds), *The Drama of the Commons*, Washington DC, 2002.

7 For the many works by H.E. Hurst and his colleagues in the Egyptian Ministry of Public Works on Nile discharges, rainfall and other hydrological phenomena, and for other works on these subjects in general, see Tvedt, *The Nile. An annotated bibliography*, which lists almost a thousand entries on the physical characteristics of the river.

8 See Honest Prosper Ngowi, Unlocking Growth and Development Potential: The Nile Basin Approach in Tanzania, in Terje Tvedt (ed), *The River Nile in the Post-Colonial Age*, London, 2010, 57–71; and Mary C. Mwiandi, The Nile Waters and the Socio-Economic Development of Kenya, in Tvedt, *Ibid.*, 93–124.

9 See Mancur Olson, *The Logic of Collective Action: Public goods and the theory of groups*, Cambridge, MA, 1965.

10 For a review, see Pamela E. Olover, 'Formal models of collective action', *Annual Review of Sociology* 19, 271–300.

11 The theory of collective goods has been developed in detail by economists. Its application to political issues stems from the work of Olson, *The Logic of Collective Action*. See also J. Barnett, *The Meaning of Environmental Security, Ecological Politics and Policy in the New Security Era*, London 2001.

12 John Waterbury (*The Nile Basin: National determinants of collective action*, New Haven 2002, 53) argues that the 'free rider' has been a problem in the Nile basin. Empirical evidence does not substantiate this general impression.

13 See for example, Waterbury, *Hydropolitics of the Nile Valley*, 1979 and Waterbury, *The Nile Basin*, 2002; Tafesse, *The Nile Question*, 2002; and especially Tvedt, *The River Nile in the Age of the British*, 2004.

14 M. Rowland, 'A framework for resolving the transboundary water allocation conflict conundrum', *Groundwater* 43:5, 700–5.

15 Elinor Ostrom, *Governing the Commons: The evolution of institutions for collective action*, Cambridge 1990.

16 These will be able to say: 'I will commit myself to follow the set of rules we have devised in all instances except dire emergencies if the rest of those affected make a similar commitment and act accordingly.'

17 This has for topographical and other geographical reasons not been an issue in the Nile basin, but in Sri Lanka the classical irrigation civilization depended on water control and – first and foremost – on the efficient water transfer among river basins. In China, the waters of the Yangtze have been diverted into the Grand Canal for more than a thousand years, and the new 'South-to-north water transfer project' will also bring Yangtze water to northern China and to a lot of other rivers, among them the Huang He.

18 Waterbury, *The Nile Basin*.

19 For a detailed empirical reconstruction of this story, see Terje Tvedt, *The River Nile in the Age of the British: Political ecology and the quest for economic power*, London and New York 2004, and Cairo 2006.

20 *Ibid*.

# Contributors

## CHAPTERS 1 AND 11

**Terje Tvedt** is Professor of Geography at the University of Bergen and Professor of Political Science at the University of Oslo. He has published extensively on water related issues, and especially on historical developments in the Nile Basin. His books include *The River Nile in the Age of the British: Political Ecology and the Quest for Economic Power* and *The River Nile. An Annotated Bibliography*. He is Series Editor for the pioneering series *A History of Water* (I.B.Tauris) and has written and co-directed two award- winning documentaries on water: *A Journey in the History of Water* and *A Journey in the Future of Water.*

## CHAPTER 2

**Pascal Nkurunziza** is Head of the Department of Earth Sciences at the University of Burundi and part-time lecturer at a number of other universities in the Great Lakes region. He holds a PhD in geology and mineralogy. His research focus is on mineralogical alteration, transport and sedimentation and, more recently, integrated water resources management (IWRM) in Burundi and in the Nile Basin.

## CHAPTER 3

**Robert Baligira** holds a PhD in hydrogeology. He has worked (especially) in the fields of water supply and sanitation, irrigation and drainage, and environmenal and water resources management and development. He was formerly Head of the Civil Engineering and Environmental Technology Department, Kigali Institute of Science Technology, Rwanda.

## CHAPTER 4

**Honest Prosper Ngowi** is a senior lecturer, researcher and consultant in economics at Mzumbe University, Tanzania. His main work has been in the fields of foreign direct investment (FDI); entrepreneurship; public private partnerships

(PPP); the informal sector; privatization; university-industry linkages; non-state actors (NSAs); program evaluations and reviews; and the global economic crisis.

## CHAPTER 5

**Raphael M. Tshimanga** is a lecturer in the Department of Natural Resources Management, Faculty of Agricultural Engineering, University of Kinshasa, Democratic Republic of Congo. He has more than six years experience in water resources management and modelling, including uncertainty analysis and stream-flow allocation, which form his current area of research interest.

## CHAPTER 6

**Mary C. Mwiandi** holds a BA in history and sociology, a postgraduate diploma in education, and an MA in history, all from the University of Nairobi. She holds a PhD in history from Michigan State University. She is currently a lecturer in the Department of History and Archaeology at the University of Nairobi, where she teaches history, and conflict and peace studies. Her research interests are in the comparative history of education, peace and conflict resolution in Africa, and social history in general.

## CHAPTER 7

**James Mulira** holds a PhD from Princeton University. He was Head of the Department of History, Development and Organizations Studies, Makerere University (2001-4) and Head of the Department of History and Government CEE, University of Nairobi (1988-95). He has taught in several universities within, and outside of, Uganda. His research interests include contemporary African history; inter-state relations and regional organizations; ethnicity and civil strife; and globalization.

## CHAPTER 8

**Yacob Arsano** is associate professor of political science and international relations and currently Dean of the College of Social Sciences at Addis Ababa University. He has published widely in hydropolitics and comparative politics. His recent publications include *Ethiopia and the Nile: Dilemmas of National and Regional Hydropolitics* (Swiss Federal Institute of Technology-Zurich, 2007).

## CHAPTER 9

**Fadwa Taha** holds a PhD in history from the University of Khartoum, where she is currently an associate professor in modern contemporary and diplomatic history and Vice Dean, Humanities, at the Graduate College. She is Visiting Professor at the Universities of Pavia and Bologna, Italy and the Institute of Arab Research & Studies, ALESCO, Cairo. She is the editor of *Adab*, the Journal of the Faculty of Arts, University of Khartoum and the author of a number of articles and books.

## CHAPTER 10

**Hosam E. Rabie Elemam** is a specialist on international waters and Nile water affairs and is a legal consultant at the Nile Water Sector, Ministry of Water Resources and Irrigation, Egypt. He acted as a legal consultant to the European Union for the Institutional Twinning project on water quality in Egypt. He worked with the United Nations-ESCWA as a lecturer on water negotiation skills. He has also been supervisor of cross-disciplinary research activity at the Middle East Research Centre, Ain Shams University, Egypt.

# Index

Abbay Basin Study Program 163
Abbay river 161, 162, 163, 167, 168, 170, 174
absolute riparian doctrine 175
African Development Fund (ADF) 150, 195
Aghakan Fund for Economic Development (AKFED) 144
'agreed curve,' Owen Falls Dam 133, 145
agriculture, plans and development in Rwanda 43–6
agro-pastoralism 110
Akagera sub-catchment 33
Akanyaru Haut sub-catchment 33
'Albert Graben' 90, 153
Albert, Lake 4, 79–80, 96
  development of semi-industrial fishing on 87
Allenby ultimatum 6
Anglo-Belgian agreement of 1894 79
Anglo-Belgian treaty, on management of river Nile 15
AQUASTAT 46
Arusha Declaration of 1967 61
Aswan Dam 5
Aswan High Dam project 7, 133, 134, 154, 189, 219, 239, 244

Bakajika Law 83, 84
Beles river, multi-purpose development project 169
biodiversity conservation 121, 152
Blue Nile 182
  water-development projects on 216
Boutros-Ghali, Boutros 237
British Nile policy 3–6
Bujagali falls 142
Bujagali 1 hydropower project 140–2
  criticism of 142–4
Bujagali 2 hydropower project 144–5
  approval of 146–7
  criticism of 145–6
Burundi
  climate change and its effects in 27–8
    fluctuation of rainfall and temperature 28–9
  demand for water in 17
  fisheries and fish culture in
    Burundi and Kagera basin 23–4
    Nile Basin Initiative (NBI) 24
  geographical and hydrological environment of 13–15
  hydroelectric potential of 19
  internal renewable sources of water in 17
  involvement in NELSAP projects 26–7
  managed and unmanaged marshes in Congo and Nile basins of 20
  and Nile basin 25–7
  and Nile treaties after independence 15–16
  water resources and utilization in 13, 16
    demand for water in sub-basin 22
    development of hydroelectric energy 18–19
    irrigation 21–2
    management of drinking water 17–18
    need for water in industrial sector 22
    supply of drinking water 17
    water pollution 23

cash crop cultivation 100, 110, 111
climate change, effects of 27–9
collective action, theories of 241–2
common property resource 242–5
Confidence-Building and Stakeholder Involvement Project 69
Congo Nile 73
challenges to development
climate change and natural resource depletion 89–90
demographic pressure 89
oil production 90
geographical and hydrographical environment 74–5
Albert, Lake 79–80
Congo basin 75–7
Congo Nile basin 77–8
Edward, Lake 78
Semliki river 78–9
Shiloango basin 75
historical context of 80
institutions, policies and cooperation
DRC and cooperative framework 85
future development plans 87–9
past development plans 86–7
post-colonial water institutions and policies 83–5
socio-economic development
agriculture 81
fishing 82
population 81
tourism 82–3
Congo Nile Development Plan 86, 87–9
Congo–Nile divide 13, 14, 15
cotton production, in Nile basin 4

*Digest of old Ethiopian judgments* 165
D–3 Project 177
drinking water
management of 17–18
supply in Burundi 17

EAC *see* East African Community (EAC)
EAGC *see* East African Governor's Conference (EAGC)
EAP *see* East Africa Protectorate (EAP)
East African Community (EAC) 23, 55, 130, 139, 140, 152, 158
East African Governor's Conference (EAGC) 117
East African Tourism and Wildlife Coordination Agency (EATWCA) 151
East Africa Protectorate (EAP) 100
'Eastern Nile' countries 216
Eastern Nile Strategic Action Program (ENSAP) 177, 230, 246
Eastern Nile Subsidiary Action Program 171, 230, 246
EATWCA *see* East African Tourism and Wildlife Coordination Agency (EATWCA)
Edward, Lake 70, 76, 77, 78, 86
fishing in 82
management of 87
Efficient Water Use for Agriculture Project 69
Egypt
cooperation with basin states
activities in basin 225–6
collective action mechanisms in basin 226–8
The Nile Basin Initiative 229
technical experience 224–5
current water situation in 217–19
development of additional water resources
desalination of seawater 222
domestic and municipal 223–4
groundwater systems 221–2
Nile water 221
rainfall and flash floods 222
reuse of agricultural drainage 223
reuse of treated wastewater 223
unconventional resources 222
differences with Ethiopia over Nile water use 9

Egypt (*cont.*)
enforcement of principle of 'exclusive user rights' for River Nile 7
irrigation economy 4
land reclamation projects 220
1959 Nile Waters Agreement and international law 233–5
opposition to dam projects on Nile 9
post-colonial developments of Nile 7
River Nile and 219–20
population growth 220
water quality 220
Shared Vision Program 229–30
subsidiary action program 230
Egyptian Revolution 4
ENSAP *see* Eastern Nile Strategic Action Program (ENSAP)
Equatorial Nile Project 80
Ethiopia
Abbay river in 162
differences with Egypt over Nile water use 9
and downstream riparian states
negotiating strategies 174–5
problem of agreement on shared waters 172–4
Ethiopian Civil Code 163–4
water rules under 164–5
hydroelectric power development 170–2
institutional development
customary water rules 165
planning and development of water sector 165–6
water in Ethiopian constitutions 164
water rules under medieval Fetha Negast 163–4
irrigable lands in water basins of 170
Nile basin initiative 175–8
shared identity of Nile 163
water planning and development in 161
Abbay (Blue Nile) master plan study 167
Gilgal Abbay project 167–8
Lake Tana project 166–7
other water development projects 169
Tana–Beles development project 168

Fincha hydropower development project 169
fishery industry, in Uganda 150–1
fish farming and aquaculture
in Kenya 120–3
in Rwanda 50
food security 35, 70, 88, 109, 119, 123, 150, 184, 207
'For The Full Utilization of the Nile Waters' Agreement 7
'Framework Agreement' 161
'Friends of Lake Victoria' non-governmental organization 122

Gabcikovo–Nagymaros dispute 234
Gezira scheme 5
area cultivated in 187
Managil extension of 189, 193
problems in 209–12
Sennar Dam and 182–4
Gilgal Abbay project 167–8
gravity irrigation 182
groundwater aquifers 222
groundwater resources
in Egypt 221–2
in Kenya 109
in Rwanda 42–3
in Sudan 205
in Tanzania 59

Hajar Asalaya Sugar Project 204
Hamdab Dam *see* Merowi Dam
Harmon doctrine 51, 232
hydraulic infrastructure, of upstream countries 9

hydroelectric dam 5, 54
hydroelectric energy, development in Burundi 18–19
hydrological monitoring network 41

independent power project (IPP) 144
Inga Falls project 76
Integrated Water Resources Management (IWRM) 218
International Bank for Reconstruction and Development (IBRD) 104
International River Network (IRN) 143
irrigation
  activities in Kenya 111–12
    fishing industry since independence 118–20
    hydropower 123
    Lake Victoria fisheries 115–18
    rice irrigation 112–13
    small-scale fish farming 120–3
    sugar production 114–15
  in Burundi 21–2
  development 4
  gravity 182
  and swamp reclamation possibilities in Kenya 108
Ituri Development Plan *see* Congo Nile Development Plan

Jebel Auliya Dam 184–5, 201
Jonglei Canal 8, 80, 200–2, 219

Kagera basin
  characteristic parameters of 53
  hydrological condition by country 52
Kagera Basin Organization (KBO) 13, 23, 54, 158
Kagera river 15, 31
  and Nile basin in Rwanda 52–3
Kanyaru river 13
Kazinga Channel 77, 78
KBO *see* Kagera Basin Organization (KBO)
Kenya
  agricultural activities 110–11
  during British rule 98–100
  development policy at independence 109–10
  economic activities in Nile basin of 106–9
  geographical and hydrological environment in 94–8
  groundwater potential in 109
  importance of Nile water agreements in 102–6
  irrigation activities in 111–12
    fishing industry since independence 118–20
    hydropower 123
    Lake Victoria fisheries 115–18
    rice irrigation 112–13
    small-scale fish farming 120–3
    sugar production 114–15
  irrigation and swamp reclamation possibilities in 108
  population distribution in 98
  socio-economic development in 93
  surface runoff in Victoria basin of 96
  'white highlands' and introduction of cash economy in 100–2
Khashm El Girba Dam 193–5, 204
Kiira hydropower dam 136–7
Kinanah Sugar Project 204
Kivu, Lake 14, 27, 41, 50, 76
Kyoga, Lake 79, 94, 96, 226

Lake Tana project 166–7
Lake Victoria Environmental Management Program (LVEMP II) 55
Lake Victoria Fisheries Organization (LVFO) 122, 158
Lake Victoria Project 63
Lutheran World Foundation (LWF) 42
LVEMP II *see* Lake Victoria Environmental Management Program (LVEMP II)

LWF *see* Lutheran World Foundation (LWF)

Mar del Plata Action Plan 232
marshland management 22
Merowi Dam 196–8, 208
Million Acre Scheme 102
Moghra aquifer 221
Mountain Gorilla tourism 83
Murchison Falls 96

Nalubale Dam *see* Owen Falls Dam
National Action Programme of Adaptation of Water Resources to Climate Changes 28
National Water Resources Plan (NWRP) 218
NBC *see* Nile Basin Commission (NBC)
NBI *see* Nile Basin Initiative (NBI)
NELSAP *see* Nile Equatorial Lakes Strategic Action Program (NELSAP); Nile Equatorial Lakes Subsidiary Action Programme (NELSAP)
Nile aquifer 221
Nile basin
  Burundi and 25–7
  distribution within Kagera basin 25
  physical characteristics of lakes located in 41
Nile Basin Commission (NBC) 157
Nile Basin Cooperative Framework 178
Nile Basin Initiative (NBI) 13, 24, 55, 85, 93, 106, 157, 176
  critical 'mass theory' and 245–6
Nile Equatorial Lakes Strategic Action Program (NELSAP) 177
Nile Equatorial Lakes Subsidiary Action Programme (NELSAP) 13, 24, 55, 230
  involvement of Burundi in 26–7
Nile, River *see* River Nile
Nile River Basin Commission (NRBC) 62
Nile-sharing agreements 8, 9
Nile treaties 9
Nile Valley Plan 6
Nile Waters Agreement 57, 60, 62, 93, 103
Nubian sandstone aquifer 221
Nyabarongo sub-catchment 33
Nyerere Doctrine 60, 61, 93, 105

Oslo Donors Conference of 2005 208
Ostrom, Elinor 242
Owen Falls Dam 5, 136
  'agreed curve' 133
  design and its implications for Nile basin 132–3
  performance of 137–40

*paysannat* system 43
Power Purchasing Agreement (PPA) 142, 143, 144
Provincial Irrigation Unit (PIU) 113
pump irrigation, in Sudan 185

Queen Elizabeth National Park 87

Rahad irrigation project 198–200
rainfall
  in Egypt 222
  in Rwanda 40
  in Sudan 205–7
rainwater
  harvesting 46
  infiltration into soil 42
REGIDESO projects 19
Regional Agriculture Program 70
Regional Power Trade Project 69
renewable energy resources 107
renewable sources of water, in Burundi 17
rice irrigation, in Kenya 112–13
Ripon Falls 131
River Nile
  in age of British 3–6
  economic transformations 6
  Egypt and 219–20

River Nile (*cont.*)
exclusive user rights for 7
geographical and hydrological characteristics 3
importance in regional development 1
international agreements on 15
major sites for hydropower development in Uganda 135
management plans during age of independence 6–11
post-colonial developments 7
Rwanda and 50–6
sharing agreements 8
sharing of water 1
tributaries in Rwanda 32
water quality of 220
river-sharing agreements 5
river system, hydraulic integrity of 4
'river war' of 1896–98 3
Roseires Dam 189, 192–3, 203, 239
'Rowland–Ostrom Framework' 242, 244
Rusizi river 14
Rusumo Falls 15
Rusumo Falls Hydroelectric Power and Multipurpose Development Project 70
Rusumo hydroelectric project 24
Ruvubu river 13, 26
Ruvyironza river 13
Rwanda
access to drinking water in 49
during colonial period 31–2
crisis after 1994 40
crisis during 1990–94 39
crops production and estimated water use in 45
development of water institutions in 36–7
first republic, 1962–73 37
second republic, 1973–94 37–8
economy 34–6
hydro-geological characteristics of 42
industrialization in 35
Nile river and 50–1
Kagera basin cooperation 53–4
Kagera river and Nile basin 52–3
Nile basin cooperation 54–6
treaties and agreements 51
water use in 31
number and location of boreholes in 44
physical characteristics 32–3
planned hydropower projects in 47
population in 33–4
role in Nile basin cooperation 31
sub-catchments of Nile-basin in 32–3
tributaries of Nile in 32
water distribution in urban areas in 49
water resources in
groundwater 42–3
rainfall 40–1
surface water 41
water resources, plans and development
agriculture 43–6
drinking water 49
fish farming and aquaculture 50
hydropower 46–9
industry 49–50
navigation 50
Rwenzori Mountains 89

SAPs *see* Subsidiary Action Programs (SAPs)
'Scramble for Africa' 15, 73
Semliki river 78–9
Sennar Reservoir 5, 182, 192, 196
Serageldin, Ismail 237
Shared Vision Coordination Project 69
Shared Vision Program (SVP) 69, 85, 229–30
Shinyanga Project 63
Sio-Malaba-Malakisi management programs 55
Socio-economic and Benefits Sharing Project 69
Sondu river, potential for hydropower generation 108

Strategic Action Program 55
Subsidiary Action Programs (SAPs) 69, 85, 230, 245
subsistence farming 110
Sudan
agro-climatic zones in 180
Anglo-Egyptian Colonial Era
flood and flush irrigation in Tokar delta and Gash river 186
Jebel Auliya Dam 184–5
1929 Nile Waters Agreement 186–7
pump irrigation 185–6
Sennar Dam and Gezira Scheme 182–4
dams in 194
domestic agricultural economy 179
geographical and hydrological position 179–81
Gezira Scheme, problems in 209–12
irrigable areas 181–2
irrigation 8, 207–8
major irrigation and water works
Gash and Tokar areas since independence 202
Guneid pump scheme 203–4
Jonglei Canal 200–2
Khashm El Girba dam 193–5
Managil extension 193
Merowi Dam 196–8
pump schemes after independence 202–3
Rahad irrigation project 198–200
Roseires Dam 195–6
sugar farms 204
water conditions before construction of Roseires Dam 192–3
position in Nile basin
cooperation with riparian countries 214–16
integration with Egypt 213
Permanent Joint Technical Commission (PJTC) 212
during post-independence period
military coup and making of Nile Waters Agreement 189–90
1959 Nile Waters Agreement 187–9
water management institutions 191–2
rainfall 182
water requirements, abundance *vs.* shortage 208–9
water resources in
groundwater 205
rainfall *vs.* Nile 205–7
Sudan Peace Agreement 8
Suez Canal 3
crisis 133–4
sugar farms 204
sugar production, in Kenya 114
surface runoff water 59
surface water, in Rwanda 41
SVP *see* Shared Vision Program (SVP)
Swynnerton Plan, for accelerating agricultural development 101

Tana–Beles development project 168
Tana, Lake 5
Tanganyika, Lake 14, 16, 28
Tanzania
colonial-era treaties 60
dependency on Nile basin water resources 59
economic policy and Nile utilization 61
groundwater in 59
hydro-meteorological survey of lakes Victoria, Kiyoga and Albert (Hydromet) 66–7
Kagera Basin Organization (KBO) 68–9
Nile Basin Initiative (NBI) 69–71
TECCONILE 67–8
*undugu* 67
involvement in cooperative efforts 65–6
Nile basin approach
1929 Nile waters agreement 61–2

Tanzania (*cont.*)
  Nile basin approach (*cont.*)
    recent Nile negotiations 62
  as Nile basin country 58–9
  participation in Nile basin arrangements 57
  projects using Nile waters in 62
    China Civil Engineering Construction Corporation (CCECC) 63–4
    Kahama water project 63
    National Contracting Company Ltd of Saudi Arabia 64–5
    Sinohydro Corporation of China 64
    use of Lake Victoria waters in mining industry 65
  sources of water in 59–60
  water administration in 60–1
Tekeze hydropower project 169
Trans-boundary Environmental Action Project 69
Transboundary Virunga Volcanoes Conservation Area 82

Uganda
  colonial history 130–1
  geographical environment in 126–30
  government's mitigation and future prospects 148–9
  historical development of 125
  hydropower development before independence 131–2
  hydropower development since independence 134–5
    Bujagali 1 hydropower project 140–4
    Bujagali 2 hydropower project 144–7
    competitive advantage in energy sector 136
    Kiira hydropower dam 136–7
    performance of Nalubale and Kiira Dams 137–40
    power sector reforms 135–6
  Medium-Term Plan (MTP) to ensure food security 150
  non-hydroelectric projects in Nile basin
    fishery industry 150–1
    irrigation 149–50
    oil sector 152–4
    tourism 151–2
  Owen Falls Dam or Nalubale Dam
    'agreed curve' 133
    design and its implications for Nile basin 132–3
    performance of 137–40
    Suez Canal crisis and 133–4
  participation in sub-basin and basin-wide programmes and projects
    basin-wide cooperation 158–9
    cooperation in Nile sub-basin of East Africa 157–8
    proposed new Nile water treaty 157
    1929 water agreement 154
    1959 water agreement 154–5
    water agreements since independence 155–7
  shortage of hydropower supply 147–8
  socio-economic development in 147–8
*Ujamaa* 61
*Umuganda* 46
United Nations Development Program (UNDP) 66
United Nations Water Conference 232

Victoria, Lake 4, 32, 96, 107, 137
  fisheries in 115–18
  as inland water fishery sanctuary 150
Virunga National Park 82–3, 87

water administration, in Tanzania 60–1
water quality management 220
water-resource management
  in Democratic Republic of Congo (DRC) 83–4
  in Kenya 107
  in Sudan 211
  in Tanzania 61, 68
Water Resources Management Project 69

water scarcity 80, 209, 217, 218, 237
water sharing, agreement between Egypt and Sudan 8
Water Users Associations 61
well-sinking projects 42
Western Rift Valley 79
'White Highlands' 100, 102
White Nile 4, 25, 28, 182
  Jebel Auliya Dam on 184–5
  at Owen Falls 134
World Meteorological Organization (WMO) 53, 66, 106